Chemistry of Oxide Superconductors

International Union of Pure and Applied Chemistry

Chemistry of Oxide Superconductors

EDITED BY

C. N. R. RAO FRS

Solid State and Structural Chemistry Unit
Indian Institute of Science
Bangalore, India

BLACKWELL SCIENTIFIC PUBLICATIONS

OXFORD LONDON EDINBURGH

BOSTON PALO ALTO MELBOURNE

and published for them by
Blackwell Scientific Publications
Editorial offices:
Osney Mead, Oxford OX2 0EL
(*Orders:* Tel: 0865 240201)
8 John Street, London WC1N 2ES
23 Ainslie Place, Edinburgh EH3 6AJ
3 Cambridge Center, Suite 208
Cambridge, Massachusetts 02142, USA
667 Lytton Avenue, Palo Alto
California 94301, USA
107 Barry Street, Carlton
Victoria 3053, Australia

First published 1988

Printed in Great Britain
at the Alden Press, Oxford

DISTRIBUTORS

USA
Blackwell Scientific Publications Inc
PO Box 50009, Palo Alto
California 94303
(*Orders:* Tel: (415) 965–4081)

Canada
Oxford University Press
70 Wynford Drive
Don Mills
Ontario M3C 1J9
(*Orders:* Tel: (416) 441–2941)

Australia
Blackwell Scientific Publications
(Australia) Pty Ltd
107 Barry Street
Carlton, Victoria 3053
(*Orders:* Tel: (03) 347 0300)

British Library
Cataloguing in Publication Data

Chemistry of oxide superconductors.
1. High temperature superconductivity
I. Rao, C.N.R. (Chintamani Nagesa
Ramachandra), *1934–* II. International
Union of Pure and Applied Chemistry
537.6′23

ISBN 0-632-02302-3

Library of Congress
Cataloging-in-Publication Data

Chemistry of oxide superconductors/
edited by C.N.R. Rao.
p. cm.
At head of title: International Union of
Pure and Applied Chemistry.
Bibliography: p.
ISBN 0-632-02302-3
1. Superconductors—Chemistry.
2. Materials at high temperatures.
I. Rao, C.N.R. (Chintamani Nagesa
Ramachandra), 1934–
II. International Union of Pure and
Applied Chemistry.
QD473.C4883 1988
537.6′23—dc19

Contents

Preface

Although superconductivity was discovered as early as 1911, the highest transition temperature achieved before 1987 was around 23 K. It appeared as though 23 K was the upper limit for the superconducting transition temperature. The situation changed suddenly, however. Oxides of the La–Ba(Sr)–Cu–O system possessing the K_2NiF_4 structure were found to show superconductivity in the 30–40 K region early in 1987. Soon after, $YBa_2CU_3O_7$ oxides possessing the defect perovskite structure as well as other rare earth cuprate derivatives with a similar formula were found to exhibit superconducting transitions at 95 ± 5 K, well above the liquid nitrogen temperature. Several other oxides showing high-temperature superconductivity have since been identified and there are indications that we may reach even higher transition temperatures in the near future. The progressive increase in superconducting transition with time over the last four decades is shown in Fig. 1 to highlight the big jump in 1987. The discovery of high-temperature oxide

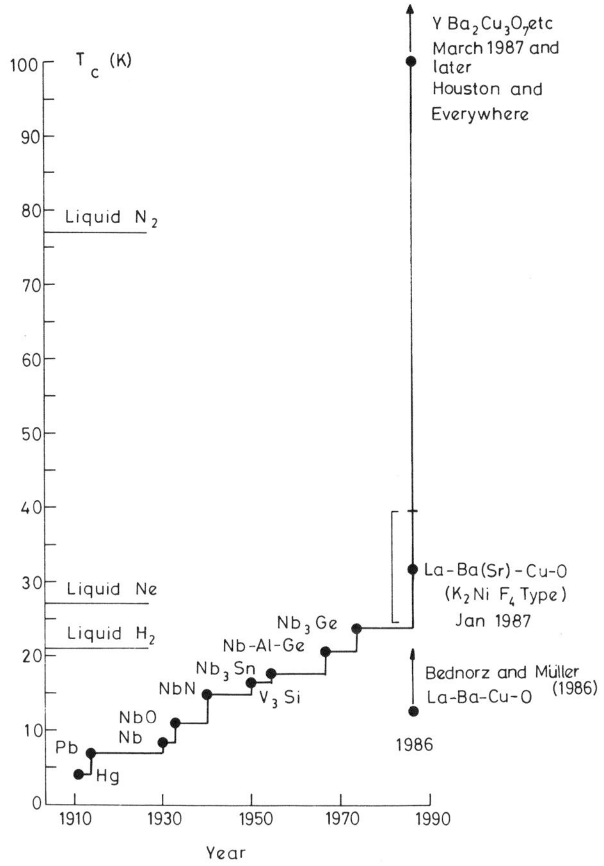

Fig. 1 Increase in the superconducting transition temperature with time since the discovery of superconductivity (modified from the figure of Muller and Bednorz, *Science*, Sept. 1987).

superconductors has created a great sensation, comparable only to that of the transistor, because of the immense technological potential of these materials in building magnets, in electronics, power and other sectors.

All the high-temperature oxide superconductors are complex chemical systems. Metal oxides have constituted a vital area of research in solid state chemistry for many years and the knowledge gained by these investigations has been of immense value to the study of the new superconductors. Many chemists have been working on the synthesis, structure, properties, theory and other important aspects of the oxide superconductors. The new superconductors have given a big boost to research in solid state chemistry.

It was suggested by Professor K. I. Zamaraev, President of the Physical Chemistry Division of the International Union of Pure and Applied Chemistry, that it would be most worthwhile to bring out a publication to highlight some of the 'chemical' contributions to high-temperature superconductivity. This volume is the result of such an effort.

I am most grateful to all the authors who so willingly contributed to this volume at such short notice. I do hope that the book gives a flavour of the chemistry of oxide superconductors and will be found useful by all those interested in oxide superconductors and in the chemistry of solids.

C. N. R. RAO
Bangalore

Chemistry of high-temperature oxide superconductors

C. N. R. Rao

Solid State and Structural Chemistry Unit, Indian Institute of Science, Bangalore-560012, India

Abstract - High-temperature superconductivity in oxides of the type $La_{2-x}Ba_x(Sr_x)CuO_4$, $Y(Ln)Ba_2Cu_3O_{7-\delta}$ and $La_{3-x}Ba_{3+x}Cu_6O_{14}$ is discussed with particular reference to the experimental findings from the author's laboratory. The role of oxygen stoichiometry in $YBa_2Cu_3O_{7-\delta}$ is examined in some detail; this oxide retains the orthorhombic structure and superconductivity upto $\delta \simeq 0.6$, beyond which it becomes tetragonal and non-superconducting. Oxygen stoichiometry in this oxide as well as in $La_{3-x}Ba_{3+x}Cu_6O_{14}$ has to be understood in terms of both structure and disorder. A transition from a high T_c superconductivity ($\sim$90K) regime to a low T_c superconductivity ($\sim$50K) regime occurs in $YBa_2Cu_3O_{7-\delta}$, accompanying the change in δ; this may be considered to be a transition from chain- to sheet-type superconductivity. There is no evidence for Cu^{3+} in these mixed valent copper oxides. Holes are on the oxygen (rather than on Cu), giving rise to O^- species. Pairing of two such O^- holes gives rise to a peroxide-like species; the reaction $2O^{2-} \rightarrow O_2^{2-} + 2e$ may indeed be of direct relevance to superconductivity in these oxides. Certain interesting features of $YBa_2Cu_3O_{7-\delta}$ such as twin or domain boundaries and microwave absorption are presented, besides important material parameters of technological relevance. Preparative aspects of these oxide materials are briefly discussed.

INTRODUCTION

The phenomenon of superconductivity has been an area of vital interest for the past several decades, ever since Kammerlingh Onnes discovered in 1911 that mercury becomes superconducting at 4.2K. Many materials, generally metals and alloys and more recently, molecular systems including organic charge-transfer compounds, have been investigated for superconductivity, but the superconducting transition temperature did not cross 23K till 1987. The average rate of increase in T_c was about 3 degrees per decade and it appeared as though 23K was the upper limit for the T_c. The highest T_c' s (in the 20K region) were exhibited by the A15 compounds such as Nb_3Sn and Nb_3Ge (ref. 1). Some of the metal oxides were known to show superconductivity, the highest T_c amongst them being exhibited by $LiTi_2O_4$ (ref. 2) and $BaPb_{1-x}Bi_xO_3$ (ref. 3) both around 13K.

Bednorz and Müller (ref. 4) recently showed the possibility of occurrence of high-temperature superconductivity in oxides of the La-Ba-Cu-O system. It was established in January 1987 that these oxides had the general formula $La_{2-x}Sr_x(Ba_x)CuO_4$ and possessed the K_2NiF_4 structure; the T_c values in these oxides were in the 20-40K region, the compositions corresponding to the maximum T_c in the Sr and Ba systems being $x = 0.2$ and 0.15 respectively (ref. 5-9). Following the heels of this discovery, superconductivity above liquid nitrogen temperature was reported in March 1987 in the Y-Ba-Cu-O system (ref. 10). It was soon found (ref. 11,12) that the oxide responsible for superconductivity in this system did not possess the K_2NiF_4 structure, but instead had the defect perovskite structure with the composition $YBa_2Cu_3O_{7-\delta}$.

The discovery of superconductivity above the liquid nitrogen temperature in oxide materials has raised much hope because of its important technological implications. Equally importantly, this has given a big boost to research in the solid state chemistry of metal oxides, an area that I have been working in for the past three decades. In what follows, some of the highlights of research on high T_c oxide superconductors will be presented, with particular reference to the results obtained in this laboratory (ref. 13). In addition, some of the special features of these ceramic oxides will be indicated. It may be remarked here that research efforts in this laboratory related to the high-temperature superconducting oxides were intiated in the early part of January 1987,

*Contribution No. 486 from the Solid State and Structural Chemistry Unit

soon after information about the $La_{2-x}Ba_x(Sr_x)CuO_4$ system became available through the American participants in the International Conference on Valence Fluctuation held in Bangalore. I am really delighted that some of the oxide systems that I have been working on for several years have gained so much importance today because of high-temperature superconductivity. I consider myself fortunate to have been a participant in this exciting area.

$La_{2-x}Ba_x(Sr_x)CuO_4$

Oxides of the general formula A_2BO_4 possess the quasi two-dimensional K_2NiF_4 structure (Fig. 1) wherein the B ions interact only in the ab plane. Structure and properties of the oxides of K_2NiF_4 structure have been examined in some detail recently by Ganguly and Rao (ref. 14). It has been known for sometime that orthorhombic La_2CuO_4 is a relatively low-resistivity material (~ 1 ohm cm at 300K with the resistivity increasing at low temperatures as shown in Fig. 2)

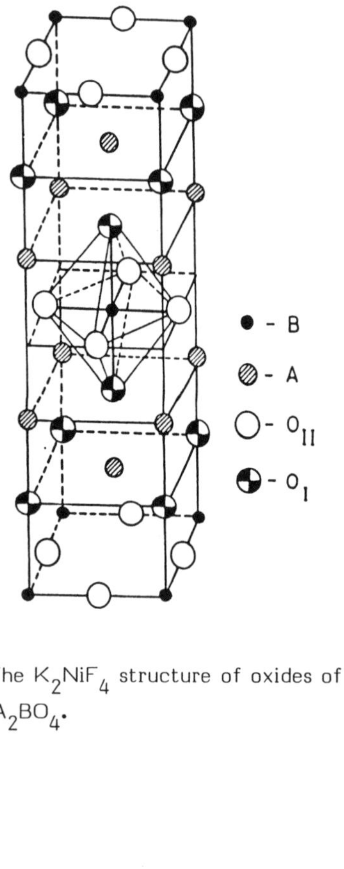

Fig. 1. The K_2NiF_4 structure of oxides of the formula A_2BO_4.

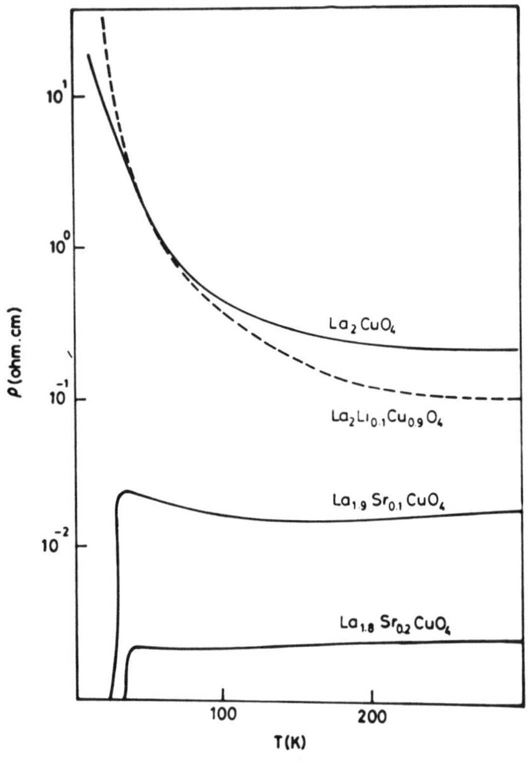

Fig. 2. Electrical resistivity behaviour of $La_{2-x}Sr_xCuO_4$ and La_2CuO_4 (From Ganguly et al, ref. 9).

and becomes antiferromagnetic at low temperatures. Substitution of La by Ba or Sr in $La_{2-x}Ba_x(Sr_x)CuO_4$ ($x > 0.05$) makes the structure tetragonal at room temperature and superconductivity manifests itself at low temperatures (ref. 8, 9). Substitution of Cu by Ni even to a small extent lowers the T_c and superconductivity is destroyed at ~ 5% substitution (ref. 9); Zn substitution has a similar effect although the ionic sizes of Zn^{2+} and Ni^{2+} differ from Cu^{2+} in the opposite directions. Superconductivity is found in the 15-40K region in oxides of the type $(La_{1-y}Ln_y)_{2-x}Ba_x(Sr_x)CuO_4$ where Ln = Pr, Nd, Gd etc. as shown in Fig. 3 (ref. 9, 15). Substitution of La by Ca lowers the T_c. These oxides, containing planar CuO_2 units, show maximum T_c around a specific value of x suggesting the importance of strong correlations (ref. 16). Electrical resistivity values of these oxides in the normal state above the T_c are in the 10^{-2} - 10^{-3} ohm. cm. range (Fig. 2) which correspond closely to Mott's minimum metallic conductivity (ref. 17).

Superconducting $La_{2-x}Ba_x(Sr_x)CuO_4$ undergoes a tetragonal-orthorhombic distortion around 180K (ref. 18,19). ESR studies have thrown light on microscopic magnetic interactions in these oxides (ref. 20). Static and dynamic aspects of the tetragonal-orthorhombic distortion in $La_{2-x}Sr_xCuO_4$ have been studied by neutron scattering and related techniques and a classical soft phonon behaviour involving CuO_6 octahedra has been observed (ref. 21).

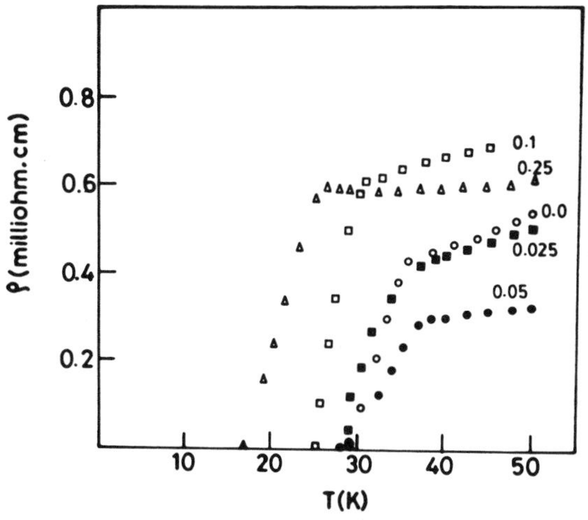

Fig. 3 (a). Resistivity of $(La_{1-y}Pr_y)_{1.8}Sr_{0.2}CuO_4$ for different values of y.

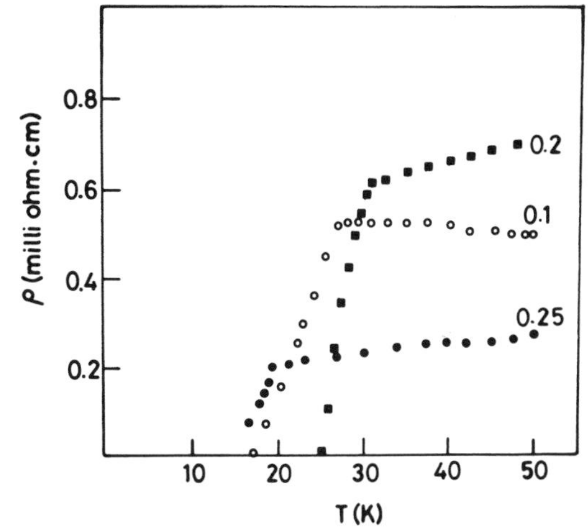

Fig. 3(b). Resistivity data of $(La_{0.9}Pr_{0.1})_{2-x}Sr_xCuO_4$ for different values of x (From Mohan Ram et al, ref. 15).

There have been several investigations of La_2CuO_4 in the last few months. Antiferromagnetism in La_2CuO_4 has been established by neutron scattering and diffraction studies, the T_c being close to 290K with a low-temperature Cu moment of 0.43 μ_B (ref. 22a). Besides establishing the magnetic structure, these studies show the occurrence of an orthorhombic-tetragonal distortion at 505K. The work of Shirane et al (ref. 22b) has shown that La_2CuO_4 is in a 2-dimensional afm quantum fluid state wherein spins are ordered instantaneously over long distances, but there is no measurable time-averaged moment. More interestingly, studies of oxygen-excess La_2CuO_4 has shown it to be superconducting with a T_c close to 40K (ref. 23). It should be noted that oxygen excess in La_2CuO_4 creates mixed valency just as substitution of La by Ba or Sr. Slight deficiency in lanthanum in the nominal composition (e.g. $La_{1.9}CuO_4$), allows the superconducting transition to be observed with ordinary ceramic preparations (ref. 24).

$YBa_2Cu_3O_{7-\delta}$ and related oxides

Wu et al (ref. 10) reported superconductivity above liquid N_2 temperature in $Y_{1.2}Ba_{0.8}CuO_4$. This composition was actually biphasic, consisting of green Y_2BaCuO_5 and a black oxide. We had initiated studies (ref. 25) on the Y-Ba-Cu-O system with compositions of the type $Y_{3-x}Ba_{3+x}Cu_6O_{14-\delta}$ by analogy with $La_{3-x}Ba_{3+x}Cu_6O_{14-\delta}$ (ref. 26). This was because Y_2CuO_4 (unlike La_2CuO_4) does not crystallize in the K_2NiF_4 structure. By comparing the x-ray diffraction patterns of $Y_{3-x}Ba_{3+x}Cu_6O_{14-\delta}$ and Y_2BaCuO_5 (Fig. 4), it was possible to identify the phase responsible for the high T_c (90-95K) superconductivity to be orthorhombic $YBa_2Cu_3O_{7-\delta}$ (ref. 11). AC susceptibility (Meissner effect) measurements have shown the superconductivity to be a bulk property (ref. 27,28). The

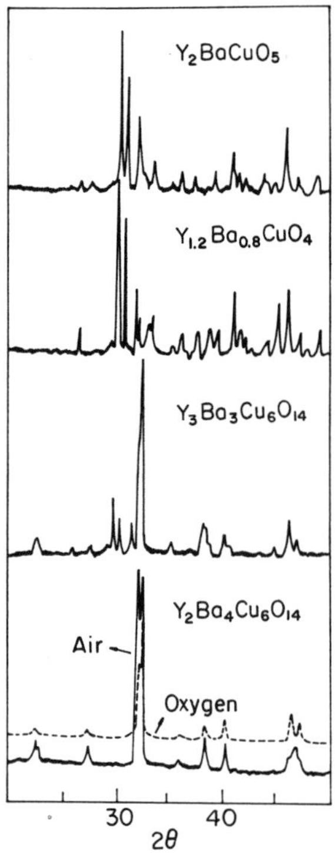

Fig. 4. X-ray powder data of $Y_{3-x}Ba_{3+x}Cu_6O_{14}$ and Y_2BaCuO_5 (From Rao et al, ref. 11).

highest Meissner effect found hitherto is ~ 80%. Other related oxides of the type $LnBa_2Cu_3O_{7-\delta}$ (Ln = Eu, Ho, Er, Dy, Gd etc.) are also high T_c superconductors, all with $T_c \simeq$ 90K (ref. 27, 29-32). Obviously, Y and Ln ions play no role in the superconductivity; they only help to keep the structure together. $LaBa_2Cu_3O_{7-\delta}$ is also superconducting and the T_c varies widely (20-90K) depending on the preparation, an aspect yet under investigation in this laboratory. The orthorhombic-tetragonal transition in $LaBa_2Cu_3O_7$ seems to occur at a low temperature. The Y/Ba or Ln/Ba ratio in $Y(Ln)Ba_2Cu_3O_{7-\delta}$ has been varied (ref. 33-37) and T_c varies somewhat with the ratio as expected.

Superconductivity in $YBa_2Cu_3O_{7-\delta}$ is very sensitive to oxygen stoichiometry (ref. 38-42). Superconducting samples of this oxide where the oxygen content is higher than 7 have also been made and they show T_c of ~ 90K upto 7.2 (ref. 43). Note that the highest theoretical composition of these oxides is $LnBa_2Cu_3O_9$. Oxygen is readily intercalated into the δ >0.5 samples and the stoichiometry reaches close to $YBa_2Cu_3O_7$ on intercalation (Fig. 5). The δ > 0.5 samples are non-super-

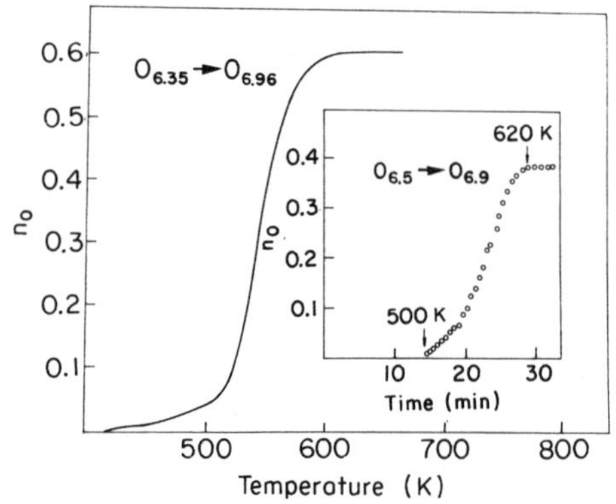

Fig. 5. TGA curves showing the oxidation of $YBa_2Cu_3O_{6.5}$ to $YBa_2Cu_3O_7$. Inset shows the rate of oxygen intercalation in a $O_{6.5}$ sample (From Rao et al, ref. 44).

conducting and tetragonal. Thermogravimetric curves show characteristic differences when δ varies between 0.0 and 1.0 (ref. 44). Structures of the orthorhombic $YBa_2Cu_3O_{7-\delta}$ and $LnBa_2Cu_3O_{7-\delta}$ (δ= 0) have been investigated in detail; the tetragonal (δ = 1.0) structure differs from this structure in an interesting manner. Besides the CuO_2 sheets, the stoichiometric (δ = 0.0) phase contains Cu-O-Cu chains (or corner-linked CuO_4 units) which are absent in the non-superconducting (δ = 1.0) sample (ref. 45-48). In Fig. 6, we compare the orthorhombic and tetragonal structures of

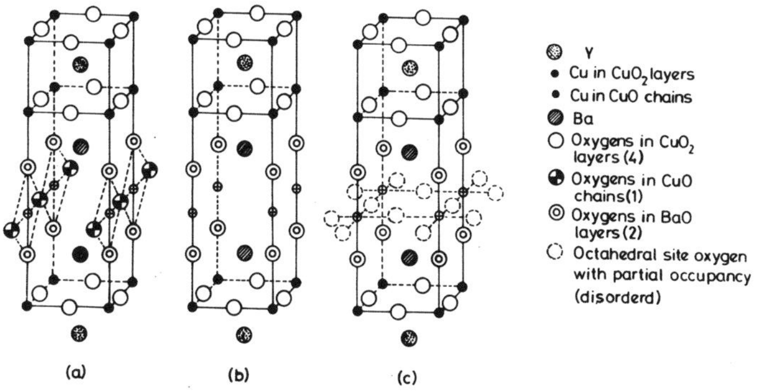

Fig. 6. Structures of $YBa_2Cu_3O_{7-\delta}$: (a) Orthorhombic structure of the superconducting phase with δ = 0.0; (b) tetragonal structure of the non-superconducting phase with δ = 1.0; (c) disordered structure of the tetragonal phase with CuO_6 octahedra where the site occupancy is small.

the δ = 0.0 and the δ = 1.0 samples. In the δ = 0.6 - 1.0 range, the tetragonal phase has very few oxygens in the Cu-O-Cu chains; this gives rise to considerable oxygen disorder and hence to distorted CuO_6 octahedra as in Fig. 6c (ref. 49). Oxygen non-stoichiometry in the $YBa_2Cu_3O_{7-\delta}$ system has therefore to be understood in terms of both disorder and structural distortion.

$YBa_2Cu_3O_{7-\delta}$ prepared in oxygen and air show thermopowers of opposite signs (ref. 27). The absolute thermopower and Hall measurements on samples with $\delta < 0.5$ show that these oxides are p-type (hole) conductors. The sign of the charge carrier may however be different at high δ values.

The variation of the superconducting transition temperature of orthorhombic $YBa_2Cu_3O_{7-\delta}$ with δ (in the δ-range 0.0 - 0.5) is most interesting. In Fig. 7, the resistivity data of a few members

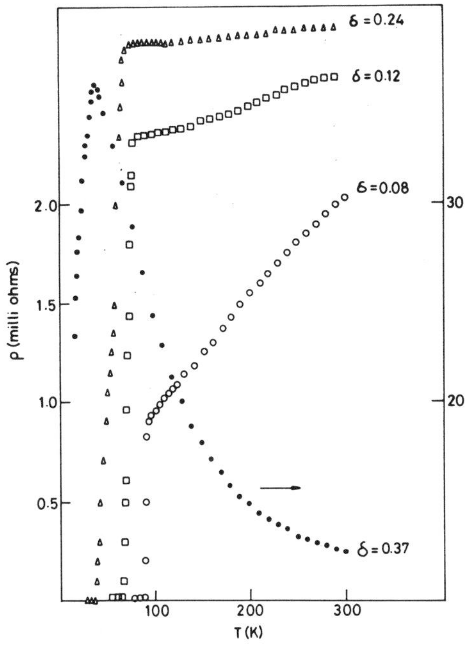

Fig. 7. Electrical resistivity behaviour of $YBa_2Cu_3O_{7-\delta}$ for different values of (Results from this laboratory).

with different δ values is shown. In Fig. 8, the T_c values from the resisitivity data (ref. 44) are plotted against δ. In Fig. 8, we also show results from the magnetic measurements of Johnston et al (ref. 40). Although the actual T_c values vary in the different sets of data (Fig. 8), we

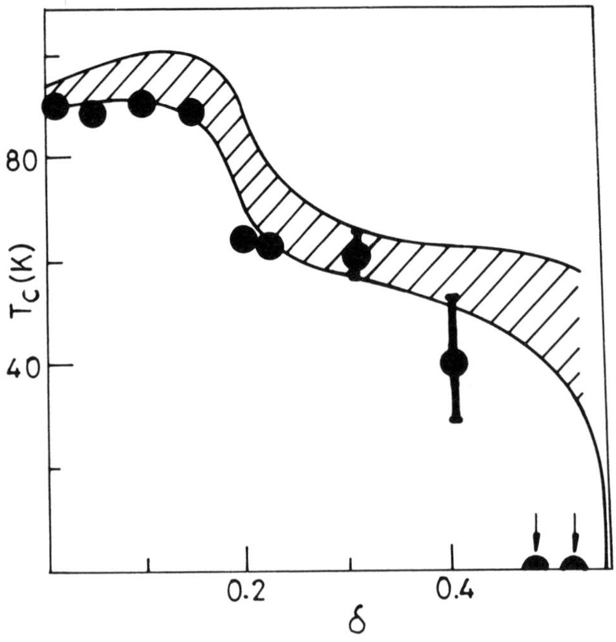

Fig. 8. Variation of T_c with δ. Magnetic measurements of Johnston et al (ref. 40) are indicated by the cross-hatched region. Our resistivity data is shown by full dark circles (From ref. 44).

see that T_c is nearly constant around 90K at low δ (0.0 - 0.2) and drops to a lower value (~50K) above δ = 0.2 showing a plateau-like behaviour. The T_c values in the plateau region are comparable to those of $La_{2-x}Ba_x(Sr_x)CuO_4$ (of the K_2NiF_4 structure). Since in this composition range, the Cu-O-Cu chains of $YBa_2Cu_3O_7$ are rather depleted of oxygen, we believe that the ~ 50K plateau is essentially characteristic of superconductivity due to CuO_2 sheets while the 90K T_c is characteristic of Cu-O-Cu chains (in the presence of CuO_2 sheets). Fig. 8 may therefore be taken to signify a transition from chain-type superconductivity to sheet-type of superconductivity in orthorhombic $YBa_2Cu_3O_7$ brought about by the change in oxygen stoichiometry.

Several workers have in recent months synthesised derivatives of $YBa_2Cu_3O_{7-\delta}$ where Cu is substituted by Zn, Ni, Co, Fe and such ions or Ba is substituted by La and other ions. Such substitution generally lowers T_c or destroys the superconductivity, the oxide generally being tetragonal or/and oxygen-deficient (see for example ref. 50); substitution of Ba by Sr also lowers the T_c. It should be noted that non-superconducting $YBa_2Cu_3O_{7-\delta}$ compositions also possess the tetragonal structure. Weakly orthorhombic or nearly tetragonal samples prepared by low-temperature methods exhibit low T_c or are non-superconducting. By removing oxygen from $YBa_2Cu_3O_7$ by the low-temperature Zr-gettered annealing technique, orthorhombic samples have been prepared upto $\delta \simeq 0.7$, all of them showing superconductivity; T_c however decreases with increrase in δ with a plateau in the δ = 0.3 - 0.5 region (ref. 51). The T_c values in this study are somewhat higher in the high range, possibly because of the ordered nature of the defects. The question then arises as to whether the orthorhombic structure is necessary for high T_c in the 90K region. A superconducting oxide (T_c = 88K) of the composition $La_{1.25}Ba_{1.75}Cu_3O_{7.1}$ with a tetragonal structure seems to have been prepared recently by Sleight and coworkers (unpublished results).

Role of oxygen

Having established the crucial role of oxygen in the superconductivity of these ceramic oxides, we have explored the mechanism of superconductivity by means of UV and X-ray photoelectron spectroscopy and Auger spectroscopy. UPS studies showed some changes in the 12 eV region in the case of $YBa_2Cu_3O_7$; furthermore, the valence band region was indicative of a low density of states and presence of strong correlations. Variable-temperature XPS studies on $La_{1.8}Sr_{0.2}CuO_4$ and $YBa_2Cu_4O_7$ in the O(1s) and Cu(2p) regions (ref. 39, 52, 53) show the presence of molecular oxygen species with a high O(1s) binding energy of ~533 eV besides features due to oxide and O^- ions at 529 and 531 eV respectively (Fig. 9). The proportion of the species responsible for the 533 eV feature increases with the lowering of temperature (Fig. 10). This oxygen species is identified as peroxide-like unit. The Cu(2p) spectrum shows the presence of a well-screened d^{10} state (Cu^{1+}) at 933 eV along with a poorly screened d^9 state (Cu^{2+}) at 942 eV. The proportion of the d^{10} state increases with the lowering of temperature. There is however no evidence for Cu^{3+} in the Cu(2p) or Auger spectra. Auger spectra also clearly show the presence of Cu^{1+} (Fig. 11). Based on these findings, it is proposed that holes are present on oxygen rather than on copper in $YBa_2Cu_3O_7$ and $La_{2-x}Ba_x(Sr_x)CuO_4$ (note that both are hole conductors). The oxygen holes (O^-) could then dimerize to give peroxide-type species (O_2^{2-}). The pair of electrons released in the

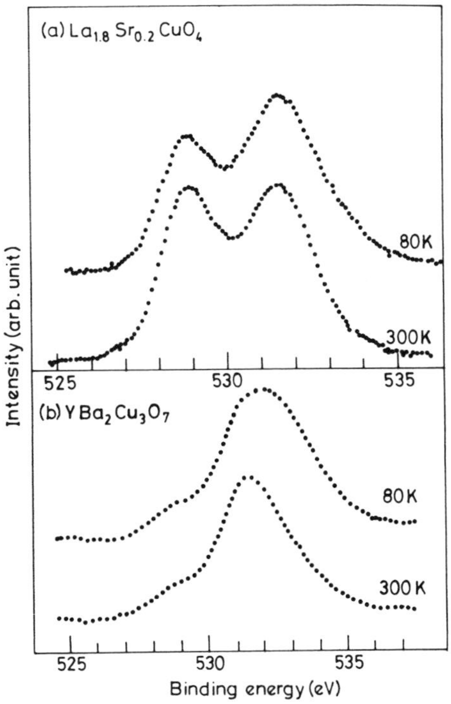

Fig. 9. O(1s) region of $La_{1.8}Sr_{0.2}CuO_4$ and $YBa_2Cu_3O_7$ showing increased intensity of the 533 eV feature due to O_2^{2-} species (From Sarma et al. ref. 52).

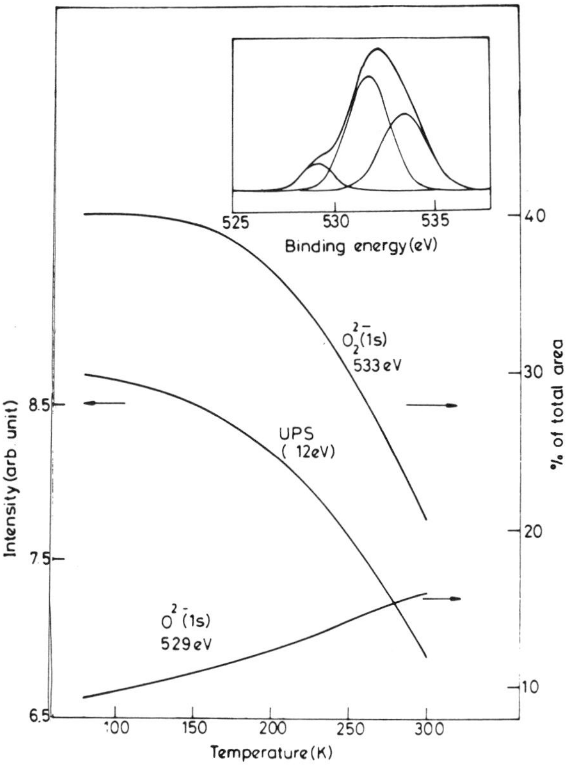

Fig. 10. Temperature-variation of O(1s) peak intensity due to O^{2-} and O_2^{2-} species (at 529 and 533 eV respectively) in $YBa_2Cu_3O_7$. Temperature-variation of the intensity of He II spectrum at 12 eV is also shown. Inset shows the O(1s) signal at 80K as consisting of three Gaussians peaking at 529, 531 and 533 eV, the one at 531 eV being due to O^- or impurity species (From Rao et al, ref. 39).

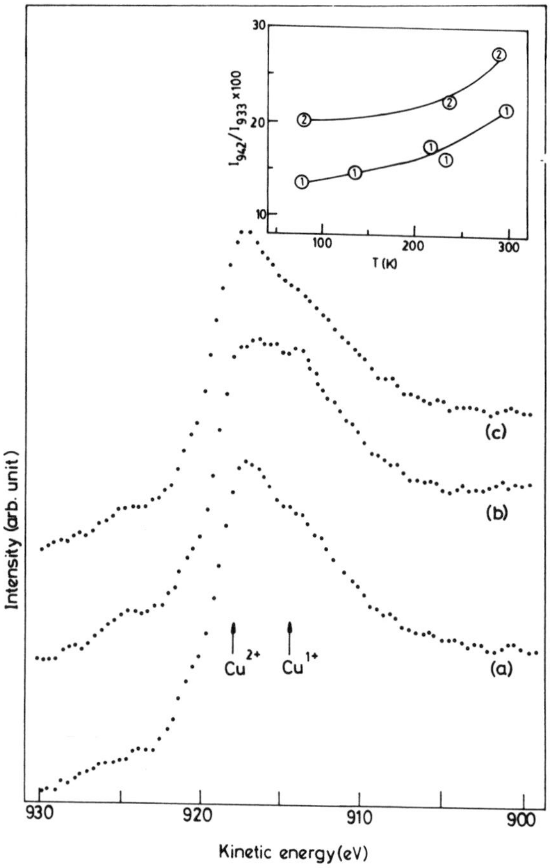

Fig. 11. Cu(L_3VV) Auger spectrum of $YBa_2Cu_3O_7$ at different temperatures. In the inset, temperature-variation of the intensity of the 942 eV feature relative to that of the 933 eV feature in the Cu($2P_{3/2}$) spectrum is shown. 1 and 2 refer to independent sets of measurements (Results from this Laboratory).

formation of such species from the oxide ions ($2O^{2-} \rightarrow 2O^-$ or O_2^{2-} + 2e) could be responsible for superconductivity. It is likely that the average charge on oxygen in these oxides is around - 1.3+0.1. Holes in the O(2p) valence bands are favoured by the Cu^{I+} (d^{10}) state, just as holes in the S(3p) or Se(4p) valence bands are generally favoured in Cu^{I+} chalcogenides (ref. 53). Since there is no real evidence for Cu^{3+} and the Cu-O bonds are highly covalent (1.8 - 1.95 Å), we may conclude that the contribution from the state with the holes on Cu is indeed negligible.

Other La-Ba-Cu-O and related systems

Besides oxides of the type $La_{2-x}Ba_xCuO_4$ (21 type) and $YBa_2Cu_3O_7$ (123 type), superconductivity has been found in $La_{3-x}Ba_{3+x}Cu_6O_{14-\delta}$ (336 type). We have found superconductivity in $La_{2.5}Ba_{3.5}Cu_6O_{14}$ and $La_2SrBa_3Cu_6O_{14}$ belonging to the $La_{3-x}Ba_{3+x}Cu_6O_{14}$ system (ref. 35b); T_c in this system increases markedly upto 70K if the samples are heated under oxygen pressure (ref. 34a). Superconductivity in this system was considered unusual since the structure was tetragonal (26) and contained no Cu-O-Cu chains. It is now known that these oxides are also orthorhombic similar to $YBa_2Cu_3O_7$ and possess the Cu-O-Cu chains (ref. 34b). $LaBa_2Cu_3O_{7-\delta}$(x = 1) is itself reported to be orthorhombic with $T_c \simeq$ 90K (34c). It appears orthorhombicity is nearly universal in all the high T_c oxides. Neutron diffraction studies are essential to decide the structures of all such oxides.

In figure 12, the various families of oxides investigated by us for superconductivity are shown. Notice that $YBa_2Cu_3O_7$ (or $LnBa_2Cu_3O_7$) is directly related to $La_{3-x}Ba_{3+x}Cu_6O_{14}$ where x = 1 gives $La_2Ba_4Cu_6O_{14}$ or $LaBa_2Cu_3O_7$. Our very initial strategy in oxide superconductor research was based on $La_{3-x}Ba_{3+x}Cu_6O_{14}$ (ref. 11, 25) and this is now fully justified in view of the relation between the structures of $La_3Ba_3Cu_6O_{14}$ and $Y(Ln)Ba_2Cu_3O_7$. We are investigating $Y_{3-x}(La_{3-x})Ba_{3+x}Cu_6O_{14}$ systems in detail in this laboratory.

Some unusual features of the oxide superconductors

Besides mixed valence, low-dimensionality and marginal metallicity (in the normal state), the new oxide superconductors exhibit some unusual features. While $La_{2-x}Sr_xCuO_4$ seems to exhibit measurable ^{18}O isotope effect, there appears to be essentially no isotope effect in the $YBa_2Cu_3O_7$ system (ref. 54). This means that the traditional BCS theory is not valid in these oxide super-

(21) $La_{2-x}Ba_x(Sr_x)CuO_4$ (SC)

(212) $La_2SrCu_2O_6$

(415) $La_4BaCu_5O_{13}$

(336) $La_{3-x}Ba_{3+x}Cu_6O_{14-\delta}$ (SC)

(123) $LnBa_2Cu_3O_{7-\delta}$ (SC)
(Ln = La, Y, Eu, Gd, Ho, Er etc)
$(Ln,Ba)_{n+1}Cu_nO_{3n+1}$

(212) $Ln_2BaCu_2O_x$
$LnBa_2Cu_2O_x$

(223) $Ln_2Ba_2Cu_3O_x$

Fig. 12. Various families of oxides investigated for superconductivity. SC stands for superconducting. Numbers in parenthesis on the left of each system are the popular designations.

conductors. While electron-phonon interaction could still play an indirect role, the actual nature of the role is not clear at present. The optical gap in $YBa_2Cu_3O_7$ seems to be ~ 200 cm^{-1} (ref. 55, 56). High-energy excitionic bands (around 0.5 eV and 1.0 eV) seem to be present in $La_{2-x}Sr_xCuO_4$ and other oxides; this aspect may be of significance and needs to be further explored.

A large number of papers on theoretical models have appeared in the literature in the last few months, but there is yet no simple model or theory to explain high T_c in the oxide superconductors. It seems best to get good experimental data and look for models later. One of the noteworthy features of the oxide superconductors is that T_c is maximum at a specific value of the nominal Cu^{3+}/Cu^{2+} ratio indicating the importance of the antiferromagnetic energy scale (ref. 16). Chemical models involving the disproportionation of Cu^{2+} to Cu^{1+} and Cu^{3+} (ref. 57, 58) may not be entirely valid since there is no real evidence for the presence of Cu^{3+}. Models based on oxygen holes are to be properly developed.

Electron microscopic studies of $YBa_2Cu_3O_7$ show the presence of domain or twin boundaries (Fig. 13) in addition to other defects (see for example, ref. 37, 44, 59-61). The nature of such domains is of interest. Noting that the coherence length is only a few angstroms in these oxides, some workers have wondered about the origin and implications of such domains. Are they due to the co-occurrence of metallic and insulating phases (with different stoichiometry) arising from phase seperation in such borderline materials? If so, what does the observed structure mean? The twin boundaries or domains such as those in figure 13 seem to be due to different orientations of the $(CuO_2)_\infty$ units (ref. 44, 61). Microstructural features such as domain or twin boundaries could as well be responsible for the transient high T_c (~ 200-300K) behaviour of some samples. Critical currents in these oxides are probably determined by the microstructure.

An interesting property of $YBa_2Cu_3O_7$ in the superconducting state is that it absorbs electromagnetic radiation over a wide range of frequencies from a few MHz to a few GHz (ref. 62). The absorption is extremely sensitive to temperature, particle size and the magnetic field and crucially depends on the presence of ambient oxygen (Fig. 14). It is suggested that Josephson junctions formed by oxygen and the superconducting grains may be responsible for this effect.

Recent specific heat measurements on $YBa_2Cu_3O_7$ show $\Delta c/\gamma T$ to be ~ 48 $mJ/mol.K^2$ giving rise to a $\Delta c/\gamma T_c$ of 1.33; there is a large temperature dependence of the Debye temperature, probably due to the large vibrational amplitudes of the loosely bound O_1 and O_2 oxygens in the

Fig. 13. Bands in electron micrographs of $YBa_2Cu_3O_7$ due to twins (From Rao et al, ref. 37).

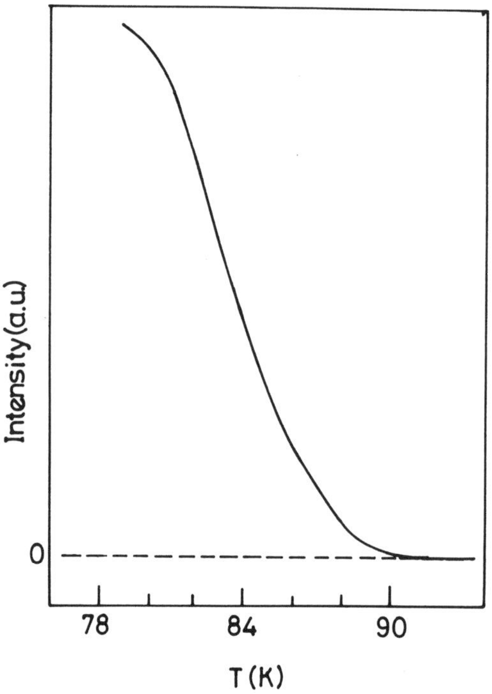

Fig. 14. Temperature-variation of the intensity of absorption recorded at 9.1 GHz (From Bhat et al, ref. 62).

Preparative aspects

Synthesis of superconducting oxides provides many opportunities and challenges. The most common method employed is the ceramic method involving the high temperature reaction of an appropriate mixture of oxides and carbonates. Citrates, nitrates and oxalates have also been employed with no definitive advantage in the case of $La_{2-x}Ba_x(Sr_x)CuO_4$ and $YBa_2Cu_3O_7$ (ref. 36). In the synthesis of $YBa_2Cu_3O_7$ and related oxides, we have found it convenient to employ BaO_2 instead of $BaCO_3$. Excess CuO sharpens the transition. Partial fluorination also seems to increase T_c.

The precursor method (ref. 64) would be an ideal alternative route for the synthesis of these oxides. Unfortunately, no single precursor compound or a precursor solid solution seems possible. Precipitation from high alkaline media has also not been successful. The sol-gel route in the traditional sense seems difficult because it is not possible to get the alkoxides of Y or Ln, Ba and Cu^{2+} to form a homogeneous mix. It has however been possible to prepare gels by adding appropriate amounts of yttrium nitrate and copper acetate to a known amount of barium hydroxide (ref. 42). It may be worthwhile examining the feasibility of using cuprous alkoxides (instead of cupric).

Heating under high oxygen pressure is found convenient to make stoichiometric or oxygen-excess samples. Thus, superconducting $La_2CuO_{4+\delta}$ and $La_{3-x}Ba_{3+x}Cu_6O_{14-\delta}$ have been made by treatment under high oxygen pressure. Plasma oxidation offers another convenient way of preparing such materials. Thus, $La_2CuO_{4+\delta}$ has been prepared by plasma oxidation; $YBa_2Cu_3O_{7-\delta}$ ($\delta \simeq 0.5$ - 1.0) is readily oxidised to the near-stoichiometric phase by plasma oxidation (ref. 42). Different workers have prepared $YBa_2Cu_3O_{7-\delta}$ by employing different treatments. The time of annealing (soaking) in oxygen varies quite markedly, from a few hours to a few days. In general, most preparations give samples with T_c in the 90-95K region. Some workers have reported much higher T_c values (>100K) in samples annealed or cycled differently (ref. 65, 66); for example, soaking samples of $YBa_2Cu_3O_7$ for extended periods in a helium or a nitrogen atmosphere has been reported to increase T_c upto 160K, but the effect does not appear to be permanent; these results are generally based on electrical resistivity measurements. The chemistry of such treatments is not clear. One possibility is that the oxygens in the Cu-O-Cu chains get perfectly ordered causing an increase in T_c. It is also worthwhile to examine Meissner effect in such samples.

Samples of $YBa_2Cu_3O_7$ prepared by low-temperature methods generally exhibit low T_c. Part of the problem could be the small particle size which in turn determines the nature of grain boundaries.

Material parameters

Since these new oxide superconductors are ceramic materials, there are inherent problems in obtaining them in the desired shape and form. However, there has already been considerable success in making wires, tapes and films. During the processing of these materials, there is a tendency to lose the labile oxygen which renders them non-superconducting; this can be solved by reheating the processed material in oxygen or by proper insulation in the initial stage itself. Another difficulty with these oxide ceramics is their chemical instability. Extended exposure to the laboratory atmosphere seems to degrade the materials. In the case of $YBa_2Cu_3O_7$, hydroxides and carbonates of the component metals and carbonization of the surface has been noticed (ref. 66, 67). Success in technological applications will depend on several factors such as material parameters and cost advantage.

Since $YBa_2Cu_3O_7$ is superconducting well above the liquid nitrogen temperature, it holds much promise. We shall briefly examine the material parameters of this ceramic oxide. Electronic properties of this oxide are anistropic (just as the structure). Some of its important properties as summarized by Malozemoff et al (ref. 68) are as follows:

1. It is a Type II Superconductor
2. Hall carrier density : $4 \times 10^{21} cm^{-3}$ (for a material of resistivity $\sim 400\ \mu\Omega$ cm just above the T_c).
3. $dH_{c2}/dT = 2T/K$
   BCS coherence length $\simeq 1.4$ nm
   London Penetration depth $\simeq 200$ nm
   Mean free path $\simeq 1.2$ nm
4. $H_c(O) \simeq 1T$
5. $H_{c2}(O) \simeq 120T$
6. Critical current density : in ceramic samples, $\sim 10^3 A/cm^2$ at 77K; in films (on $SrTiO_3$) $\sim 10^5 A/cm^2$ at 77K; at 4.2K $\sim 10^6 A/cm^2$ (in crystals and films)

Depairing current density $10^7 - 10^8 A/cm^2$ (estimated)

Concluding Remarks

The tremendous possibilities of applications of $YBa_2Cu_3O_7$ and other yet-to-be-discovered high T_c ceramic oxides make this area of research most exciting. Clearly, the high-temperature oxide superconductors constitute the most sensational discovery since the transistor. It has given a great boost to research in solid state chemistry and there is little doubt that within a short time frame, some technological applications will become a reality even with the existing materials. The future offers unlimited vistas and opportunities in metal oxide research as well as in electromagnetic technology. Synthesis of materials with T_c's close to room temperature is no longer a remote possibility. Some of the oxides showing indications of T_c around 300K have been found, but alas they are all unstable. A stable room-temperature superconductor is probably lying around somewhere in the laboratory of a solid state chemist.

Acknowledgements

The author thanks the Department of Science and Technology, Government of India, the University Grants Commission and the US National Science Foundation for support of this research. He also expresses his deep sense of gratitude to his colleagues and students for their enthusiastic participation in the research activities related to oxide superconductors.

References

1. C.N.R. Rao and J. Gopalakrishnan, New Directions in Solid State Chemistry, Cambridge University Press, 1986; C.N.R. Rao, K.J. Rao and J. Gopalakrishnan, Annual Reports, Royal Society of Chemistry, London, Part C, 1985
2. D.C. Johnston, H. Prakash, W.H. Zachriasen and R. Viswanathan, Mat. Res.Bull. 8, 777 (1973)
3. A.W. Sleight, J.L. Gilson and F.E. Bierstedt, Solid State Commun. 17, 27 (1975)
4. J.G. Bednorz and K.A. Muller, Z. Phys. B64, 187 (1986)
5. R.J. Cava, R.B. van Dover, B. Batlogg and E.A. Reitman, Phys. Rev. Lett. 58, 408 (1987)

6. C.W. Chu et al., Phys. Rev. Lett. 58, 405 (1987)
7. S. Uchida, H. Takagi, K. Kitazenwa and S. Tanaka, Jap. J. Appl. Phys. 26, L1 (1987)
8. C.N.R. Rao and P. Ganguly, Curr. Sci. 56, 47 (1987)
9. P. Ganguly, R.A. Mohan Ram, K. Sreedhar and C.N.R. Rao, Solid State Commun. 62, 807 (1987)
10. M.K. Wu, J.R. Ashburn, C.J. Torng, P.H. Hor, R.L. Meng, L. Gao, Z.H. Huang, Y.Q. Wang and C.W. Chu, Phys. Rev. Lett. 58, 908 (1987)
11. C.N.R. Rao, P. Ganguly, A.K. Raychaudhuri, R.A. Mohan Ram and K. Sreedhar, Nature, 326, 856 (1987)
12. T. Siegrist, S. Sunshine, D.W. Murphy, R.J. Cava and S.M. Zahurak, Phys. Rev. B35, 7137 (1987)
13. C.N.R. Rao, (Proc. Adriatico Conference, Trieste), Int. J. Mod. Phys. B1, No. 3 (1987)
14. P. Ganguly and C.N.R. Rao, J. Solid State Chem. 53, 193 (1984)
15. R.A. Mohan Ram, P. Ganguly and C.N.R. Rao, Phase Transitions, 10, 107 (1987)
16. K. Sreedhar, T.V. Ramakrishnan and C.N.R. Rao, Solid State Commun, 63, 835 (1987)
17. C.N.R. Rao and P. Ganguly, in The Metallic and the Non-metallic States of Matter (Eds. P.P. Edwards and C.N.R. Rao), Taylor and Francis, London, 1987
18. P. Day, M. Resseinsky, K. Prassides, W.I.F. David, O. Moze and A. Soper, J. Phys. C Solid State, 20, L429 (1987)
19. D.M. Paul, G. Balakrishna, N.R. Bernhoeft, W.I.F. David and W.T.A. Harrison, Phys. Rev. Lett., 58, 1976 (1987)
20. H. Thomann, D.C. Johnston, P.J. Tindall, D.P. Goshorn and R.A. Klemm, Phys. Rev. Lett., In Print
21. R.J. Birgeneau, C.Y. Chen, D.R. Gabbe, H.P. Jenssen, M.A. Kastner, C.J. Peters, P.J. Picone, T. Thio, T.R. Thurston and H.L. Tuller, To be published.
22. (a) S. Mitsuda, G. Shirane, S.K. Sinha, D.C. Johnston, M.S. Alvarez, D. Veknin and D.E. Moncton, To be published
(b) G. Shirane, Y. Endoh, R.J. Birgeneau, M.A. Kastner, Y. Hidaka, M. Oda, M. Suzuki and T. Murakami, To be published
23. J. Beille, R. Cabanel, C. Chaillout, B. Chevalier, G. Demazeau, F. Deslands, J. Etorneau, P. Le Jay, C. Michel, J. Provost, B. Raveau, A. Sulpice, J. Tholence and R. Tournier, CR Acad Sci Paris, 18, 304 (1987)
24. S.M. Fine, M. Greenblatt, S. Simizu and S.A. Friedberg, ACS Symposium Series 351, 1987
25. P. Ganguly, A.K. Raychaudhuri, K. Sreedhar and C.N.R. Rao, Pramana-J. Phys. 27, L229 (1987)
26. L. Er-Rakho, C. Michel, J. Provost and B. Raveau, J. Solid State Chem, 37, 151 (1981)
27. A.K. Raychaudhuri, K. Sreedhar, R.A. Mohan Ram, P. Ganguly and C.N.R. Rao, Phil. Mag. Lett., 55, 29 (1987)
28. R.J. Cava, B. Batlogg, R.B. van Dover, D.W. Murphy, S. Sunshine, T. Siegrist, J.P. Remeika, E.A. Rietman, S. Zaurak and G.P. Espinosa Phys. Rev. Lett. 58, 1676 (1987)
29. R.A. Mohan Ram, N.Y. Vasanthacharya and C.N.R. Rao, J. Solid State Chem. 69, 186 (1987)
30. J.M. Tarascon, W.R. McKinnon, L.H. Greene, G.W. Hull and E.M. Vogel, Phys. Rev. B36, 226 (1987)
31. J.M. Tarascon et al. Proc. MRS Meeting, Anaheim, 1987.
32. G. Xiao, F.H. Streitz, A. Gavrin and C.L. Chien, Solid State Commun, 63, 817 (1987).
33. C.C. Toradi, E.M. McCarron, M.A. Subramanian, H.S. Horowitz, J.B. Michel, A.W. Sleight and D.E. Cox, ACS Symposium Series 351, 1987.
34. (a) D.B. Mitzi, A.F. Marshall, J.Z. Sun, D.J. Webb, M.R. Beasley, T.H. Geballe and A. Kapitulnik, To be published.
(b) C.U. Segre, B. Dabrowski, D.G. Hinks, K. Zhang, J.D. Jorgensen, M.A. Beno and I.K. Schuller, Nature, 329, 227 (1987).
(c) A. Maeda et al. Jap. J. Appl. Phys. 26, L1366 (1987).
35. (a) R.A. Mohan Ram, K. Sreedhar, A.K. Ray Chaudhuri, P. Ganguly and C.N.R. Rao, Phil. Mag. Lett., 55, 257 (1987).
(b) A.K. Ganguly, L. Ganapathi, K. Sreedhar, R.A. Mohan Ram, P. Ganguly and C.N.R. Rao, Pramana-J. Phys. 29, L335 (1987).
36. V. Bhat, A.K. Ganguly, K.S. Nanjunda Swamy, R.A. Mohan Ram, J. Gopalakrishnan and C.N.R. Rao, Phase Transitions, 10, 87 (1987).
37. C.N.R. Rao, P. Ganguly, K. Sreedhar, R.A. Mohan Ram and P.R. Sarode, Mat. Res. Bull., 22, 849 (1987)
38. C.N.R. Rao and P. Ganguly, Jap. J. Appl. Phys., 26, L882 (1987)
39. C.N.R. Rao, P. Ganguly, J. Gopalakrishnan and D.D. Sarma, Mat. Res. Bull., 22, 1159 (1987)
40. D.C. Johnston, A.J. Jacobson, J.M. Newsam, J.T. Lewandowski, D.P. Goshorn, D. Xie and W.B. Yelon, ACS Symposium Series 351, (1987)
41. D.W. Murphy, S.A. Sunshine, P.K. Gallagher, H.M.O'Bryan, R.J. Cava, B. Batlogg, R.B. van Dover, L.F. Scheemeyer and S.M. Zahurak, ACS Symposium Series 351, 1987
42. J.M. Tarascon, P. Barboux, B.G. Bagley, L.H. Greene, W.R. McKinnon and G.W. Hull, ACS Symoposium Series 351, 1987
43. S.W. Keller, K.J. Leary, T.A. Faltens, J.N. Michaels and A.M. Stacey, ACS Symposium Series 351, 1987
44. C.N.R. Rao, L. Ganapathi and R.A. Mohan Ram, Mat. Res. Bull., In Print
45. W.I.F. David, W.T.A. Harrison, J.M.F. Gunn, O. Moze, A.K. Soper, P. Day, J.D. Jorgensen, M.A. Beno, D.W. Capone, D.G. Hinks, I.K. Schuller, L. Soderholm, C.U. Segre, K. Zhang

and J.D. Grace, Nature, 327, 310 (1987)
46. P. Bordet, C. Chaillout, J.J. Capponi, J. Chenavas and M. Marezio, Nature, 327, 687 (1987)
47. A. Santoro, S. Miraglia, F. Beech, S.A. Sunshine, D.W. Murphy, L.F. Schneemeyer and J.V. Waszczak, Mat. Res. Bull., 22, 1009 (1987)
48. F. Beech, S. Miraglia, A. Santoro and R.S. Roth, Phys. Rev., B35, 8778 (1987)
49. J.D. Jorgensen, M.A. Beno, D.G. Hinks, L. Soderholm, K.J. Volin, R.L. Hitterman, J.D. Grace, I.K. Schuller, C.U. Segre, K. Zhang and M.S. Kleefisch, Phys. Rev., B35, 7915 (1987)
50. (a) J. Thiel, S.N. Song, J.B. Ketterson and K.R. Poeppelmeier, ACS Symposium Series 351, 1987
(b) K. Fueki, K. Kitazawa, K. Kishio, T. Hasegawa, S. Uchida, H. Takagi and S. Tanaka, ACS Symposium Series 351, 1987
51. R.J. Cava, B. Batlogg, C.H. Chen, E.A. Rietman, S.M. Zahurak and D. Werder, Nature, 329, 423 (1987)
52. (a) D.D. Sarma, K. Sreedhar, P. Ganguly and C.N.R. Rao, Phys. Rev. B36, 2371 (1987)
(b) D.D. Sarma and C.N.R. Rao, J. Phys. C. Solid State, 20, L659 (1987)
53. C.N.R. Rao, P. Ganguly, M.S. Hegde and D.D. Sarma, J. Am. Chem. Soc., In Print
54. B. Batlogg, R.J. Cava, A. Jayaraman, R.B. van Dover, G.A. Kowrokis, S. Sunshine, D.W. Murphy, L.W. Rupp, H.S. Chen, A. White, K.T. Short, A.M. Mujsce and R.A. Rietman, Phys. Rev. Lett., 58, 2333 (1987)
55. M. Cardona, L. Genzel, R. Liu, A. Wittlin, H. Mattausch, F. Garcia-Alvarado and F. Gracia-Gonazalez, Solid State Commun, In Print
56. L. Genzel, A. Wittlin, J. Kuhl, H. Mattausch, W. Bauhofer and A. Simon, Solid State Commun., 63, 843 (1987)
57. A.W. Sleight, ACS Symposium Series 351, 1987
58. A. Simon, Angew. Chem. Intnl. Ed. (Engl) 26, 579 (1987); L. Pauling, Phys. Rev. Lett. 59, 225 (1987)
59. G.N. Subbanna, P. Ganguly and C.N.R. Rao, Mod. Phys. Lett B, 1, 155 (1987)
60. Y. Matsui, E.T. Muranachi, A. Ono, S. Horiuchi and K. Kato, Jap. J. Appl. Phys., 26, L777 (1987)
61. B. Raveau, C. Michel and M. Hervieu, ACS Symposium Series 351, 1987
62. S.V. Bhat, P. Ganguly, T.V. Ramakrishnan and C.N.R. Rao, J. Phys. C. Solid State, 20, L559 (1987)
63. K. Kadowaki et al, To be published
64. C.N.R. Rao and J. Gopalakrishnan, Accts. Chem. Res., 20, 228 (1987)
65. I.K. Gopalakrishnan, J.V. Yakhmi and R.M. Iyer, Nature, 327, 604 (1987)
66. (a) D.N. Matthews, A. Bailey, R.A. Vaile, G.J. Russell and K.N.R. Taylor, Nature, 328, 786 (1987)
(b) R.N. Bhargava, S.P. Herko and W.N. Osborne, Phys. Rev. Lett., 59, 1468 (1987)
67. B.G. Hyde, J.G. Thompson, R.L. Withers, J.G. FitzGerald, A.M. Stewart, D.J.M. Bevan, J.S. Anderson, J. Bitmead and M.S. Paterson, Nature, 327, 402 (1987)
68. A.P. Malzemoff, W.J. Gallagher and R.E. Schwall, ACS Symposium Series 351, 1987

"It is a capital mistake to theorize before you have all the evidence. It biases the judgement"

Sherlock Holmes

The role of crystal chemistry for generation of high T_c oxide superconductors

Bernard Raveau, Claude Michel and Maryvonne Hervieu

Laboratoire de Cristallographie et Sciences des Matériaux, ISMRa, Campus 2, Boulevard du Maréchal Juin, Université de Caen, 14032 Caen Cedex, France

Abstract - The factors which govern the high critical temperature of the mixed-valence copper oxides are examined in the light of the recent experimental results. The mixed-valence of copper, the influence of the low dimensionality of the structure, the deviation from oxygen stoichiometry are particularly discussed for the following phases : $La_{2-x}Sr_xCuO_{4-y}$, La_2CuO_4, $YBa_2Cu_3O_{7-\delta}$, $YBa_2Cu_{3-x}Pd_xCu_3O_{7-\delta}$, $LaBa_2Cu_3O_{7-\delta}$. The HREM observations show the complex the crystal chemistry of the "123" oxides.

INTRODUCTION

It is well known since the discovery of the superconductivity by Kamerlingh Onnes (ref. 1) in mercury, that this phenomenon will mainly appear in compounds characterized by a metallic conductivity. In this respect, the evidence for superconductivity in oxides with metallic properties is not unexpected. It has indeed been observed the last fifteen years that the bronzes Rb_xWO_3 and Cs_xWO_3 (ref. 2-3), the oxide $LiTi_2O_4$ (ref. 4-5) and $Ba(Pb_{1-x}Bi_x)O_3$ (ref. 6) were superconductors with critical temperatures ranging from 0.6K to 13.7K. The copper oxides $La_{2-x}A_xCuO_{4-x/2+\delta}$ (ref. 7) whose metallic properties have been observed as soon as 1982 (ref. 8) belong to this family of superconductors (ref. 9). But the surprising feature which was observed by Bednorz and Muller in those oxides concerns their high critical temperature, close to 35K. The crystal chemistry of these materials is rather complex in spite of the apparent easiness of their synthesis. We analyze here, the different factors which govern the superconductivity in mixed-valence copper oxides related to the pervoskite and we present the recent developments which were carried out in Caen in this field.

The first factor which governs the superconductivity in oxides deals with the mixed-valence of the metallic element which participates to the edification of the host lattice MO_n. This property is not in fact specific to superconductivity since it is well known, from solid state chemists, that mixed-valence is necessary for the obtention of metallic properties of oxygen bronzes which are explained in terms of a band structure resulting from the overlapping of the d(M) orbitals and p(O) orbitals. The mixed-valence Cu(II)-Cu(III) which characterizes copper oxides differs from that observed in other bronzes such as W(V)-W(VI) or Ti(IV)-Ti(III) oxides in two points : copper takes generally two sorts of coordinations simultaneously in those metallic compounds contrary to tungsten or titanium and one observes a delocalization of the holes over the CuO_n framework, whereas a delocalization of electrons is generally observed in other bronzes. Clearly, the mixed-valence Cu(II)-Cu(III) is formal like in other bronzes, but what seems to be specific to copper is the fact that the holes tend to spend much more time on oxygen than on copper leading to the configuration $3d^9\underline{L}$ ($\underline{L}$ = ligand hole) rather than $3d^8$ for Cu(III) (ref. 10). This particular behavior of copper and especially its ability to take several coordinations is closely related to its Jahn Teller effect. The latter is also observed for manganese but, in that case, the mixed-valence Mn(IV)-Mn(III) does not involve a delocalization of the electrons, so that Mn^{3+} and Mn^{4+} ions can be identified, leading to magnetic properties. Nickel, should exhibit intermediate properties for the mixed-valence Ni(III)-Ni(II) owing to its position in the periodic table. The stabilization of Cu(III) is not a trivial problem : we have previously shown (ref. 7, 11) that the introduction of basic cations and especially of alkaline earth ions may induce the formation of Cu(III). Nevertheless the presence of basic cations is not sufficient for the stabilization of Cu(III), the nature of the structure must be considered. It is the case of the perovskite structure in which the possibility of formation of anionic vacancies (ref. 12) allows additional oxygen to enter into the lattice, i.e. a partial oxidation of Cu(II) into Cu(III).

The second factor which governs the superconductivity of copper oxides concerns the low dimensionality of the structure. The importance of low dimensionality in those superconductors has been pointed out for the first time by the theory of Labbé and Bok (ref. 13). Many other theories do not take into account this structural feature for the explanation of superconductivity in copper oxides. However the experimental results are clearly in favour of the great influence of this factor upon superconductivity in these oxides. Among the mixed-valence copper oxides actually known, two oxides exhibit a three-dimensional framework : the oxide $La_4BaCu_5O_{13+\delta}$ (ref. 14) has its host-lattice formed of corner-sharing CuO_6 octahedra and CuO_5 pyramids (Fig. 1) whereas in the oxide $La_{8-x}Sr_xCu_8O_{20}$ (ref. 15), the CuO_6 octahedra, CuO_5 pyramids and CuO_4 square planar groups share their corners (Fig. 2). Both structures derive from the perovskite by elimination of rows of oxygen atoms parallel to [001] and both compounds are characterized by a metallic conductivity, but they do not exhibit any sign of superconductivity down to 4K. On the opposite the oxides $La_{2-x}A_xCuO_{4-x/2+\delta}$ (A = Ca, Sr, Ba) (ref. 7, 16) and $YBa_2Cu_3O_{7-\delta}$ (ref. 17-18) which exhibit superconductivity below about 40K (ref. 9-19, 20) and 90K (ref. 21-22) respectively can be described as layer structures. The first family belongs to the K_2NiF_4 structure (Fig. 3) and corresponds to the intergrowth of insulating SrO-type layers and of single oxygen-deficient perovskite layers known for their metallic or semi-metallic conductivity (ref. 23). The oxygen deficient perovskite $YBa_2Cu_3O_{7-\delta}$ called "123" exhibits also two-dimensional character ; the structure consists of triple layers of polyhedra, $[Cu_3O_7]_\infty$, whose cohesion is ensured by "yttrium" planes (Fig. 4). However the bidimensional character of

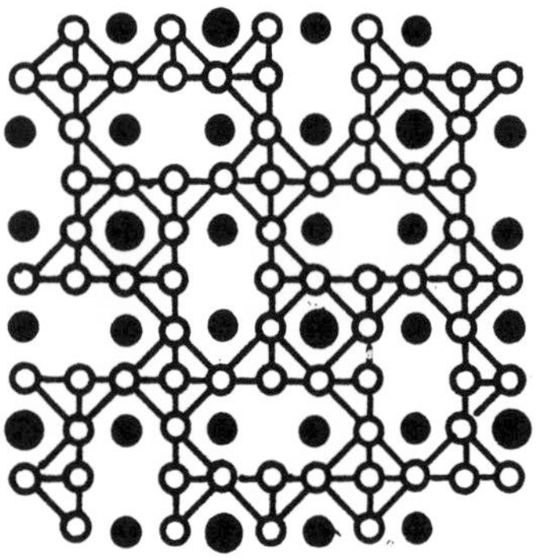

Fig. 1. Projection onto [001] of $Ba_4LaCu_5O_{13}$

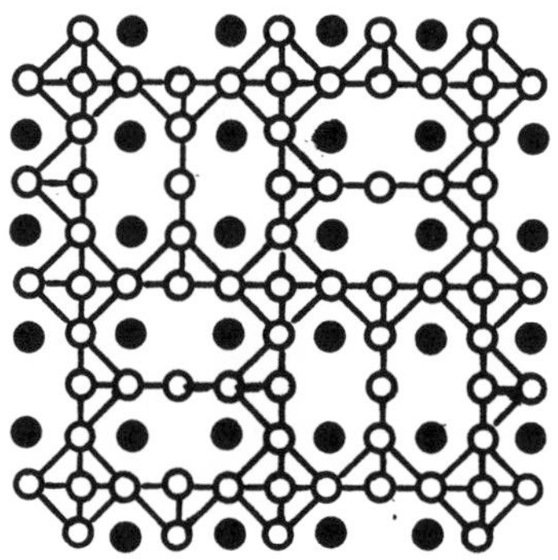

Fig. 2. Projection onto [001] of $La_{6.4}Sr_{1.6}Cu_8O_{20}$ (x = 1.6).

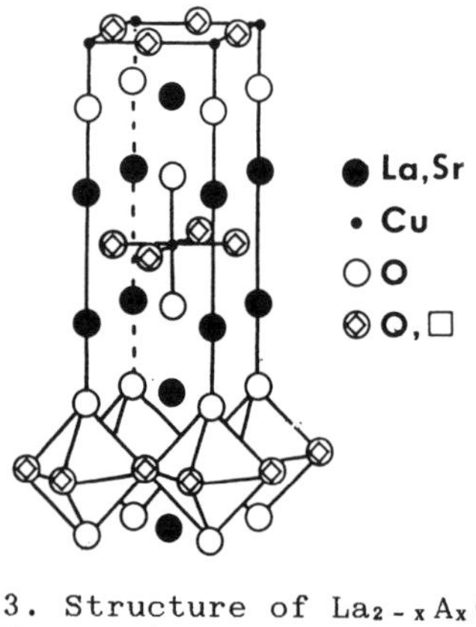

Fig. 3. Structure of $La_{2-x}A_xCuO_{4-x/2+\delta}$ (K_2NiF_4 type).

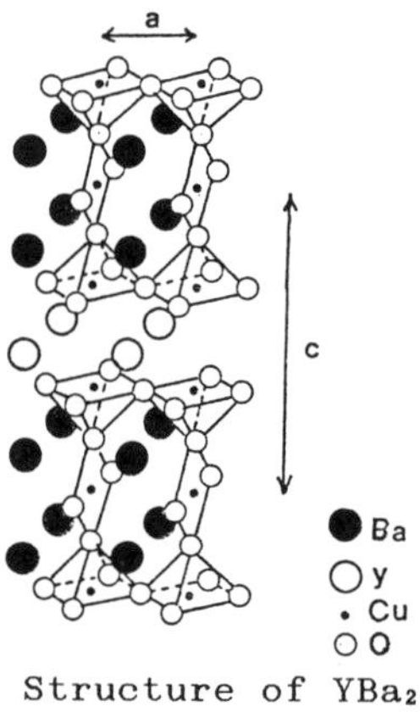

Fig. 4. Structure of $YBa_2Cu_3O_7$.

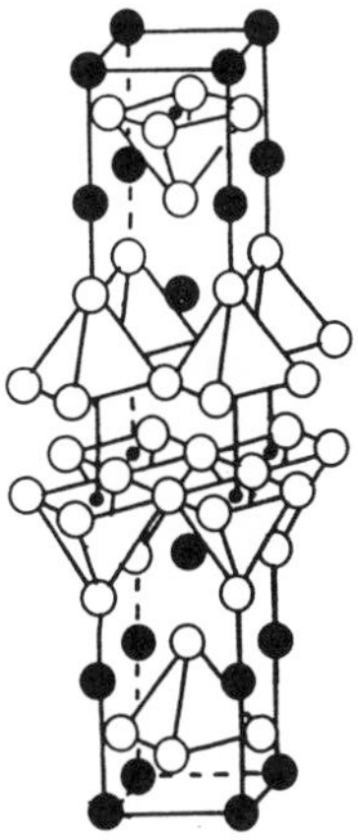

Fig. 5. Structure of $La_2SrCu_2O_6$ ($x = 0$).

this structure is less pronounced compared to the first one, in spite of its higher T_c : each triple layer is indeed built up from two single $[CuO_{2.5}]_\infty$ layers of corner-sharing pyramids connected through $[CuO_2]_\infty$ files of corner sharing CuO_4 square planar groups. Thus it seems that the unidimensional character of the $[CuO_2]_\infty$ rows could play a role in the increase of T_c. The fact that the mixed-valence oxide $La_{2-x}Sr_{1+x}Cu_2O_{6+x/2+\delta}$ (ref. 24) does not exhibit any superconductivity is in agreement with this point of view. The two dimensional character of this structure (Fig. 5) is intermediate between that of $La_{2-x}A_xCuO_{4-x/2+\delta}$ and that of $YBa_2Cu_3O_{7-\delta}$. One can recognize double oxygen-deficient perovskite layers which are intergrown with stoichiometric insulating SrO-type layers. However the absence of $[CuO_2]_\infty$ rows could explain the absence of superconductivity in this oxide which exhibits metallic or semi-metallic properties (ref. 8). The recent study of the substitution of palladium for copper in $YBa_2Cu_3O_{7-\delta}$ brings an interesting information about the role played by the $[CuO_2]_\infty$ rows. One can indeed isolate the isostructural oxide $YBa_2Cu_{3-x}Pd_xP_{7-\delta}$ (ref. 25) in which Pd(II) partially replaces copper in an ordered manner, i.e. in the CuO_4 groups of the $[CuO_2]_\infty$ rows so that the structure can be described as $[Cu_{3-x}Pd_xO_7]_\infty$ triple layers built up from two pyramidal $[CuO_{2.5}]_\infty$ layers linked through $[Cu_{1-x}Pd_xO_2]_\infty$ rows. Contrary to $YBa_2Cu_3O_{7-\delta}$ this oxide is no more a superconductor : a large resitive transition is observed arround 70K, but the diamagnetism smaller than 1 % indicates that it is likely due to $YBa_2Cu_3O_{7-\delta}$ as an impurity. Clearly, Pd(II) kills the superconductivity in spite of its $4d^8$ configuration similar to that of Cu(III) ; this points out the particular configuration of Cu(III) which would be $3d^9\underline{L}$ rather than $3d^8$ (ref. 10) and suggests that the $[CuO_2]_\infty$ rows of square planar groups in $YBa_2Cu_3O_{7-\delta}$ would be mainly occupied by Cu(III) in agreement with the model recently developed by Labbé (ref. 26). Moreover, it appears that the existence of the mixed-valence Cu(II)-Cu(III) in the pyramidal layers of $YBa_2Cu_{3-x}Pd_xO_{7-\delta}$ does not induce any superconductivity. Thus $[Cu^{III}O_2]_\infty$ rows of square lanar groups seem to be mainly responsible for superconductivity in $YBa_2Cu_3O_{7-\delta}$. Nevertheless the interaction between these rows and the $[CuO_{2.5}]_\infty$ layers may also play a role in the superconducting properties of this oxide.

The low dimensionality is not the only structural parameter which governs the superconductivity. No superconductivity was indeed observed for the oxides $La_{2-x}A_xCuO_{4-y}$ with $x < 0.10$, in spite of the mixed-valence of copper and of the bidimensionality of the structure. Several authors have explained this phenomenon in terms of symmetry of the K_2NiF_4-type structure : one observes indeed an orthorhombic symmetry for $x < 0.10$, whereas the symmetry of the superconducting phase is tetragonal or pseudo-tetragonal. The recent discovery of superconductivity in La_2CuO_4 (ref. 27) shows that the symmetry is not much involved but that the deviation from stoichiometry can drastically influence the superconductivity. Stoichiometric La_2CuO_4 was claimed by many authors not to be a semiconductor up to february 1987 in agreement with the only presence of Cu(II). However the application of an oxygen pressure of about 500 bars, allows the superconductivity to be obtained, in the bulk with a $T_c \approx 37K$. This phenomenon is easily explained by the fact that the non superconducting oxide which involves only Cu(II) contains Schottky defects symmetrical on all sites according to the formula $La_{2-2\epsilon}Cu^{II}_{1-\epsilon}O_{4-4\epsilon}$, whereas an intercalation of oxygen on the anionic sites allows the partial oxydation of Cu(II) into Cu(III), i.e. the mixed-valence of copper to be realized leading to the superconductor $La_{2-2\epsilon}Cu^{II,III}_{1-\epsilon}O_4$. This effect of non stoichiometry is enhanced by changing the molar ration La:Cu as shown from Fig. 6. An excess of copper oxide (x > 1) allows the formation of additional lanthanum and oxygen vacancies ($La_{2-\delta}Cu^{II}O_{4-3\delta/2}$) ; the presence of oxygen vacancies favours the introduction of oxygen, even under normal pressure leading to the limit formulation $La_{2-\delta}Cu^{II,III}O_4$, in which mixed valence and bidimensionality explain the superconductivity. An excess of lanthanum oxide (x < 1) is also favourable to the appearance of superconductivity, due to the formation of additional anionic vacancies ($La_2Cu^{II}_{1-\delta}O_{4-\delta}$), leading after oxidation to the limit formulation $La_2Cu^{II,III}_{1-\delta}O_4$, also characterised by the mixed-valence of copper. However in this latter case, the presence of a greater number of vacancies on the copper sites, tends to destroy the superconductivity in agreement with the decrease of Tc observed for x < 1 (Fig. 7).

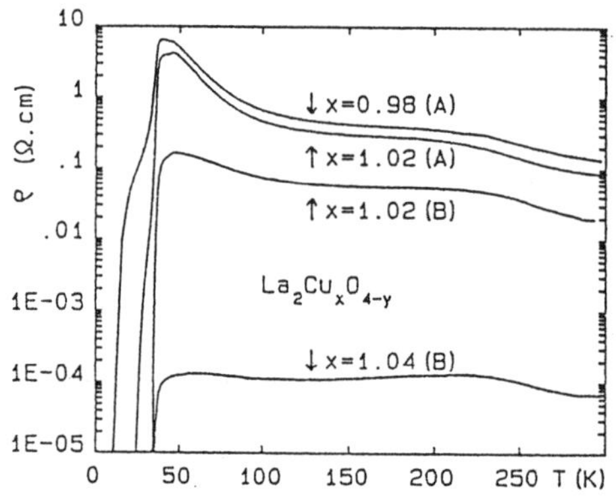

Fig. 6. The logarithm of the resistivity of $La_2Cu_xO_{4-y}$ versus T for x = 0.98, 1.02 and 1.04. The sample A was annealed at 450°C under a pressure of 1 bar of oxygen and B at 900°C under 500 bars of oxygen.

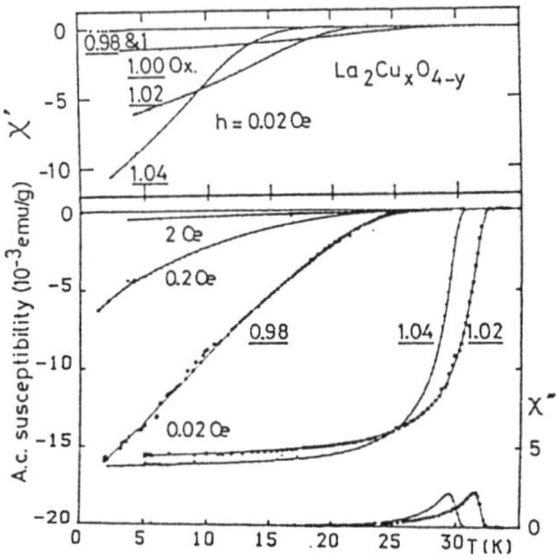

Fig. 7. The ac susceptibility X' versus T for different X values after annealing at 450°C under 1 bar oxygen The amplitude of the ac field is 0.02 oe.

The disappearance of superconductivity in $YBa_2Cu_3O_{7-\delta}$ during the transition from the orthorhombic to the tetragonal symmetry is another proof of the importance of the $[CuO_2]_\infty$ rows in the superconducting properties of this compound. The structure of the tetragonal semiconducting phase (Fig. 8) obtained by quenching the compound from 950°C to room temperature can be

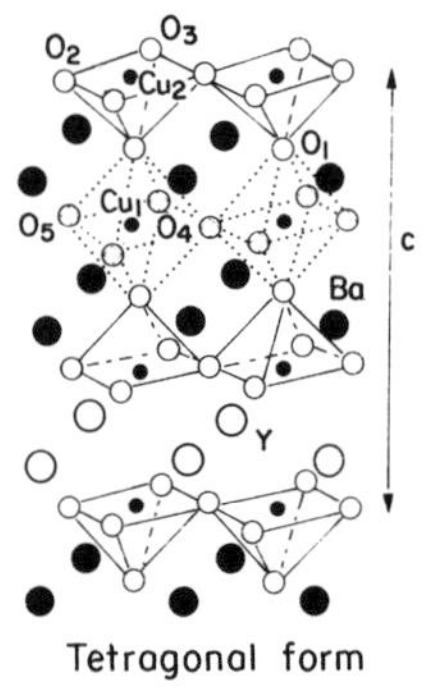

Fig. 8. Structure of the tetragonal semiconductor $YBa_2Cu_3O_{6.5}$.

described, from the X-ray and neutron diffraction studies (ref. 28-29), as formed from triple layers $[Cu_3O_{7-\delta}]_\infty$, where cohesion is ensured by planes of yttrium ions. The triple layers are built up from two $[CuO_{2.5}]_\infty$ single layers linked through an oxygen deficient layer $[CuO_{2-\delta}]_\infty$. Thus the structure of the tetragonal form differs mainly from the orthorhombic one only by the nature of the $[CuO_{2-\delta}]_\infty$ layer which does not present $[CuO_2]_\infty$ rows any more, but can be described as an oxygen-deficient octahedral layer whose anionic vacancies are statistically distributed in the basal plane of the octahedra. However the powder neutron diffraction study does not reflect the actual structure : one observes a high background which suggests the presence of an amorphous phase, and high values of the anisotropic thermal factors of the oxygen atoms of the basal plane. The electron microscopy study (ref. 28) of this "phase" confirms its inhomogeneous character. The electron diffraction study of quenched tetragonal samples of nominal composition $YBa_2Cu_3O_{6.50}$ shows the existence of an amorphous part and, for the crystallized grains, the presence of diffusion streaks along c and variation of parameters from one crystal to the other. The high resolution electron microscopy study shows the existence of crystals systematically coated with an amorphous layer of 40 to 80 Å (Fig. 9), and variations of contrast from one zone of a crystal to the other, which result from a local variation of the oxygen content as shown from a careful simulation of the images (Fig. 10). Clearly, the chemical analysis does not reflect the composition of the crystals, as shown here where the refinements

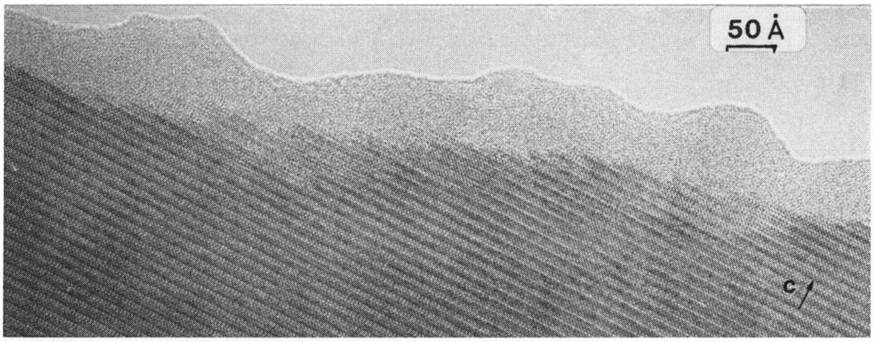

Fig. 9. Typical HREM image of a crystal coated with an amorphous layer

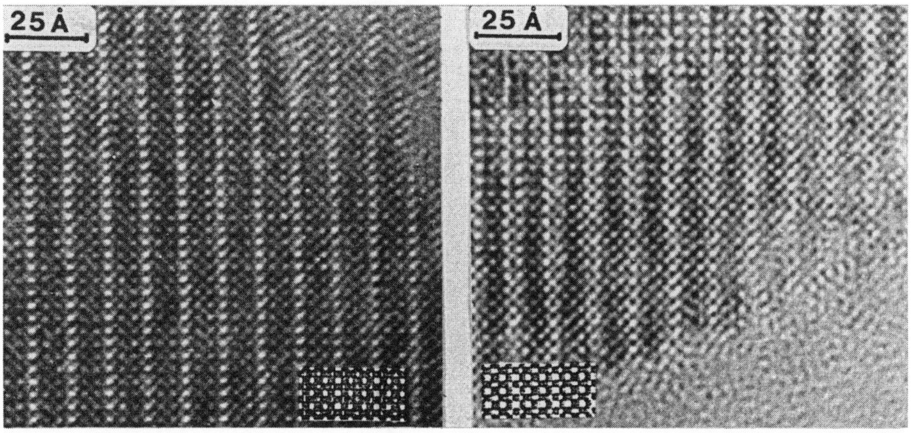

Fig. 10. [100] HREM image showing different oxygen content areas : poor (a) and rich (b). Inset images calculated with D_f = -150 Å, t = 30 Å correspond to the formulation $YBa_2Cu_3O_{6.2}$ (a) and $YBa_2Cu_3O_7$ (b).

by neutron diffraction converge for the composition $YBa_2Cu_3O_{6.25}$ (ref. 27). From these studies it appears that such crystals cannot involve only Cu(II) and Cu(I), since this would lead for a part of Cu(II) to a coordination smaller than four, which is not likely. A more probable interpretation which agrees with HREM observations consists in considering the existence of two sorts of zones in the crystals : $Ba_2YCu_3O_7$ zones (Fig. 4) involving the mixed-valence Cu(II)-Cu(III) and $Ba_2YCu_3O_6$ zones whose structure has previously been established (ref. 30) (Fig. 11) in which Cu(II) and Cu(I) exhibit the pyramidal and the twofold coordination respectively. Thus the crystal of composition $YBa_2Cu_3O_{6.25}$ would better be formulated :

$$(YBa_2Cu^{II}{}_2Cu^{III}O_7)_{0.25}(YBa_2Cu^{II}{}_2Cu^{I}O_6)_{0.75}.$$

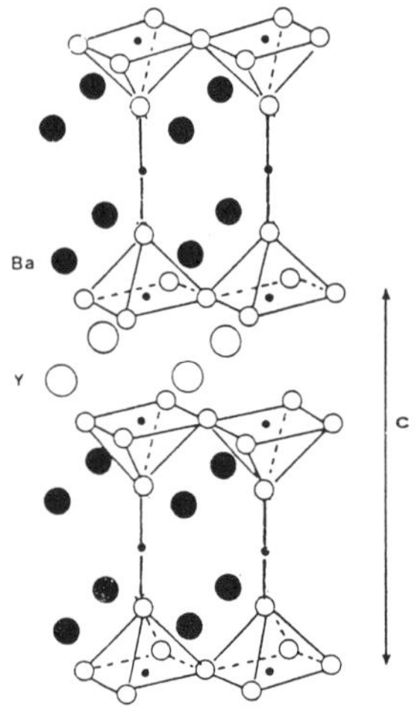

Fig. 11. Structure of $YBa_2Cu_3O_6$

The crystal chemistry of the orthorhombic superconductor $YBa_2Cu_3O_{7-\delta}$ is also very complex in spite of the apparent simplicity of the structure established by neutron diffraction. A systematic microtwinning of the crystals was observed by X-ray diffraction (ref. 17, 18-31) and electron microscopy (ref. 32-35). The twinning boundaries are spaced of about 500 to 2000 Å (Fig. 12).

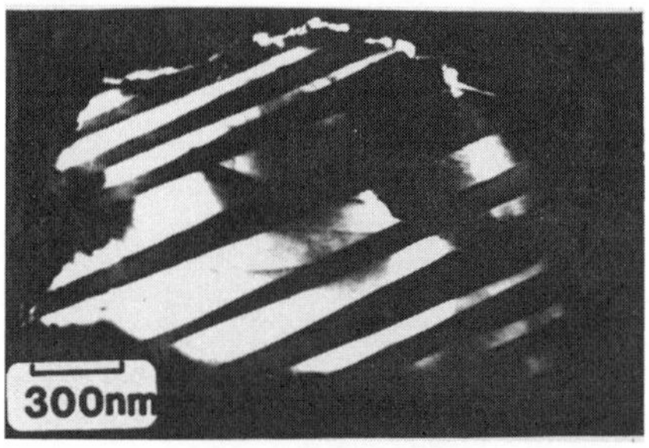

Fig. 12. Dark field image of a twinned crystal.

This twinning phenomenon results from the orthorhombic-tetragonal phase transition. Several models can be proposed for the "structure" of the boundaries which can involve either CuO_6 octahedra (Fig. 13a) or CuO_5 pyramids (Fig. 13b) or CuO_4 tetrahedra (Fig. 13c) ; the two first models change the ratio Cu(III)/Cu(II) without introducing any distortion of the structure,

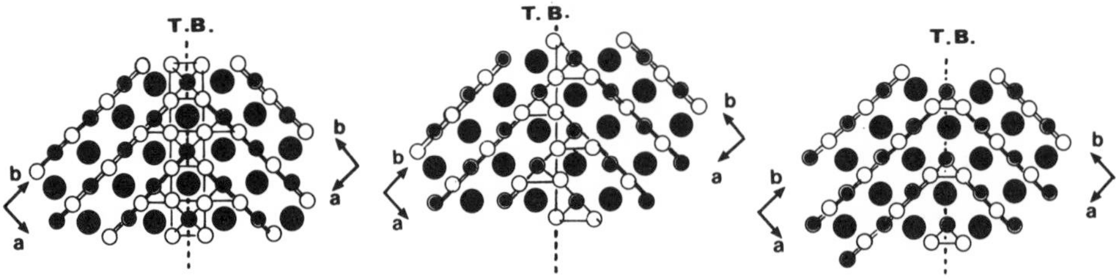

Fig. 13. Models proposed for the "structure" of the boundaries.

whereas the third model does not change the stoichiometry, i.e. Cu(III)/Cu(II), but should lead to a distortion of the polyhedra at the junction and is less favourable from the energetic point of view. The systematic HREM investigation of superconducting ceramics shows that many crystals exhibit oriented domains and defects. Such phenomena cannot be detected and proved by the observation alone. A careful simulation of the images, taking into account the defocus series, the thickness of the crystals, is of course necessary ; attention must be drawn on the fact that variations of contrast which are due to the variations of the oxygen stoichiometry are, most of the time, much more sensitive to the small displacements of the cations than to the oxygen content itself. Oriented domains (Fig. 14) are often observed

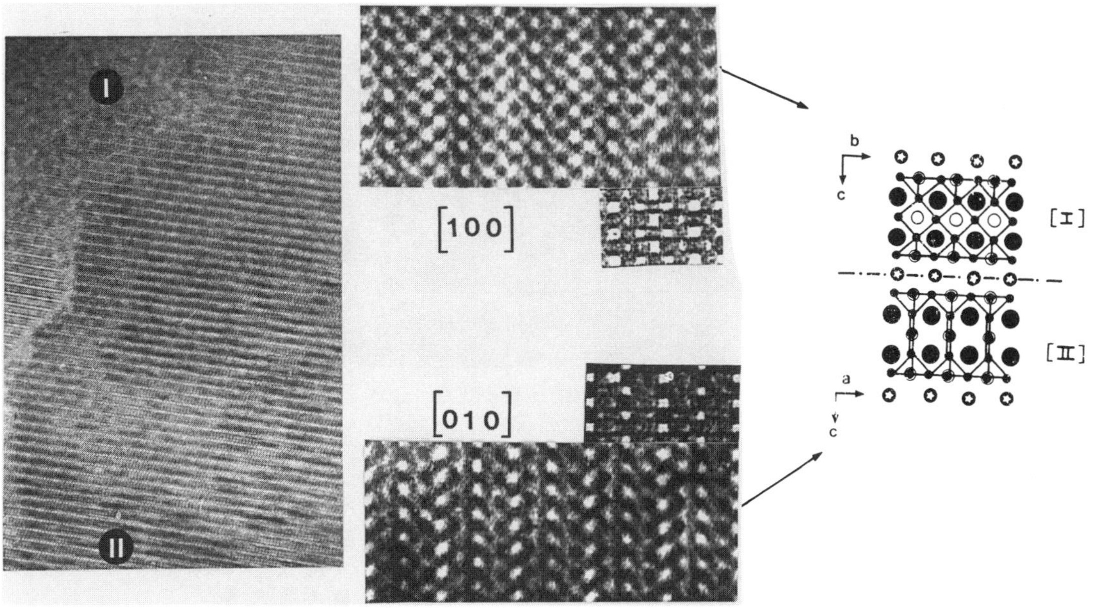

Fig. 14. A change in the direction of CuO_4 square planar groups from one layer to the other results in oriented domains.

which result from the fact that the CuO_4 square planar groups belonging to the $[CuO_2]_\infty$ planes are twinned of 90° from one region of the crystal to the next one. This implies the coincidence of "a_I" and "b_{II}" in the (001) domain boundary, leading for several crystals to a misorientation of the domains and even to microcracks due to the different values of the "a" and "b" parameters. Defects, of 40 to 50 Å diameter are often observed (ref. 36), which show a drastic variation of contrast (Fig. 15) ; a simulation of these defects allowed an excess of oxygen to be detected, leading to $YBa_2Cu_3O_8$-type regions (Fig. 16). In the same way oxygen deficient defects corresponding to $SrCuO_2$-type arrangement (ref. 37) characterized by double rows of edge-sharing CuO_4 square planar groups were also observed (Fig. 17). These examples of the local variation of the structure and of the oxygen stoichiometry suggest that superconductivity of the ceramics $YBa_2Cu_3O_{7-\delta}$ cannot be understood by only taking into account the structure of the pure oxide.

The importance of crystal chemistry in the generation of the superconducting properties is alike well illustrated in the $LaBa_2Cu_3O_{7-\delta}$-$La_3Ba_3Cu_6O_{14+\delta}$ system, which appears more complex than the yttrium one as well by the behavior of $LaBa_2Cu_3O_{7-\delta}$ oxide as by the existence of an homogeneity range. In this system, the essential features of the structure of the triple-layer perovskites were previously determined for $La_3Ba_3Cu_6O_{14+\delta}$ (ref. 45) : mixed-valence of copper, bidimensional character ($c = 3a_p$) and particularly, the ability of lanthanum ions to substitute partly the barium ions. The first results obtained for the superconductor $LaBa_2Cu_3O_{7-\epsilon}$ showed that this oxide had a particular behavior (ref. 37) : a tetragonal symmetry (from X-ray diffraction) and a lower critical temperature (T_c . 70-75K) were observed. It appears in fact that this tetragonal symmetry is not compatible with an ordering of the oxygen to build O-Cu-O chains up. One of the main points was first to determine the nature of this feature : artefact or a different origin of the superconductivity. The HREM investigation of a "tetragonal superconducting" oxide $LaBa_2O_{6.7}$ by electron diffraction was carried out (ref. 39-40). Two sorts of crystals were mainly observed : crystals with a rigourous tetragonal symmetry and crystals characterized by a distortion of

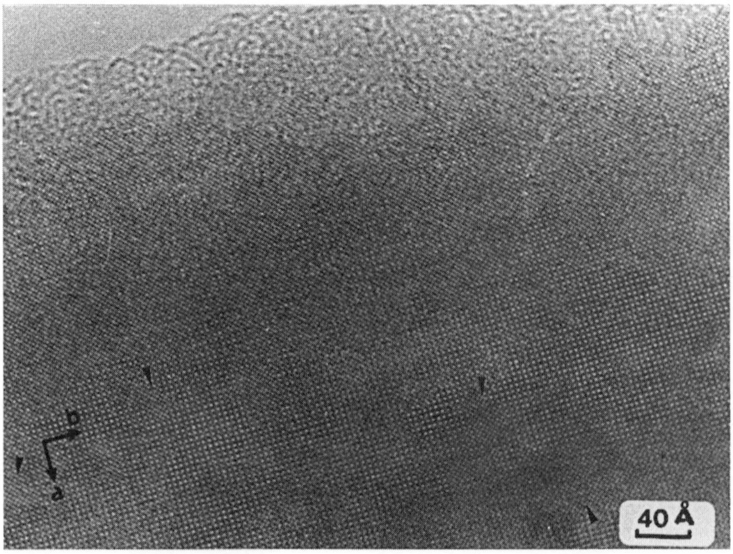

Fig. 15. Variation of contrast are often observed in domains of 40 to 50 Å diameter [001] image 3.

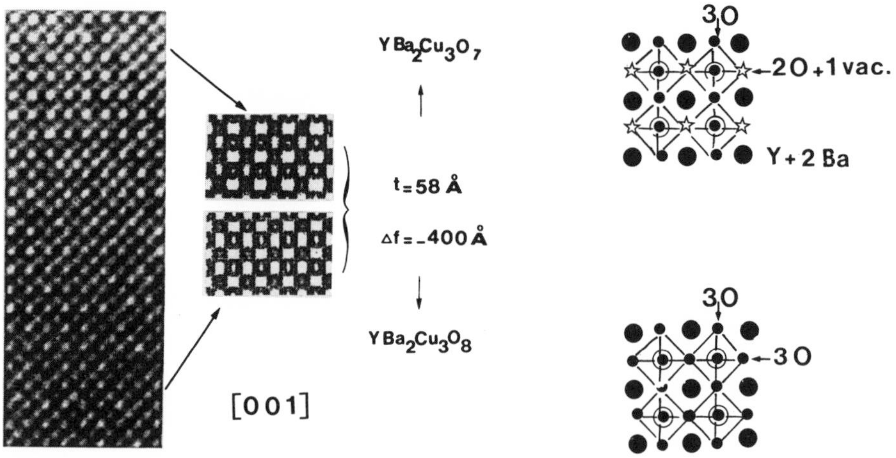

Fig. 16. Enlargement of such a domain and calculated corresponding image.

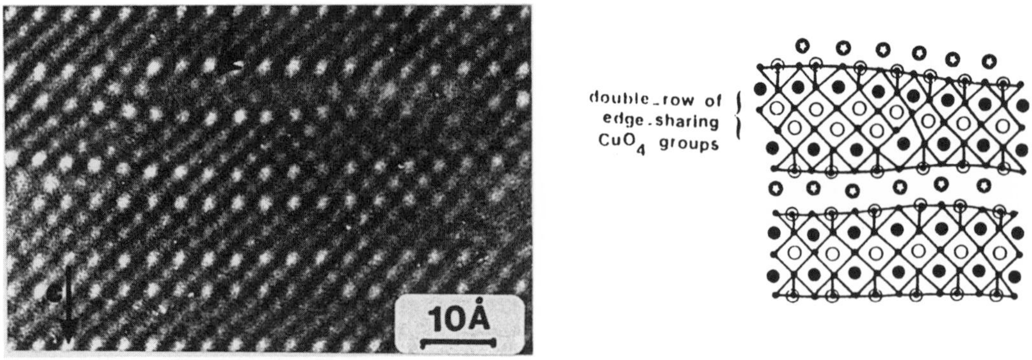

Fig. 17. Image of a defect corresponding to $SrCuO_2$ type arrangement and idealized model.

the tetragonal cell, likely orthorhombic. All crystals exhibit numerous domains which result from different structural phenomena. The first type of domains correspond to the setting up of the superstructure ($3a_p$) along three perpendicular equivalent orientations of the perovskite subcell (Fig. 18).

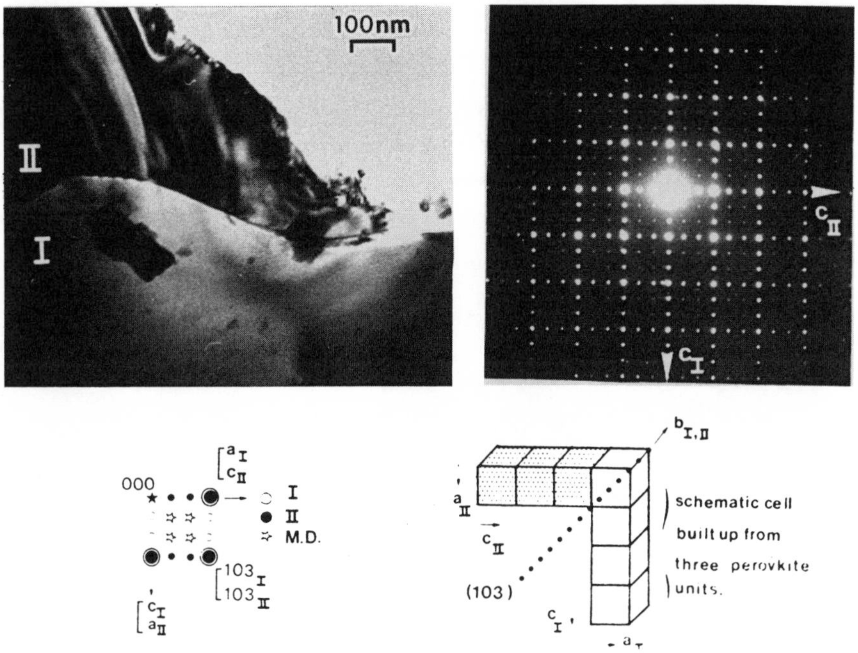

Fig. 18. Low resolution image and electron diffraction pattern of two domains corresponding to the setting up of the superstructure ($3a_p$) along two perpendicular directions of the perovskite subcell (idealized drawing of the E.D. pattern and of the schematic cells).

This behaviour was not observed in $YBa_2Cu_3O_{7-\delta}$ owing to the parameter ratios which are significantly different from 1 (a/b, 3a/c and 3b/c). However those phenomena are often observed in the oxygen deficient perovskites (ref. 41-44) where they generally appear as localized domains. The second type of domains, encountered in numerous crystals, can be described as microtwinning domains resulting from an orthorhombic distortion of the cell : the observation of the splitting of the reflections, except hh0, showed that the amplitude of the distortion varies from one crystal to the other (Fig. 19). Such a result is consistent with the "apparent tetragonal symmetry" deduced fromm the X-ray diffraction patterns, where the components of the various distortions do not appear. From these results a systematic study of the experimental, and particularly of the thermal conditions of synthesis, was undertaken in order to stabilize one of the orthorhombic forms (ref. 40). Thus, a sample sintered at 1020°C under oxygen, slowly cooled (45°C h^{-1}) in an oxygen flow and then annealed in an oxygen flow at 450°C during one week was indexed in an orthorhombic cell. The isotypism of this oxide with the superconductor $YBa_2Cu_3O_7$ was confirmed by a preliminary X-ray diffraction study (ref. 40) ; a more complete study, by neutron diffraction is in progress. These results suggest that the oxygen content and the type of ordering of the oxygen vacancies vary from one crystal to the other. Then, the problem of the superconductivity in the "tetragonal $LaBa_2Cu_3O_7$" phase appears to be the result of a percolation of various distorted crystals in the bulk. This hypothesis is also supported by the results of the magnetic measurements : a small diamagnetic volume (10 %) is observed in the "tetragonal" phase while more than 90 % diamagnetism are observed in the orthorhombic one. Moreover the existence of a wide domain of homogeneity, from $LaBa_2Cu_3O_{7-\delta}$ to $La_3Ba_3Cu_6O_{14+\delta}$, shows the specific behavior of lanthanum with respect to other rare earths and Y. This particular property is likely due to its size intermediate between that of barium and other lanthanides. Its size smaller than barium allows an order to be established, as in $La_4BaCu_5O_{13+\delta}$ (ref. 14) or $LaBa_2Cu_3O_{7-\delta}$ (ref. 37-39) but is close enough to that of barium for a partial occupation of the same sites, as in $La_3Ba_3Cu_6O_{14+\delta}$ (ref. 45). The existence of oriented domains in the

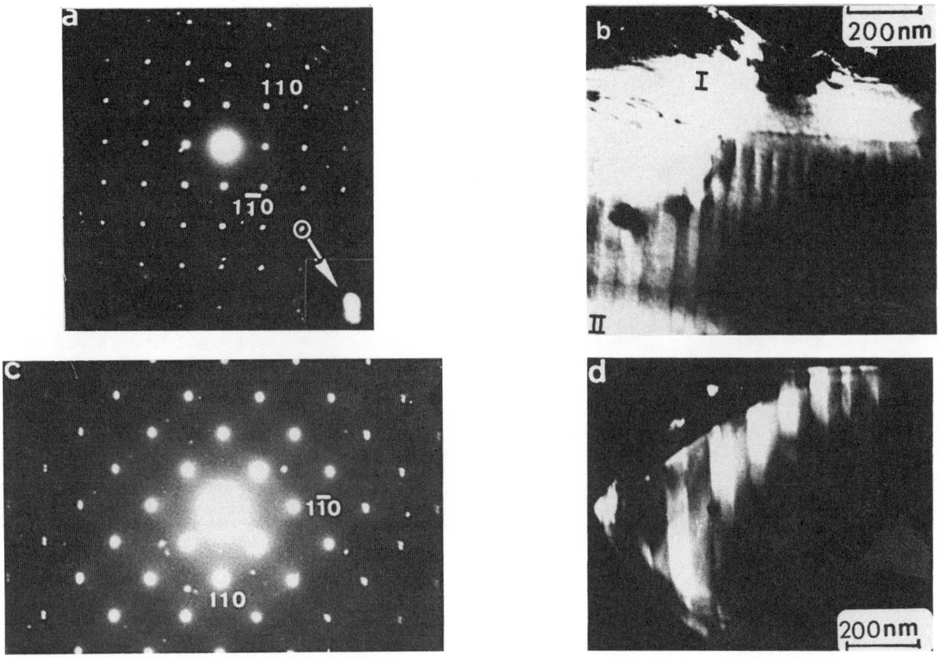

Fig. 19. Electron diffraction patterns and corresponding low resolution image of two crystals showing that the amplitude of the distorsion varies from one crystal to the other (weak : a, b and strong c, d).

lanthanum oxides (ref. 39) may be related to the size effect which involves a quasi-cubic subcell. These two features, the tendancy to oxygen disordering added to the partial mixing of La and Ba, would imply a very complex microstructural state of the crystals which cannot be determined by X-ray and neutron powder diffraction methods alone.

REFERENCES

1. H. KAMERLINGH ONNES, Akad. Van Weterschappen, 14, 818 (1911).
2. R.K. STANLEY, R.G. MORRIS and W.G. MOULTON, Phys. Rev., B20, 1903-1914 (1979).
3. M.R. STOKAN, W.G. MOULTON and R.C. MORRIS, Phys. Rev., B20, 3670 (1979)
4. A. DESCHANVRES, B. RAVEAU and Z. SEKKAL, Mat. Res. Bull., 6, 699 (1971).
5. D.C. JOHNSTON, H. PRAKASH, W.H. ZACHARIASEN and R. VISWANATHAN, Mat. Res. Bull., 8, 777 (1973).
6. A.W. SLEIGHT, J.L. GILLSON and F.E. BIERSTEDT, Solid State Commun, 17, 27 (1975).
7. N. NGUYEN, J. CHOISNET, M. HERVIEU and B. RAVEAU, J. Solid State Chem., 39, 120-127 (1981).
8. N. NGUYEN, C. MICHEL, F. STUDER, B. RAVEAU, Mat. Chem., 7, 413-428 (1982)
9. J.G. BEDNORZ and K.A. MULLER, Z. Phys. B, 64, 189-193 (1986).
10. A. BIANCONI, J. BUDNICK, A.M. FLANK, A. FONTAINE, P. LAZARDE, A. MARCELLI, H. TOLENTINO, B. CHAMBERLAND, G. DEMAZEAU, C. MICHEL and B. RAVEAU, Phys. Letters A, submitted.
11. L. ER-RAKHO, C. MICHEL, J. PROVOST and B. RAVEAU, J. Solid State Chem., 37, 151-156 (1981).
12. B. RAVEAU, Proc. Indian Natn. Sci. Acad., 52A, 1, 67-101 (1986).
13. J. LABBE and J. BOK, Europhysics Letters, to be published.
14. C. MICHEL, L. ER-RAKHO, M. HERVIEU, J. PANNETIER and B. RAVEAU, J. Solid State Chem., 68, 143-152 (1987).
15. L. ER-RAKHO, C. MICHEL and B. RAVEAU, J. Solid State Chem., in press.
16. C. MICHEL and B. RAVEAU, Rev. Chim. Miner., 21, 407-425 (1984).
17. Y. LEPAGE, W.R. McKINNON, J.M. TARASCON, L.H. GREENE, G.W. HULL and D.M. HWANG, Phys. Rev., B35, 7245-7248 (1987).
18. J.J. CAPPONI, C. CHAILLOUT, A.W. HEWAT, P. LEJAY, M. MAREZIO, N. NGUYEN, B. RAVEAU, J.L. SOUBEYROUX, J.L. THOLENCE and R. TOURNIER, Europhysics Lett., 12, 1301-1307 (1987).

19. R.J. CAVA, R.B. VAN DOVER, B. BATTLOGG and E.A. RIETMAN, Phys. Rev. Lett., 58, 408-410 (1987).
20. C.N.R. RAO and P. GANGULY, Current Science, 56, 47 (1987).
21. C. MICHEL, F. DESLANDES, J. PROVOST, P. LEJAY, R. TOURNIER, M. HERVIEU and B. RAVEAU, C.R. Acad. Sci., 304 II, 1059-1061 (1987).
22. R.J. CAVA, B. BATTLOG, R.B. VANDOVER, D.W. MURPHY, S. SUNSHINE, T. SIEGRIST, J.P. REMEIKA, E.A. REITMAN, S. ZAHURAK and G.P. ESPINOSA, Phys. Rev. Lett., 58, 1676-1679 (1987).
23. N. NGUYEN, F. STUDER, B. RAVEAU, J. Phys. Chem. Solids, 44, 389-400, (1983).
24. N. NGUYEN, L. ER-RAKHO, C. MICHEL, J. CHOISNET and B. RAVEAU, Mat. Res. Bull., 15, 891-897 (1980).
25. Y. LALIGAN, G. FEREY, M. HERVIEU, B. RAVEAU, A. SULPICE, R. TOURNIER, J. Phys., submitted.
26. J. LABBE, Seminaire de l'Ecole Normale Supérieure de Paris, 30 Octobre 1987.
27. J. BEILLE, R. CABANEL, C. CHAILLOUT, B. CHEVALLIER, G. DEMAZEAU, F. DESLANDES, J.. ETOURNEAU, P. LEJAY, C. MICHEL, J. PROVOST, B. RAVEAU, A. SULPICE, J.L. THOLENCE, R. TOURNIER, C.R. Acad. Sci., 304 II, 1097-1102 (1987).
28. B. DOMENGES, M. HERVIEU, V. CAIGNAERT, B. RAVEAU, J. THOLENCE and R. TOURNIER, J. Microsc. Spectrosc. Elect., submitted.
29. F. ISUMI, H. ASANO, T. ISHIGAKI, E. TAKYAMA-MUROMACHI, Y. UCHIDA and WATANABE, J. Appl. Phys., V26, 7, 1214-1217 (1987).
30. G. ROTH, B. RENKER, G. HEGER, M. HERVIEU, B. DOMENGES and B. RAVEAU, Z. Phys., submitted.
31. G. ROTH, D. EWERT, G. HEGER, M. HERVIEU, C. MICHEL, B. RAVEAU, F. D'YVOIRE and A. REVCOLEVSKI, Z. Phys., submitted.
32. M. HERVIEU, B. DOMENGES, C. MICHEL, G. HEGER, J. PROVOST and B. RAVEAU, Phys. Rev., B36, 3920 (1987).
33. M. HERVIEU, B. DOMENGES, C. MICHEL, J. PROVOST and B. RAVEAU, J. Solid State Chem., submitted.
34. G. VAN TENDELOO, H.W. ZANDBERGER and S. AMELINCKX, J. Solid State Com., submitted.
35. H.U. ZANDBERGER, G. VAN TENDELOO, T. KABE and S. AMELINCKX, Phys. Statu Solidi (A), to be published.
36. M. HERVIEU, B. DOMENGES, C. MICHEL and B. RAVEAU, Europhysics Lett., 4 (2), 205-210 (1987).
37. B. DOMENGES, M. HERVIEU, C. MICHEL and B. RAVEAU, Europhysics Lett., 4 (2), 211-214 (1986).
38. C. MICHEL, F. DESLANDES, J. PROVOST, P. LEJAY, R. TOURNIER, M. HERVIEU and B. RAVEAU, C.R. Acad. Sci., 304 II, 1169-1172 (1987).
39. M. HERVIEU, B. DOMENGES, J. PROVOST, F. DESLANDES and B. RAVEAU, Angewandte Chemie, submitted.
40. M. HERVIEU, B. DOMENGES, F. DESLANDES, C. MICHEL and B. RAVEAU, Angewandte Chemie, submitted.
41. M. HERVIEU, N. NGUYEN, V. CAIGNAERT and B. RAVEAU, Physica Statu Solidi, 2, 83, 473-483 (1984).
42. V. CAIGNAERT, M. HERVIEU, N. NUGYEN, B. RAVEAU, J. Solid State Chem., 62, 281-289 (1986).
43. V. CAIGNAERT, M. HERVIEU and B. RAVEAU, Mat. Res. Bull., 21 (10), 1147- (1986).
44. V. CAIGNAERT, M. HERVIEU, J.M. GRENECHE, B. RAVEAU, Chem. Scripta, 27, 283-294 (1987).
45. L. ER-RAKHO, C. MICHEL, J. PROVOST and B. RAVEAU, J. Solid State Chem., 37, 151-156 (1981).

Ba(Pb, Bi)O_3 superconductors and their relationship to the copper oxide based superconductors

Arthur W. Sleight

Central Research and Development Department, E. I. du Pont de Nemours and Company, Experimental Station, Building 356, Wilmington, Delaware 19898, USA

Abstract - The Ba(Pb,Bi)O_3 system is reviewed in light of evidence that the origin of superconductivity in this system is essentially the same as that in the recently discovered copper oxide based systems. Common features of the two systems include proximity to a metal-insulator boundary, very high covalency of certain bonds, a conduction band which is σ^* in character, and mixed valency. A mechanism for superconductivity based on disproportionation of Bi^{IV} or Cu^{II} is consistent with available data.

INTRODUCTION

The purpose of this paper is to review the $BaPb_{1-x}Bi_xO_3$ system and point out common features with the recently discovered copper oxide based systems. The highest T_c obtained in the $BaPb_{1-x}Bi_xO_3$ system is about 13 K, and this remains the record high T_c for any material not containing a transition element. Thus, this material has been considered anomalous since its discovery (ref. 1), and there have been numerous proposals for a mechanism to explain the unexpectly high T_c. There are several reasons for believing that the mechanism for superconductivity is the same in $BaPb_{1-x}Bi_xO_3$ phases as in the copper oxide based systems.

This paper is not intended to be a comprehensive review of the $BaPb_{1-x}Bi_xO_3$ system. There are two recent reviews that serve that function. One review (ref. 2) is from the University of Tokyo group which has been very active in this area for many years. The other review (ref. 3) is from the Russian literature.

$BaPb_{1-x}Bi_xO_3$ SYSTEM

$BaPbO_3$

This compound has a typical orthorhombic distortion of the perovskite structure. Crystals have been grown hydrothermally; they are black and possess metallic conductivity (ref. 4). Superconductivity is reported for $BaPbO_3$ with a T_c of 0.5 K (ref. 5). The structure of $BaPbO_3$ is a clear indication that Pb is present in the tetravalent state. Thus the metallic properties might be considered surprising. One might well have expected a gap between an oxygen 2*p* valence band and a lead 6*s* conduction band. Indeed, $BaSnO_3$ is a white insulator. However, in $BaPbO_3$ it appears that the valence and conduction bands have just overlapped (Fig. 1). Thus, we may term this compound a semimetal or a zero gap semiconductor. It could also be that the metallic properties of $BaPbO_3$ are due to some slight oxygen deficiency which would give electron carriers in the Pb 6*s* band. This possibility cannot be completely excluded. However, if $BaPbO_3$ is slightly oxygen deficient, this could be the result of an overlap of the O 2*p* and Pb 6*s* bands. Whatever the intrinsic properties of stoichiometric $BaPbO_3$, we may be certain that it is right on the boundary between localized and delocalized properties. Metallic properties are also observed for PbO_2 whereas the $A_2^{III}Pb_2^{IV}O_7$ pyrochlores are semiconducting.

Indicating that the oxidation state of Pb is tetravalent in PbO_2, $BaPbO_3$ and $A_2Pb_2O_7$ pyrochlores does not suggest anything about the real charge on Pb.

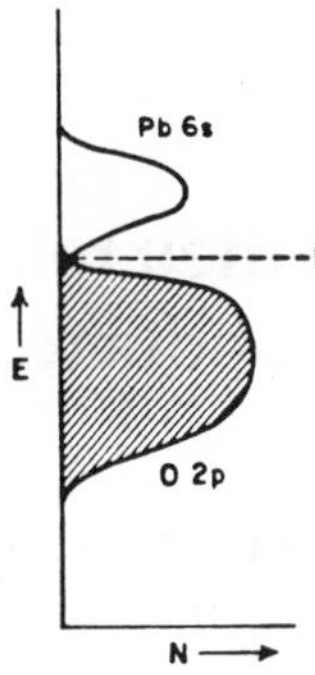

Fig. 1. Schematic density of states *vs.* energy for $BaPbO_3$; E_F is the Fermi level.

The Pb-O bond is highly covalent. This covalency increases with the Pb oxidation state, and it increases due to the presence of Ba. Thus, the real charge on Pb in $BaPbO_3$ is closer to zero than to four, but this is no reason to suggest using other oxidation state representations such as Pb^{II} and O^{-I}. Real charges and oxidation states are two very different quantities. For more discussion on this subject see ref. 6.

$\underline{BaBiO_3}$

This compound also possesses a perovskite-type structure. However, neutron diffraction studies clearly show two types of Bi present in equal amounts (refs. 7-10). The structure is shown in Fig. 2. Based on the Bi-O distances, one may conclude (ref. 9 and 11) that these two bismuth cations are Bi^{III} and Bi^V. This is not surprising since Bi^{III} and Bi^V are the only two expected oxidation states for Bi. A Bi^{IV} species at such high concentrations would have been surprising.

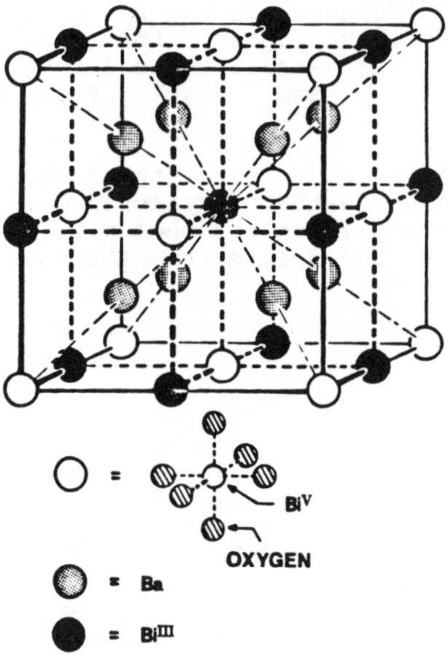

Fig. 2. The ordered perovskite structure of $Ba_2Bi^{III}Bi^VO_6$.

Crystallographic studies (ref. 9) indicate that $BaBiO_3$ is monoclinic at room temperature and below, but it has two phase transitions above room temperature. A first-order monoclinic-to-rhombohedral transition occurs at about 405 K. There also appears to be a second-order transition to cubic symmetry at about 800 K. However, throughout the entire stability range of $BaBiO_3$, there are two types of Bi best described as Bi^{III} and Bi^V.

The situation for $BaBiO_3$ is then an analogy of the situation found for $A^I_2SbX^{-I}_6$ phases. Despite the fact that such $A^I_2SbX^{-I}_6$ phases are roughly isostructural with $A^I_2Pt^{IV}X^{-I}_6$ and $A^I_2Pb^{IV}X^{-I}_6$ phases, careful crystallographic studies (ref. 12) show two types of Sb in equal amounts. Mössbauer studies (ref. 13) of these $A^I_2SbX^{-I}_6$ phases confirm that these two types of Sb must be regarded as Sb^{III} and Sb^{V}. The disproportionation of Sb^{IV} into Sb^{III} and Sb^{V} or of Bi^{IV} into Bi^{III} and Bi^{V} may be equally well described as a charge density wave which is commensurate with the lattice. Covalency is high, and real charges of the cations are much less than their oxidation state. Thus, the real charges difference between Bi^{III} and Bi^{V} is much less than two.

Lone pair cations such as Sb^{III} and Bi^{III} frequently take on a very asymmetric environment. This is caused by the antibonding nature of the filled 5*s* or 6*s* orbitals, respectively. Hybridization of this filled *s* level with the unfilled *p* levels allows the antibonding electron density to be pushed out of the bonds. The lone pair of electron is pushed off center, and strong bonds form on the opposite side. Consequently lone pair cations tend not to be found at inversion centers. Nonetheless, there are several known cases where Bi^{III} and Sb^{III} are found at inversion centers. Thus, the presence of an inversion center for Bi^{III} in $Ba_2Bi^{III}Bi^{V}O_6$ cannot be used as evidence to suggest that Bi^{III} is not actually present in $BaBiO_3$.

Since $BaBiO_3$ does not melt congruently, crystals were grown hydrothermally (ref. 1). The crystals are lustrous with a gold appearance, but like fools gold (FeS_2) they are actually semiconducting from 4 K to the decomposition temperature of about 1100 K (ref. 9). The activation energy remains remarkably constant at about 0.2eV from 100 to 473 K (Fig. 3). This activation energy can presumably be related to electron hopping involving the reaction $Bi^{III} + Bi^{V} \rightarrow 2Bi^{IV}$. Below 100 K, $BaBiO_3$ becomes so insulating we were unable to determine resistance values.

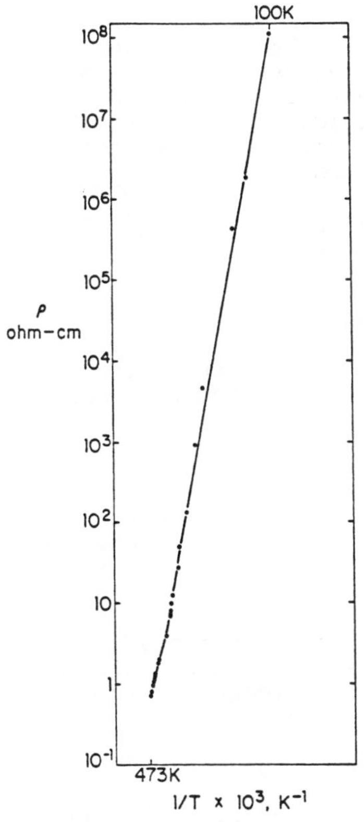

Fig. 3. Electrical resistivity for $BaBiO_3$ as a function of temperature.

<u>$BaPb_{1-x}Bi_xO_3$ (0.35<x<1)</u>

Over this entire region, the materials remain semiconducting with a gold color. The symmetry is, however, orthorhombic instead of monoclinic. (Cell dimensions for the entire $BaPb_{1-x}Bi_xO_3$ system are given in Fig. 4.) There is only one crystallographic site for the Pb and Bi cations, but this does not imply that Bi^{IV} disproportionation into Bi^{III} and Bi^{V} has not occurred. It is in fact clear from the semiconducting properties that Bi^{IV} has largely disproportionated in this region. However, the resulting Bi^{III} and Bi^{V} have

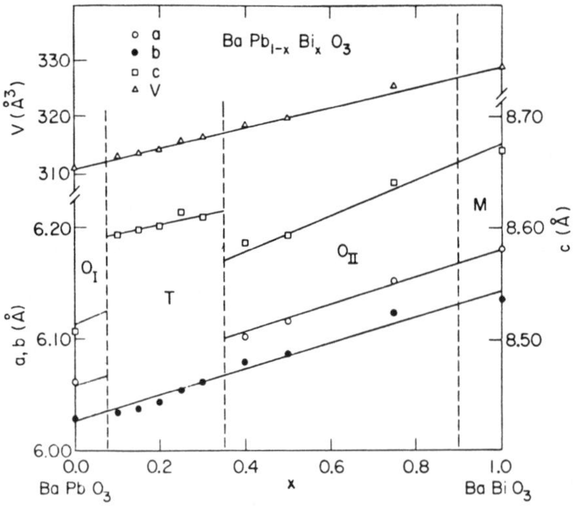

Fig. 4. Unit cell edges and volumes in the system $BaPb_{1-x}Bi_xO_3$. The broken lines indicate schematically the four structural regions described in the text. The solid lines are a visual guide.

not taken on long range ordering in the lattice because Bi and Pb randomly occupy the octahedral sites. This disorder between Pb and Bi prevents the long range order of Bi^{III} and Bi^{V}. There must of course be a two phase region between monoclinic or rhombohedral $BaBiO_3$ and the orthorhombic $Ba(Pb,Bi)O_3$ phases. However, the extent of this two phase region as a function of temperature has not been determined.

One question that needs to be addressed is why $BaPb_{0.6}Bi_{0.4}O_3$ is not metallic even if all the Bi is present as a mixture of Bi^{III} and Bi^{V}. After all, $BaPbO_3$ has metallic properties even though it formally has no 6*s* electrons for the conduction band. If the metallic properties of $BaPbO_3$ are caused by overlap of the Pb 6*s* and O 2*p* bands, this same overlap could result in metallic properties for $BaPb_{0.6}Bi_{0.4}O_3$. Note however (Fig. 4) that the lattice expands considerably as Bi is substituted for Pb. This, plus the dilution effect, could well lead to sufficient band narrowing to eliminate the overlap of the Pb 6*s* and O 2*p* bands.

$BaPb_{1-x}Bi_xO_3$ (0>x>0.35)

This is the region of greatest interest. Metallic and superconducting properties exist from x = 0 to x about 0.35. The color is now black to blue-black. The critical temperature first increases slowly (T_c = 0.5 K for x = 0), achieves a peak value of about 13 K at x about 0.27 and then rapidly decreases as x increases further. Determination of T_c as a function of x is difficult because the transitions as measured by flux exclusion are always broad. Thus a plot of T_c versus x may not be very meaningful. Presumably, the problem of obtaining sharp transitions is at least partly due to the distribution of Bi and Pb on the octahedral sites.

Solid solutions are always frustrating systems to understand because the structure at the atomic level is intrinsically disordered and undefined. In the $Ba(Pb,Bi)O_3$ system, any member of the solid solution will necessarily have regions which are Bi rich and Pb rich relative to the average value of x. The fact that it is impossible for x to be uniform on an atomic scale may cause some diffusness of certain properties, including superconducting properties. The best properties, e.g. sharpest superconducting transition, are obtained in rapidly quenched samples of the $BaPb_{0.73}Bi_{0.27}O_3$ composition. Slower cooler results in more segregation of Pb and Bi which ultimately results in phase separation. Another method to sharpen the transition is to replace some Ba with K (ref. 1). Initially this substitution, i.e. $Ba_{1-y}K_yPb_{1-x}Bi_xO_3$, was intended as an independent way of varying electron concentration. However, it appears more likely that the beneficial effect of K is that it promotes a more random distribution of Bi and Pb. The more random this distribution becomes, the more homogeneous the materials become, and thus the transition becomes sharper.

The crystal symmetry in the metallic region is orthorhombic or tetragonal. At low values of x, including x = 0, the symmetry is clearly orthorhombic. For samples with optimum superconducting properties, the symmetry as determined by high resolution x-ray powder diffraction (refs. 10, 14, 15) is tetragonal. It also appears that samples of a given x value may be either tetragonal or orthorhombic depending on thermal treatment which could effect oxygen content or the distribution of Bi and Pb on the octahedral sites. To add even further confusion, some samples in this region appear to be mixtures of tetragonal and orthorhombic phases. Of course, this complication is to be expected near the boundaries between the tetragonal and orthorhombic regions. Interpretation of powder diffraction patterns becomes very complex because the cell edges of the tetragonal and orthorhombic phases are so similar. Even high resolution techniques are of little value if there is intrinsic line broadening.

Presumably then, there is a two phase region near x = 0.35. This is not surprising because there normally is a discontinuity between metallic and semiconducting regions of a compositional phase diagram. A postulated phase diagram for the $BaPb_{1-x}Bi_xO_3$ system is shown schematically in Fig. 5. This then implies that the best superconductors in this system are not thermodynamically stable phases.

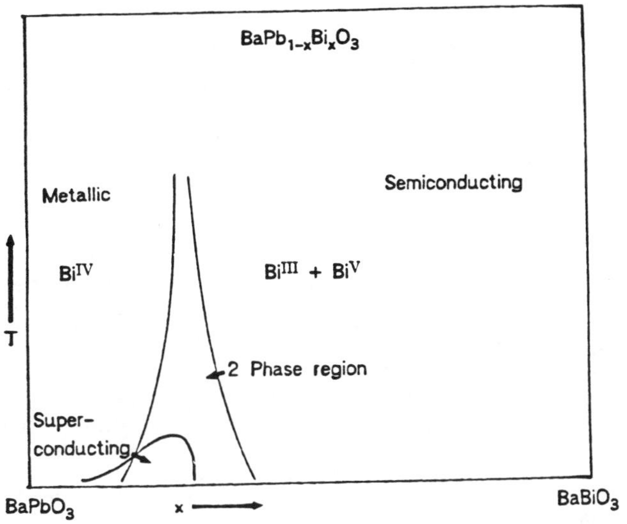

Fig. 5. Schematic phase diagram for the BaPbBiO system. The two phase region is not normally observed.

Crystals of the $BaPb_{1-x}Bi_xO_3$ can be grown hydrothermally. In fact, electrical resistivity data on such cyrstals were reported in our first publication (ref. 1) on the superconducting compositions. Although transitions measured by the electrical resistance of crystals are generally sharp, this does not imply that the crystal is more homogeneous or a better defined material than the polycrystalline samples. There will still be Pb and Bi rich regions on an atomic scale, and this may lead to a diffuse transition as measured by flux exclusion.

The first observation of superconductivity in the $BaPb_{1-x}Bi_xO_3$ system was by electrical resistance measurement on a sintered pellet of $BaPb_{0.75}Bi_{0.25}O_3$. Above T_c, the resistivity versus tempeature behavior was that of a semiconductor rather than a metal. However, we had previously seen such behavior in other oxide superconductors, such as the tungsten bronzes. We therefore assumed that this semiconductor-to-superconductor transition was due to insulating grain boundaries. However, such behavior has been reported (ref. 16) for $BaPb_{1-x}Bi_xO_3$ crystals when x exceeds 0.3. The flux exclusion in these samples is low. Therefore, the semiconductor-to-superconductor transition is likely related to inhomogenieties in both crystals and polycrystalline samples. The Pb rich regions are superconducting, and the Bi rich regions are semiconducting.

MECHANISM

Common features of the $Ba(Pb,Bi)O_3$ system and copper oxide based superconductors include perovskite type structures where all oxygen forms two highly covalent bonds to either Bi or Cu, a conduction band which is sigma antibonding in character, proximity to a metal-insulator boundary, a low density of states at the Fermi level, and mixed valency where three consecutive oxidation states pertain (Bi^{III}, Bi^{IV}, Bi^{V} or Cu^{I}, Cu^{II}, Cu^{III}). A feature of the copper oxygen systems not present in the $Ba(Pb,Bi)O_3$ system is magnetism for insulating compositions. However, this magnetism must be surpressed before superconductivity can occur.

$Ba(Pb,Bi)O_3$ System

In the superconducting region of this system, we may think of bismuth being in the Bi^{IV} state. We may also think of the 6*s* electron associated with Bi^{IV} as being delocalized in a band composed of Pb 6*s*, Bi 6*s*, and O 2*p* states. This is a stable situation only as long as Bi^{IV} is dilute in a $BaPbO_3$ mattrix. When the Bi^{IV} cations begin to strongly sense each other through Bi-O-Bi linkages, the disproportionation reaction ($2Bi^{IV} \rightarrow Bi^{III} + Bi^{V}$) becomes favored. The superconducting transition temperature reaches its maximum just as the nondisproportionated state (Bi^{IV}) becomes essentially degenerate with the disproportionated state ($Bi^{III} + Bi^{V}$). The disproportionation reaction can be considered a mechanism for superconductivity because it is spin pairing in nature:
$2\ (Bi^{IV}) \rightarrow\ (Bi^{III}) + Bi^{V}$. Since Bi^{III}-O distances are somewhat longer than Bi^{V}-O distances, the tendency to disproportionate will be coupled to breathing mode type displacements of oxygen. Thus, we would expect phonons to be involved in this mechanism. Indeed, a substantial suppression of T_c has been found when ^{16}O is replaced with ^{18}O in $BaPb_{0.8}Bi_{0.2}O_3$ (ref. 17).

Copper Oxide Based Superconductors

The disproportionation analogy for the copper oxides would be $2Cu^{II} \rightarrow Cu^{I} + Cu^{III}$. Although this reaction is not well established for Cu^{II}, it is well known for Ag^{II} and Au^{II}. In fact, AgO is really $Ag^{I}Ag^{III}O_2$. Apparently Au^{II} always disproportionates into Au^{I} and Au^{III}. Furthermore, we know that Cu^{I}, Cu^{II} and Cu^{III} can all exist in the 1:2:3 structure, i.e. $YBa_2Cu^{I}Cu_2^{II}O_6$ and $YBa_2Cu^{III}Cu_2^{II}O_7$. Thus, it is entirely reasonable to suggest that the disproportionation of Cu^{II} could be the spin pairing reaction leading to superconductivity *providing* that Cu^{II} is in square planar rather than octahedral coordination. If the d_{z^2} and $d_{x^2-y^2}$ levels of Cu^{II} become degenerate, the disproportionation reaction is no longer simple spin pairing in nature.

REFERENCES

1. A. W. Sleight, J. L. Gillson, and P. E. Bierstedt, Solid State Commun., 17, 27-28 (1975).
2. S. Uchida, K. Kitazawa and S. Tanaka, Phase Transitions, 8, 95-127 (1987).
3. A. M. Gobovich and D. P. Moiseev, Soviet Phys. Usp., 29, 1135 (1986).
4. R. D. Shannon and P. E. Bierstedt, J. Amer. Cer. Soc., 53, 635 (1970).
5. V. V. Bagotko and Yu. N. Venetsev, Sov. Phys. Solid State, 22, 705-706 (1980).
6. A. W. Sleight, Superconductivity: Synthesis, Properties and Processing, ed. W. Hatfield, Marcel Dekker, Inc. (1988).
7. D. E. Cox and A. W. Sleight, Solid State Commun., 19, 969-973 (1976).
8. G. Thornton and A. J. Jacobson, Acta Cryst., B34, 351-354 (1978).
9. D. E. Cox and A. W. Sleight, Acta Cryst., B35, 1-10 (1979).
10. D. E. Cox and A. W. Sleight, In Proceedings of the Conference on Neutron Scattering, R. M. Moon (Ed.) Gathinburg, National Technical Information Service, Springfield, VA, USA. 45-54 (1976).
11. N. K. McGuire and M. O'Keeffe, Solid State Commun., 52, 433-434 (1984).
12. S. L. Lawton and R. A. Jacobson, Inorg. Chem., 5, 743-749 (1966).
13. G. Longworth and P. Day, Inorg. Nucl. Lett., 12, 451-453 (1976).
14. A. W. Sleight and D. E. Cox, Solid State Commun., 58, 347-350 (1986).

15. M. Oda, Y. Hidaka, A. Katsui and T. Murakami, Solid State Commun., 60, 897-900 (1986).
16. H. Takogi, M. Naito, S. Uchida, K. Kitazawa, S. Tanaka, and A. Katsui, Solid State Commun., 55, 1019-1022 (1985).
17. H.-C. L. Loye, K. J. Leary, S. W. Keller, W. K. Ham, T. A. Faltens, J. N. Michaels, and A. M. Stacy, Science, in press.

Chemistry and high T_c superconductivity in the La–Ba–Cu–O system

Shu Li and Martha Greenblatt+

Department of Chemistry, Rutgers, The State University of New Jersey, New Brunswick, NJ 08903

Abstract - A systematic investigation of the chemistry and superconductivity in the La-Ba-Cu-O system was carried out, using X-ray powder diffraction, resistivity and TGA measurements. A phase diagram for the La-Ba-Cu-O quasi-ternary system at 950°C has been determined and the relationship between the phases in this system is discussed. A series of A-cation (ABO_3-type perovskite) deficient compounds of the type of $(LaBa_2)_xCu_3O_{7.0}$ ($0.67 < x < 0.96$) were prepared. Cation deficiencies appear to stabilize the structure but depress T_c. Partial substitution of La and Ba in $LaBa_2Cu_3O_{7+\delta}$ by other rare earth and alkali earth cations, respectively, were also investigated. An enhancement in T_c, compared with $La_{1+x}Ba_{2-x}Cu_3O_{7+\delta}$, was found in $LaBa_{2-x}Ca_xCu_3O_{7+\delta}$ ($0 < x < 1.0$), providing further evidence for the importance of reduced lattice dimensions in stabilizing the superconducting phase of La-123.

INTRODUCTION

Since the discovery of the high T_c superconductor $YBa_2Cu_3O_{7-\delta}$ (often referred to as the 123 compound) which was shown to be an oxygen deficient perovskite-type compound[1], the substitution of yttrium by other elements has become a very active field in the chemistry of high T_c superconducting materials. Yttrium can be substituted for by all of the rare-earth elements except cerium and terbium,[2] to form the same type of perovskite-like structure with the stoichiometry of $ReBa_2Cu_3O_{7-\delta}$ (Re = Rare-earth). However, some of these rare-earth analogues show superconducting properties different from that of $YBa_2Cu_3O_{7-\delta}$. For example, $PrBa_2Cu_3O_{7-\delta}$ can be prepared with the identical structure of $YBa_2Cu_3O_{7-\delta}$ but it does not show superconductivity, possibly, because of the mixed-valence of praseodymium cations. 123 compounds containing the heavier rare-earths (Er-Lu) and yttrium can display bulk superconductivity at 90 K, while the lanthanum analogue, $LaBa_2Cu_3O_{7+\delta}$ becomes superconducting at about 55-75 K[3,4,5]. Moreover, it is very difficult to prepare the La analogue in pure phase form; $BaCuO_2$ is always present as an impurity phase together with nominal $LaBa_2Cu_3O_{7+\delta}$. Recently, the T_c for $LaBa_2Cu_3O_{7+\delta}$ was reported to be 90 K, but the Meissner effect measurement indicated only about 30% bulk superconductivity.[6,7]

It is known that lanthanum can form various cuprate phases with perovskite-related structures, e.g. La_2CuO_4, $La_2SrCu_2O_7$[8] and $LaCuO_3$[9]. Yttrium and other heavy rare earths do not form these phases. The structure of $LaBa_2Cu_3O_{7+\delta}$ was recently determined by neutron diffraction techniques[4]. Compared with $YBa_2Cu_3O_{7-\delta}$, some notable differences in the oxygen content, oxygen distribution, symmetry of the crystal lattice, and the disordering between the La^{3+} and Ba^{2+} cations in the lattice were found in the $LaBa_2Cu_3O_{7+\delta}$ system. In general, the La-Ba-Cu-O perovskite-type structure appears to be more stable than the Y-Ba-Cu-O system; therefore, variations in the $LaBa_2Cu_3O_{7+\delta}$ stoichiometry might be more feasable.

The copper chemistry in the La-Ba-Cu-O system is also very interesting. Several La-Ba-Cu-O compounds show very high formal oxidation states for copper. For example, $La_4BaCu_5O_{13.4}$ has a copper valence of 2.50. However, it is metallic and not superconducting. In $La_{1+x}Ba_{2-x}Cu_3O_{7+\delta}$ the copper valence does not change with x, however, T_c decreases with increasing x for $0.2 < x < 0.5$ (i.e. the range of superconductivity).[7,4] Although the existance of Cu^{3+} species in these high T_c superconducting materials is arguable,[10] the significance of some form of mixed valent copper (or oxygen) is apparent.

In this work, we report on a systemetic study of the La-Ba-Cu-O system. We have investigated the phase diagram at 950°C and will discuss the relationship between the various phases present. The dependence of the T_c on the formal copper valence, oxygen content, impurity phases, cation vacancies, and the disordering of trivalent and divalent A-cations in the structure were examined.

EXPERIMENTAL

All the samples were prepared using conventional solid state reaction procedures. Stoichiometric mixtures of La_2O_3, $BaCO_3$ and CuO were calcined at 950°C for 24 hrs with repeated grindings, then quenched to room temperature and pressed into pellets at 80,000 psi. The pellets were then sintered at 950°C for 24 hrs, followed by oxygen annealing at 450°C-500°C for another 24 hrs. Samples were identified by powder x-ray diffraction on a SCINTAG PAD IV system with Si as an internal standard. The lattice parameters were calculated using a least-square refinement of the observed x-ray diffraction data. The oxygen stoichiometry of the samples was determined using a Du Pont 951 thermogravimetric analyzer. The standard four-probe technique was used to measure the resistivity of all samples. Indium leads were attached to rectangularly-shaped pellets using either ultrasonic soldering or silverprint.

RESULTS AND DISCUSSION

1. The phase diagram of $LaO_{1.5}$-BaO-CuO system and its basic properties.

An effort to find new compounds in the La-Ba-Cu-O system resulted in the phase diagram, as shown in Fig 1. The $LaO_{1.5}$-BaO-CuO system includes several previously known phases:

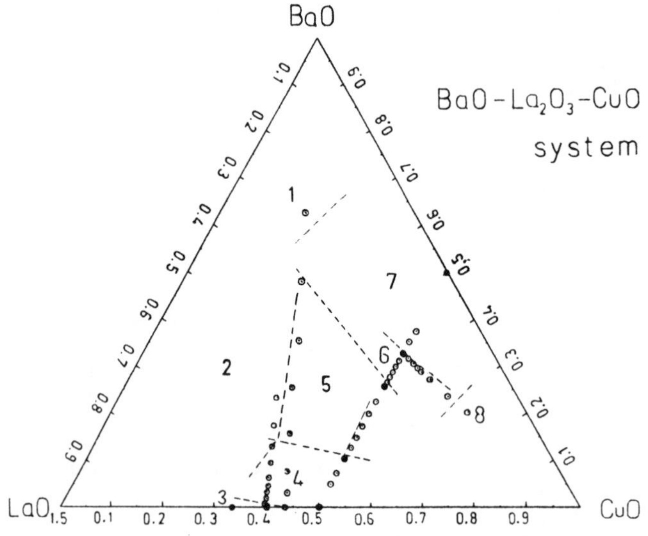

Fig. 1 Phase diagram of $LaO_{1.5}$-BaO-CuO system
1) $BaCuO_2$ + unknown; 2) $La_{2-x}Ba_xCuO_4$ + unknown;
3) $La_{2-x}Ba_xCuO_4$; 4) $La_4BaCu_5O_{13+\delta}$ + $La_{2-x}Ba_xCuO_4$;
5) $La_4BaCu_5O_{13+\delta}$ + $La_3Ba_3Cu_6O_{14+\delta}$ + $La_{2-x}Ba_xCuO_4$;
6) $LaBa_2Cu_3O_{7+\delta}$-type (La-123); 7) La-123 + $BaCuO_2$;
8) La-123 + CuO

La_2CuO_4, $La_4BaCu_5O_{13+\delta}$,(11) $La_3Ba_3Cu_6O_{14+\delta}$,(12) $La_{1+x}Ba_{2-x}Cu_3O_{7+\delta}$, $LaCuO_3$, and $BaCuO_2$. All of these phases, except $BaCuO_2$, may be considered to form oxygen deficient perovskite-related structures. However, the compounds differ in oxygen content and oxygen distribution with formation of different coordination around the copper, lanthanum and barium cations in the various structures as shown in Fig. 2. These differences are undoubtedly important in determining the electrical properties of these compounds. $LaCuO_3$ has a rhombohedrally distorted perovskite structure, and it only forms at high oxygen pressure; therefore, it will not appear in the phase diagram. $La_4BaCu_5O_{13+\delta}$ may be prepared as a homogeneous pure phase only in a very small region of the phase diagram. It forms in a tetragonal perovskite-like structure with space group P4/mmm, with a = 8.648Å, c = 3.859Å. However, $La_4BaCu_5O_{13+\delta}$ is found together with $La_{2-x}Ba_xCuO_4$ and $La_{1+x}Ba_{2-x}Cu_3O_{7+\delta}$ $(0 < x \leq 0.5)$ in a large region in the middle of the phase diagram. It disappears as the ratio of Ba/La exceeds one in the nominal starting compositions. Raveau et al have reported detailed studies of the structural and basic conduction properties of this compound in the temperature range from 100 K to 300 K.(13) We measured the temperature dependence of the resistivity for $La_4BaCu_5O_{13+\delta}$ down to 4 K and the data are shown in Fig. 3.

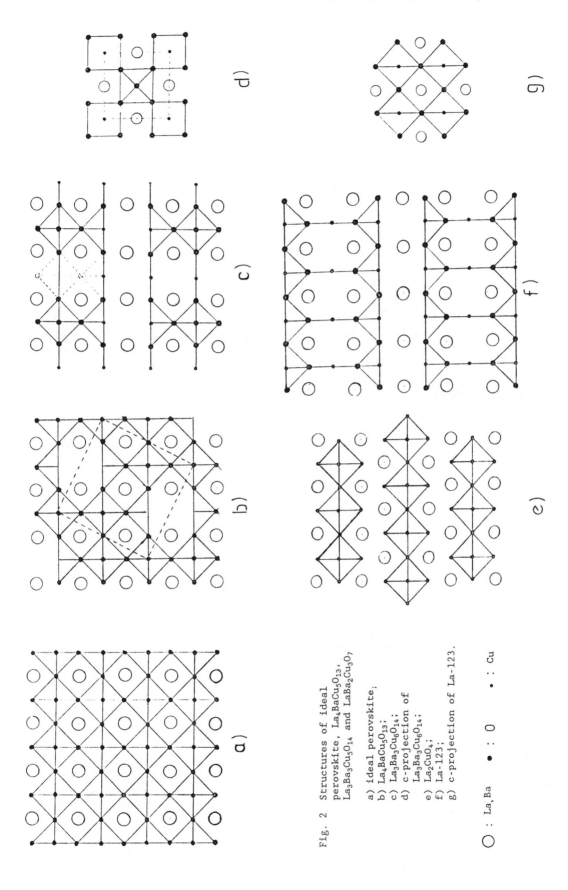

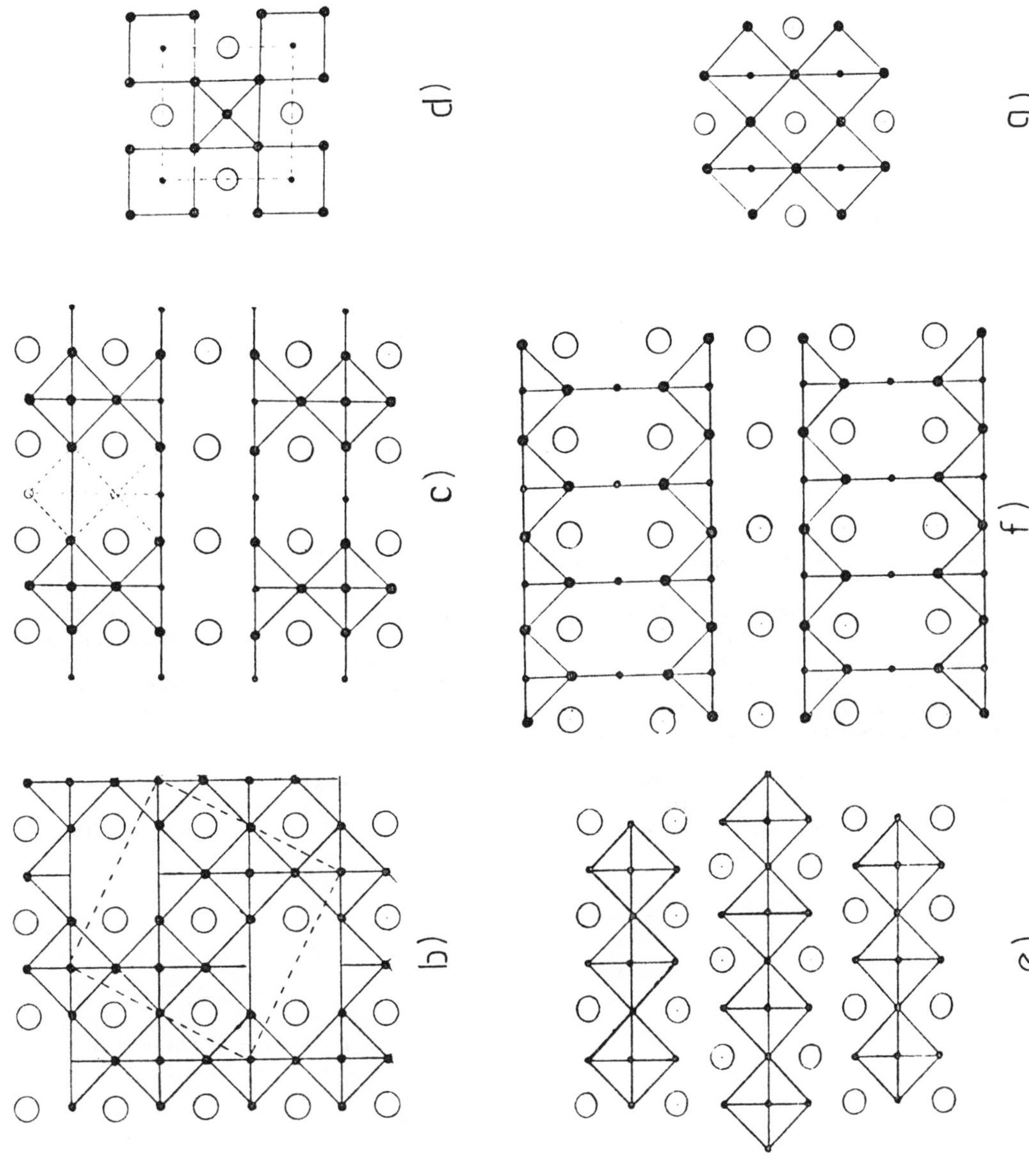

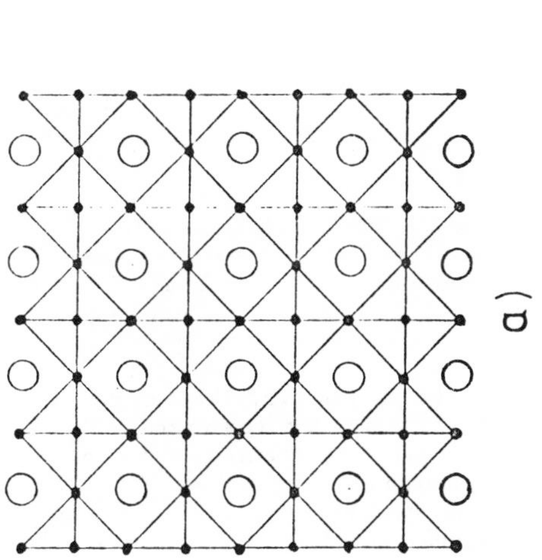

Fig. 2 Structures of ideal perovskite, $La_4BaCu_5O_{13}$, $La_3Ba_3Cu_5O_{14}$ and $LaBa_2Cu_3O_7$

a) ideal perovskite;
b) $La_4BaCu_5O_{13}$;
c) $La_3Ba_3Cu_6O_{14}$;
d) c-projection of $La_3Ba_3Cu_6O_{14}$;
e) La_2CuO_4;
f) La-123;
g) c-projection of La-123.

○ : La, Ba ● : O • : Cu

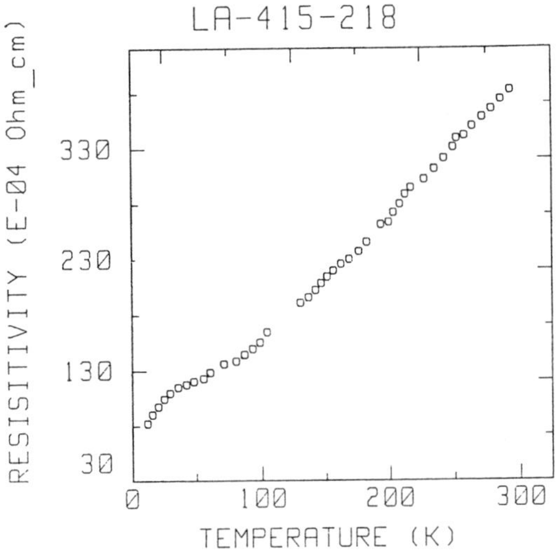

Fig. 3 Resistivity vs. temperature of $La_4BaCu_5O_{13+\delta}$

At temperatures above 60 K, typical metallic behavior is observed. At lower temperatures, the ρ vs. T relationship deviates from linearity. Recently, Rao et al reported an anomaly at about 100 K for a sample of $La_4BaCu_5O_{13+\delta}$,(14) but this anomaly is not seen in our data (Fig. 3). Partial substitution by other rare-earth ions including Pr, Nd and Sm for La ($La_{4-x}Re_xBaCu_5O_{13+\delta}$, $0 < x < 1.6$), as well as strontium or calcium for barium, ($La_4Ba_{1-x}M_xCu_5O_{13+\delta}$, (M = Ca or Sr, $0 < x < 0.3$) were also investigated. Although single phase solid solutions were formed, these substitutions have little or no effect on the temperature dependence of the resistivity.

As the Ba/La ratio is increased from 0.25 (Fig. 1), $La_3Ba_3Cu_6O_{14+\delta}$ appears as a competing phase. The higher the Ba/La ratio, the more $La_3Ba_3Cu_6O_{14+\delta}$ is in the mixture. Interestingly, this mixture of $La_4BaCu_5O_{13+\delta}$ and $La_3Ba_3Cu_6O_{14+\delta}$ exhibits an X-ray diffraction pattern very similar to that of $YBa_2Cu_3O_{7-\delta}$ (Fig. 4).

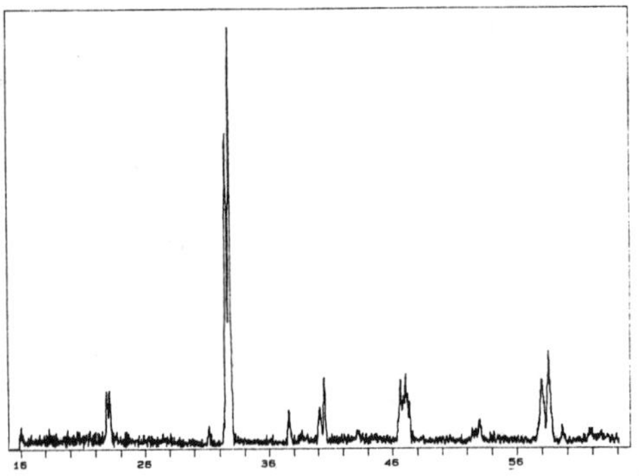

Fig. 4 The X-ray powder diffraction pattern of a mixture of $La_4BaCu_5O_{13+\delta}$ and $La_3Ba_3Cu_6O_{14+\delta}$, quite similar to that $YBa_2Cu_3O_{7-\delta}$

When the Ba/La ratio equals one, pure $La_3Ba_3Cu_6O_{14+\delta}$ is formed. The electrical and structural properties of this compound have been investigated by Raveau.(12,15) A recent neutron diffraction study of $La_3Ba_3Cu_6O_{14+\delta}$ indicates a tetragonal lattice with a = 3.90Å and c = 11.69Å,(4) compared to the lattice parameters of a = 5.52Å and c = 11.72Å reported by Raveau.(12)

$La_3Ba_3Cu_6O_{14+\delta}$ is semiconducting and the copper valence and the oxygen distribution can be varied by intercalation of oxygen into the structure. Fig. 5 shows the temperature dependence of the resistivity for $La_3Ba_3Cu_6O_{14+\delta}$ and our data are consistent with those previously reported.(15) When 1 < Ba/La < 2, the high T_c superconducting phase $La_{1+x}Ba_{2-x}Cu_3O_{7+\delta}$, La-123 is formed. It has either tetragonal symmetry with a = 3.923Å and c = 11.78Å or a slight orthorhombic distortion from tetragonal symmetry.(4,6) However, it is very difficult to prepare $LaBa_2Cu_3O_{7+\delta}$ as a pure phase. Even when $BaCuO_2$ is not detectable by X-ray diffraction, it has been shown to be present by other techniques, such as Raman spectroscopy.(7) We propose that at the preparative temperatures necessary for its formation, pure $LaBa_2Cu_3O_{7+\delta}$ tends to decompose by the following reaction.

$LaBa_2Cu_3O_{7+\delta}$ --------> $((3 - x)/3)La_{1+y}Ba_{2-y}Cu_3O_{7+\delta} + xBaCuO_2$

(where $y = x/(3 - x)$)

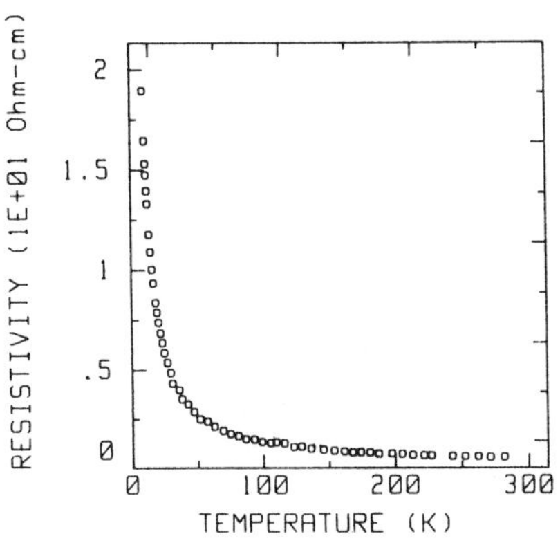

Fig. 5 Temperature dependence of the resistivity for $La_3Ba_3Cu_6O_{14+\delta}$

Therefore, the compound responsible for the high T_c superconductivity is $La_{1+y}Ba_{2-y}Cu_3O_{7+\delta}$ and not the compound with exact 1-2-3 stoichiometry. Thus, $La_{1.15}Ba_{1.85}Cu_3O_{7+\delta}$ was prepared without any $BaCuO_2$ impurity and this compound is superconducting at about 60 K, as shown in Fig. 6. Others have found similar results in the $La_{1+x}Ba_{2-x}Cu_3O_7$ system,[4]

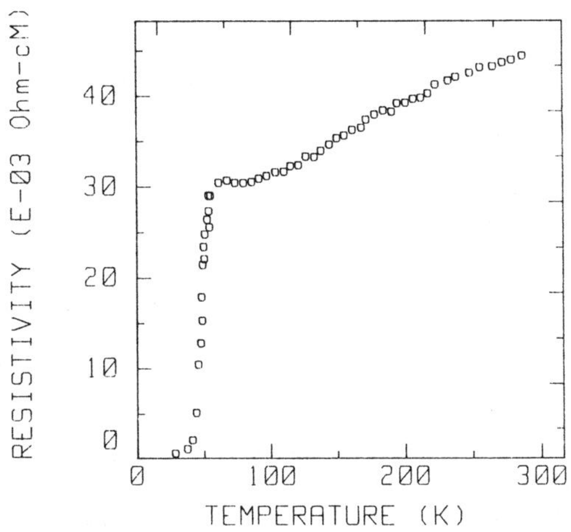

Fig. 6 Resistivity vs. temperature behavior of $La_{1.15}Ba_{1.85}Cu_3O_{7.3}$

with superconductivity for $0.2 < x < 0.5$ and decreasing values of T_c with increasing x.

The reason for the instability of $LaBa_2Cu_3O_{7+\delta}$ is not quite clear. It may be related to the A-cation size effect in the perovskite-type structure. It is well known that the stability of perovskite-type structure is affected by the A-cation size. In the case of pure $LaBa_2Cu_3O_{7+\delta}$, perhaps, the A-cations (33% La^{3+} + 67% Ba^{2+}) are too large, and the lattice is stabilized by reducing the Ba/La ratio, and hence the unit cell volume. Actually, the La-123 structure can form in a relatively large region of the phase diagram if the volume of the latice is reduced. We have prepared a series of A-cation deficient single phase compounds, $(LaBa_2)_xCu_3O_{7+\delta}$ $(0.67 < x < 0.96)$, which have smaller unit cells than near stoichiometric La-123. Table 1 summarizes the lattice parameters of $(LaBa)_xCu_3O_{7+\delta}$ phases. Substitution of La^{3+} by other rare-earth elements or the substitution of Ba^{2+} by Sr^{2+} and/or Ca^{2+} also facilitates the formation of pure compounds of the La-123 structure, as indicated by X-ray powder diffraction. This will be discussed in the following section.

The compounds, $La_{2-x}Ba_xCuO_4$ with the K_2NiF_4-type structure, also exist in a large area of the phase diagram as pure phases or as a competing phases. In addition to $BaCuO_2$, there is evidence of other binary Ba-Cu-O compounds, however, these could not be identified unambiguously, although Wang et al recently reported the existence of Ba_2CuO_3 and Ba_3CuO_4.[16]

A comparison of the phase diagram of the Y-Ba-Cu-O[16] and the La-Ba-Cu-O systems shows

Table 1. Lattice parameters of $(LaBa_2)_xCu_3O_{7.0}$ compounds

Sample	Stoichiometry(x)	a(Å)	c(Å)	T_c(K)
La-123	1.00	3.923	11.782	--
301	0.96	3.914(1)	11.771(2)	55.3
302	0.91	3.917(1)	11.752(2)	53.2
303	0.83	3.917(1)	11.748(2)	47.5
304	0.75	3.916(1)	11.745(2)	34.8(not sc)
305	0.67	3.907(2)	11.723(4)	no

some interesting differences. Generally, lanthanum and yttrium have quite different oxide chemistry, probably, partly due to the great differences in their effective ionic radii. Phases such as $Y_2Cu_2O_5$, Y_2BaCuO_5 and $YBa_3Cu_3O_7$ do not form in the La-Ba-Cu-O system. In the La-Ba-Cu-O system, only compounds with perovskite-related structures are found. Therefore, if the perovskite-type structure is essential for high T_c, the La-Ba-Cu-O system may be a worthwhile target, and should be further explored.

2. Superconductivity and copper chemistry in La-Ba-Cu-O system

The oxidation state of copper in $La_4BaCu_5O_{13.4}$ is 2.50, in $La_3Ba_3Cu_5O_{14.5}$ it is 2.35 and in $La_{1+x}Ba_{2-x}Cu_3O_{7+\delta}$ it is a constant 2.35. In the $(LaBa_2)_xCu_3O_{7.0}$ (0.67 x < 0.96) phases studied here, the the formal valence of copper changes from 2.4 to 3.1. However, $La_4BaCu_5O_{13.4}$ is a typical p-type metal; $La_3Ba_3Cu_6O_{14.5}$ is semiconducting and only $La_{1+x}Ba_{2-x}Cu_3O_{7+\delta}$ ($0.2 < x < 0.5$) shows T_c to be compositional dependent, but not copper valence dependent. This suggests that the copper valence is not the only factor in determining the high T_c superconductivity.

In $(LaBa_2)_xCu_3O_{7.0}$, the formal copper valence can be varied because of the Ba/La deficiency in the lattice. The T_c of the samples is a function of the nominal copper valence (Table 1, Fig. 7 and Fig. 8). Interestingly, $(LaBa_2)_{0.67}Cu_3O_{7.0}$ indicates an extraordinarily high copper valence; it may contain only Cu^{3+} species, but it is only

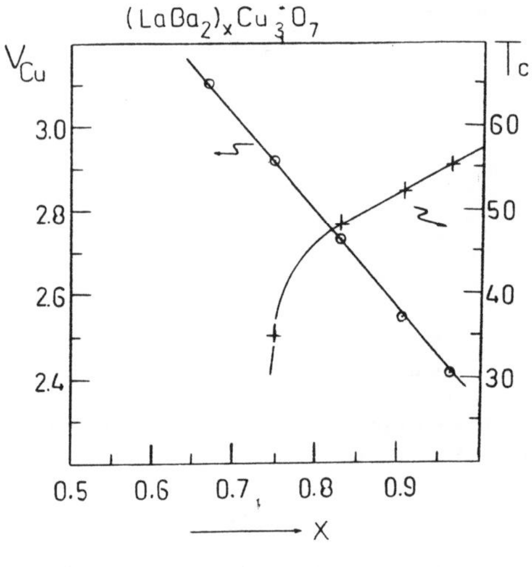

Fig. 7 Relationship between the formal copper valence and T_c for $(LaBa_2)_xCu_3O_{7.0}$ compounds

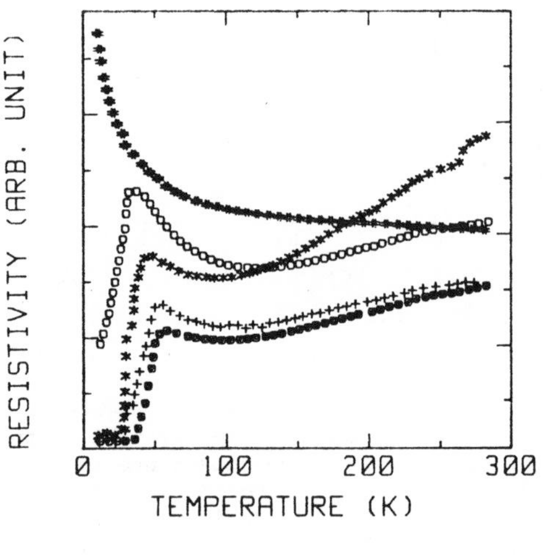

Fig. 8 Temperature dependence of the resistivity for $(LaBa_2)_xCu_3O_{7.0}$ •: x = 0.96; +: x = 0.91; *: x = 0.83; o: x = 0.75; #: x= 0.67

semiconducting. $(LaBa_2)_{0.75}Cu_3O_{7.0}$, displays a formal copper valence of 2.9 and a transition in the ρ vs. T at about 35 K (Fig. 8) but the resistivity does not go to zero down to 4 K. The phases with larger x values and correspondingly lower copper valences show superconductivity at higher T_c (Fig. 8).

All these compounds have the same seven oxygen content as determined by TGA. This oxygen stoichiometry is relatively high, compared with other high T_c compounds of the same class. The copper valence varies with the amount of deficiencies of the Ba/La cations in $(LaBa_2)_xCu_3O_{7.0}$. In $YBa_2Cu_3O_{7-\delta}$ the formal oxidation state of copper can vary only with the change in oxygen content. The conduction band of high T_c superconductors results from the overlap of Cu $d_{x^2-y^2}$ and oxygen 2p-π orbitals. If both the copper formal valence (or possibly the oxygen valence) and oxygen content change, it is not clear which factor has a major effect on T_c in these materials. In $(LaBa_2)_xCu_3O_{7.0}$, there appears to be a relationship between the formal oxidation state of copper and T_c, since the oxygen content is constant at 7.0. But, both oxygen content and oxygen ordering are important factors for superconductivity as shown in the $YBa_2Cu_3O_{7-\delta}$ system.[17,18] Moreover, lattice distortions due to A-cation deficiencies may also affect the electrical properties significantly by oxygen and possibly Cu^{3+} ordering.

3. Substitution for Ba^{2+} in the La-123 compound

Since we noted that a smaller lattice facilitates the stability of the La-123 structure, Ca-substituted La-123 compounds were prepared as $LaBa_{2-x}Ca_xCu_3O_{7+\delta}$ $(0 < x < 1)$. The lattice parameters of these compounds are summarized in Table 2.

Table 2. Lattice parameters and T_c of $LaBa_{2-x}Ca_xCu_3O_{7+\delta}$.

Sample	x	a(Å)	C(Å)	T_c(K)
La-123	0	3.923	11.782	--
336	0.2	3.914	11.73(1)	60.0
337	0.4	3.893(0)	11.70(0)	75.5
338	0.6	3.883(2)	11.66(1)	77.0
339	0.8	3.877(1)	11.62(0)	78.4
340	1.0	3.868(1)	11.62(1)	80.5

The lattice parameters decrease uniformly with increasing x in $LaBa_{2-x}Ca_xCu_3O_{7+\delta}$, relative to "pure" La-123 (Table 2). Calcium substitution eliminates the $BaCuO_2$ impurity from the product as indicated by X-ray powder diffraction. T_c increases from 60 K up to 80 K as the calcium content increases. Fig. 9 illustrates the temperature dependence of the resistivity of the $LaBa_{2-x}Ca_xCu_3O_{7+\delta}$ samples. In $La_{1+x}Ba_{2-x}Cu_3O_7$, the La^{3+}, Ba^{2+}

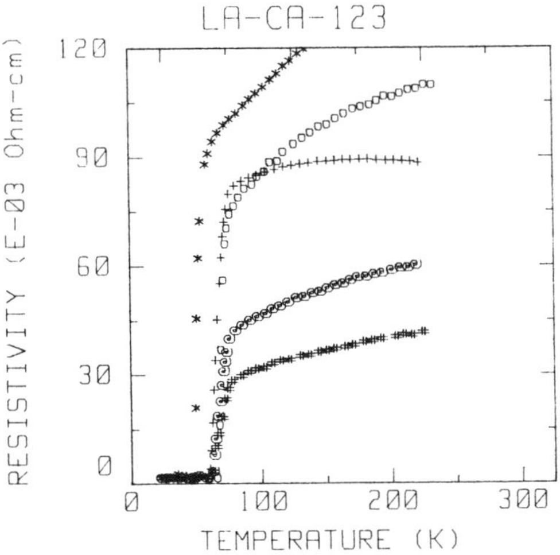

Fig. 9 Superconducting properties of $LaBa_{2-x}Ca_xCu_3O_{7+\delta}$ compounds. #: = 1.0; @: x = 0.8; +: x = 0.6; o : x = 0.4; *: x = 0.2

disordering is believed to affect the T_c. In the Ca^{2+} substituted phases, cation ordering similar to that found in $YBa_2Cu_3O_7$, may be occurring. It is noteworthy that the calculated lattice parameters of $LaBa_2CaCu_3O_{7+\delta}$ are very close to those of of $YBa_2Cu_3O_{7-\delta}$, further confirming that critical lattice dimensions are important for high superconducting transition temperatures in oxocuprates.

CONCLUSION

We have established the phase diagram of the La-Ba-Cu-O system in which compounds only with perovskite-type structures were found. The phase diagram is very different from that of Y-Ba-Cu-O. $(LaBa_2)_xCu_3O_7$ $(0.67 < x < 0.96)$ phases were prepared with a large range of cation deficiencies. With careful oxygen annealing of these samples, the oxygen content in each was found to be constant at 7. Thus, the oxidation state of copper varied with x.

A clear relationship between the formal copper valence and T_c has been demonstrated. There is evidence that $LaBa_2Cu_3O_{7+\delta}$ is not stable. Cation deficiencies appear to reduce the lattice volume and stabilize the structure as demonstrated above. This is also confirmed by the formation of Ca substituted $LaBa_{2-x}Ca_xCu_3O_{7+\delta}$, where the T_c is enhanced with increasing x and the lattice parameters are comparable to those of Y-123 at x = 1.

ACKNOWLEDGEMENT

We greatfully acknowledge helpful discussions with Dr. S. Fine and Dr. K.V. Ramanujachary. This work was supported by the Office of Naval Research and by the National Science Foundation Solid State Chemistry Grants DMR-84-04003 and DMR-84-08266.

REFERENCES

1. R.M. Hazen, L.W. Finger, R.J. Angel, C.T. Prewitt, N.L. Ross, H.K. Mao, C.G. Hadidiacos, P.H. Hor, R.L. Meng and C.W. Chu, Phys. Rev. B., 35, 7238 (1987).
2. Shigetoshi Ohshima and Tokuo Wakiyama, Jap. J. Appl. Phys. Lett., 26, L815 (1987).
3. I. Kirshner and Bankuti; Phys. Rev. B., 36, 2313 (1987).
4. C.U. Segre, B. Dabrowski, D.G. Hinks, K. Zhang, J.D. Jorgensen, M.A. Beno and I.K. Schuller; Nature, 329, 227 (1987).
5. Sung-Ik Lee, John P. Golben, Sang Young Lee, Xia-Dong Chen, Yi Song, Tae W. Noh, R.D. McMichale, J.R. Gains, D.L. Cox and Bruce R. Patton, Phys. Rev. B., 36, 2417 (1987).
6. Atsataka Maeda, Tomoaki Yabe, Kunimitsu Uchinokula and Shoji Tanaka, Jap. J. Appl. Phys. Lett., 26, L1368 (1987).
7. S.A. Sunshine, L.F. Schneemeyer, J.V. Waszczak, D.W. Murphy, S. Miraglia, A. Santoro and F. Beech, a preprint, submitted for publication.
8. N. Nguyen, C. Michel, F. Studer, and B. Raveau, Mater. Chem., 7, 413 (1982).
9. M. Arjomand and David J. Machin, J. Chem. Soc. Dalton, 1061 (1975).
10. Zhi-xun Shen, J.W. Allen, J.J. Yeh, J.S. Kang, W. Ellis, W. Spice, I. Lindau, M.B. Maple, Y.DD. Dalichaouch, M.S. Torikachvili and J.Z. Sun, to be published.
11. C. Michel, L. Er-Rakho, M. Hervieu, J. Pannetier and B. Raveau, J.Solid State Chem., 68, 143 (1987).
12. J. Provost, F. Studer, C. Michel and B. Raveau, Synthetic Metals, 4, 147 (1981).
13. C, Michel, L. Er-Rakho and B. Raveau, Mater. Res. Bull., 20, 667 (1985).
14. P. Ganguly, R.A. Mohan Ram, K. Sreedhar and C.N.R. Rao, Solid State Commun., 62, 807 (1987).
15. J. Provost, F. Studer, C. Michel and B. Raveau, Synthetic Metals, 4, 157 (1981).
16. G. Wang, S.J. Hwu, S.N. Song, J.B. Ketterson, L.D. Marks, K.R. Peoppelmeier and T.O. Mason, Adv. Ceram. Mater., 2, 313 (1987).
17. J.D. Jorgensen, B.W. Veal, W.K. Kowk, G.W. Crabtree, A. Umezawa, L.J. Nowicki and A.P. Paulikas, Phys. Rev. B., 36, 5731 (1987).
18. Izumi Nakai, Katsuhiro Imai, Takuji Kawashima and Ryozo Yoshizaki, Jap. J. Appl. Phys. Lett., 26 L1244 (1987).

Synthesis, structural chemistry and properties of $YBa_2Cu_3O_{7-x}$

A. J. Jacobson, J. M. Newsam, D. C. Johnston*, J. P. Stokes, S. Bhattacharya, J. T. Lewandowski, D. P. Goshorn, M. J. Higgins and M. S. Alvarez.

Exxon Research and Engineering Company
Annandale, NJ 08801, USA

Abstract The solid state syntheses of $YBa_2Cu_3O_{7-x}$ samples for $0.00 \leq x \leq 1.00$ and the use of various characterization techniques including thermogravimetric analysis, X-ray and neutron powder diffraction and magnetization measurements are described. Ultrasonic propagation, resistivity and magnetic susceptibility anomalies in $YBa_2Cu_3O_{7-x}$; $x \approx 0.0$ are reported. Data pertaining to the T - x phase diagram for $YBa_2Cu_3O_{7-x}$, and the dependence of the magnetic, structural and superconducting properties of $YBa_2Cu_3O_{7-x}$ on x are discussed.

INTRODUCTION

The occurrence of superconductivity at relatively high temperatures in copper-containing oxides (refs. 1-6) adopting structures related to K_2NiF_4 or perovskite, $CaTiO_3$, has focussed widespread interest on these rather common structure types (refs. 7-10). Phases related to $YBa_2Cu_3O_{7-x}$, the 90K superconducting material (refs. 4-6) that adopts a tripled oxygen-deficient perovskite structure (refs. 11,12,13 and refs. cited), in particular, have been subjected to close scrutiny. In several senses these particular materials are special. They have enabled superconducting onset temperatures to be raised from 23K to at least 90K, and perhaps much higher. However, the structural chemistries of these phases, which at first sight appear complex, can be rationalized readily in the context of the wealth of information that is available for related perovskite materials. This correspondence, which has not gone unrecognized (refs. 14,15), is discussed below.

There is already a large amount of literature data on $YBa_2Cu_3O_{7-x}$ materials, and the major features of their behavior are now well established. We focus here mainly on our own results, but also cite from the growing literature base. First, we discuss some general characteristics of oxygen ion deficient perovskites, approaches to the synthesis of non-stoichiometric $YBa_2Cu_3O_{7-x}$ phases, and data obtained by thermogravimetric analysis (tga) and powder diffraction techniques. We then outline anomalies in the ultrasonic propagation, resistivity and magnetic susceptibility data for $YBa_2Cu_3O_{7-x}$; $x \approx 0.0$. Finally, we describe and discuss the manner in which the magnetic, structural and superconducting properties of $YBa_2Cu_3O_{7-x}$ evolve with oxygen content.

THE PEROVSKITE STRUCTURE

As a class of materials, compounds with the perovskite structure (Fig. 1), of general composition ABO_3, display an interesting range of magnetic, electronic, dielectric and catalytic properties. The structure can accommodate a wide range of chemical substitution and is tolerant of large deviations from stoichiometry on all of the A, B and anion (O) sublattices (refs. 16,17). For example, the A sites are partially occupied in the perovskite tungsten bronzes, A_xWO_3 (ref. 18), while one third of the B sites are empty in the hexagonal but perovskite related compound $Ba_3W_2O_9$ (ref. 19). By far the most common type of nonstoichiometry, however, and one of primary interest in relation to the high temperature superconducting oxides is that involving anion vacancies which can occur in ABO_{3-x} perovskites for $0.0 \leq x \leq 1.0$. Unlike other non-stoichiometric oxides, the oxygen ion vacancies cannot be eliminated from the structure by crystallographic shear (ref. 17) as a consequence of the coordination requirements of the A cation and the fixed A/B ratio. Instead, the perovskite structure basis is maintained but with oxide vacancies which are

*Present Address: Department of Physics and Ames Laboratory (USDOE), Iowa State University, Ames, IA 50011

ordered or disordered depending on synthesis conditions and composition. Ordered vacancies define a new structure type but one which is related to the stoichiometric equivalent in having a similar arrangement of A and B cations.

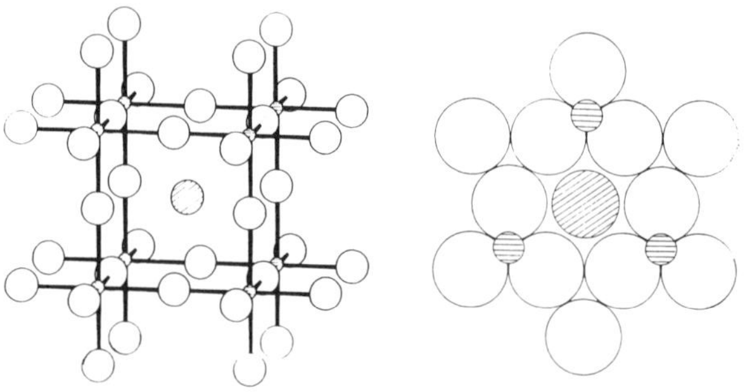

Fig. 1. The ABO_3 perovskite structure (left) and its 111 projection showing AO_3 close-packing (right). (⊘ = A, ⊖ = B, ○ = O).

Oxide vacancies alter the local B cation coordination. In a general way, the vacancy ordering scheme adopted by a specific system depends on the ability of different B cations in different oxidation states to exist in different coordinations. Two examples illustrate this effect. In the compound brownmillerite (ref. 20), $Ca_2Fe_2O_5$ ($CaFeO_{2.5}$), all of the iron atoms are present as $Fe^{3+}(3d^5)$ and the vacancies are ordered in rows along the - perovskite 110 direction, in layers that alternate along 001. The result is that half of the iron atoms are tetrahedrally coordinated while the other half are octahedral. The two different types of iron atom are thus arranged in alternating layers along 001. In the related compound Ca_2FeCoO_5 (ref. 21) the vacancy distribution is the same but the $Co^{3+}(3d^6)$ cations populate all of the octahedrally coordinated sites because of their strong preference for this particular coordination. In many cases, simple crystal field arguments for predicting site preferences provide a useful guide to the general features of vacancy order. The manganese analog of the iron system, $Ca_2Mn_2O_5$, is an example (refs. 22,23). The $Mn^{3+}(3d^4)$ cation has a preference for trigonal prismatic or square pyramidal coordination and the ordered vacancy structure adopted enables all of the manganese ions to be square pyramidally coordinated.

For most systems, the oxygen non-stoichiometry spans a broad range of composition at high temperatures. However, as the temperature is lowered discrete phases with narrower ranges of composition appear as a consequence of, for example, vacancy ordering. This general type of behavior is illustrated by $BaBiO_{3-x}$ (ref. 24). Thermodynamic measurements show that above 600°C this material is continuously non-stoichiometric for $0.0 \leq x \leq 0.5$. On cooling below 600°C, however, line phases separate with compositions corresponding to x = 0.0, 0.2 and 0.45. In some systems, several ordered intermediates have been identified and characterized, primarily by electron diffraction and high resolution lattice imaging. For example, in the $CaMnO_{3-x}$ system referred to above (refs. 22,23), five distinct ordered structures have been found, each one corresponding to a different composition and all based on arrangements of different numbers of MnO_6 octahedra and MnO_5 square pyramids (ref. 25).

These examples represent somewhat idealized behavior. The non-stoichiometric behavior of most systems is often much more complex and the appearance or absence of long range vacancy ordering is very strongly dependent on synthesis conditions. For example, in $Ca_xLa_{1-x}FeO_{3-y}$ ($2/3 < x < 1$) disordered intergrowths of $Ca_2Fe_2O_5$ (as above) and $Ca_{2/3}La_{1/3}FeO_{2.67}$ (which has ordered A cations, and hence a tripled unit cell), are observed. However, under different synthesis conditions the same compositions yield an apparently single phase that has a complex microdomain structure (refs. 26,27). In systems where the B cations can adopt more than one oxidation state, synthesis is complicated by changes in composition which occur at constant oxygen partial pressure when annealing temperatures are changed. Unfortunately, annealing at constant composition requires much more detailed information about the thermodynamics than is usually available.

In this context, the structural chemistry of $YBa_2Cu_3O_{7-x}$ that is described in some detail below has many resemblances to the other oxygen defect perovskites. At x = 0, the structure is based on perovskite cation positions but with vacancies ordered to give square pyramidal and square planar copper ions (see Fig. 13 below), coordinations consistent with formal oxidation states close to 2+ and 3+, respectively. Further oxide vacancies are introduced by forming linearly coordinated Cu^+ (Fig. 13 below). One unusual structural feature of this phase, worth noting here because there are very few other examples, is that the perovskite cation framework is stable to a composition (x=1.0) corresponding to ABO_2. The structure at this composition and for other values of x is discussed in more detail below.

SYNTHESIS

Samples of $YBa_2Cu_3O_{7-x}$ were prepared from appropriate proportions of $BaCO_3$, Y_2O_3 (pre-fired at 1000°C to remove any hydroxide or carbonate contamination) and CuO. In a typical synthesis, the reactants were ground together and heated in an alumina crucible at 850°C(48h), 950°C(24h) and, finally, at 500°C(24h). Samples were reground well between treatments. This conventional solid state synthesis route works well and gave, under the above conditions, a material with x = 0.04, and with no detectable impurities. Manual regrindings are tedious for large amounts of material but are, however, necessary to ensure the preparation of uniformly black, homogeneous samples. Such material can be fully oxidized to x = 0.00 at 1 atm. oxygen pressure at 400-425°C. Although $YBa_2Cu_3O_{7-x}$ materials with $x < 0.0$ have been reported by other groups (refs. 28,29), we have not observed such compositions under the preparative conditions discussed here.

After extended exposure to the atmosphere samples develop a light grey coloration, reflecting surface degradation to barium carbonate and/or hydroxides. The presence of such degradation is also evident in the thermogravimetric analysis data (tga - see below) as a low temperature peak in the temperature derivative of the weight loss centered at c.a. 250°C.

In order to investigate the low temperature properties of the $YBa_2Cu_3O_{7-x}$ system as a function of x, it was necessary to prepare samples with preselected oxygen contents. The tga data, and diffraction experiments of other groups (refs. 30,31,13 and refs. cited), demonstrate that desired stoichiometries can be attained by defining an appropriate elevated temperature and oxygen partial pressure and then quenching to room temperature. However, treatment of relatively large volume samples to partially reducing conditions at an elevated temperature followed by 'rapid' quenching to room temperature (or below) is not generally a satisfactory procedure and can give inhomogeneous samples. However, samples small enough to attain thermal equilibrium rapidly can apparently be prepared in homogeneous form by the quenching method. Our initial investigations of the dependence of the superconducting transition temperature and magnetic properties on oxygen content, x, were, for example, based on samples generated 'in situ' in a George Associates Faraday balance (ref. 13). Successive values of x were obtained by heating the previously measured sample (in-situ) up to 460-530°C, waiting until the desired weight change was achieved, dropping the temperature to prevent further weight loss, and removing the evolved oxygen under vacuum. The homogeneity of these samples was confirmed by comparison of the results with data from other samples prepared by annealing in sealed tubes.

Further series of materials with various preselected oxygen contents were prepared by sealed tube equilibrations (refs. 13,32). The x = 1.00 material is prepared simply by heating any $x \neq 1.00$ composition in an inert atmosphere for several hours at $T \simeq 840°C$. Intermediate compositions were then synthesized by mixing appropriate proportions of x = 0.04 and x = 1.00 materials (the latter prepared, specifically, by reduction of $YBa_2Cu_3O_{6.96}$ in He at 840°C for 48h, followed by furnace-cooling in He), sealing the mixtures in silica tubes under vacuum, equilibrating at 650-670°C for 16h, and finally, furnace-cooling.

The sealed tube method affords a simple and accurately reproducible route to intermediate compositions, $0.0 < x < 1.0$. The method is also applicable to single crystals and to sintered pellets, although the restricted oxygen mobilities in these latter cases require much longer equilibration times. The technique also allows the approach to equilibrium with respect to oxygen content to be explored (ref. 32). Thus, a series of samples with targeted compositions close to x = 6.75 was prepared by equilibration in H-tubes. Appropriate amounts of samples with x = 0.10 and x = 1.00 were placed separately into each of the two H-tube arms, the tubes were sealed under vacuum and then heated at a temperature of 150°C, 250°C, or 350°C for 66h. After equilibration under these conditions the contents of the two arms of the H-tubes were separately analysed by tga, powder X-ray diffraction (PXD) and magnetic and Meissner Effect (ME) measurements (see below). The samples in both arms of the H-tubes after treatment at 150°C and 250°C were identical to the starting materials. However, the tga and ME data for the 350°C samples from both sides are closely similar. The tga data (Fig. 2) gave oxygen contents of 6.79 for both samples. The Meissner effect data show nearly identical onset temperatures and similar diamagnetic susceptibilities. PXD data showed both samples to be orthorhombic with similar lattice constants (ref. 32). These experiments show that a temperature of 350°C is sufficient to achieve equilibration for powder samples ($\simeq 1\mu m$ crystallite sizes). Further experiments at temperatures between 250°C and 350°C and at other compositions are being pursued to help define the lower temperature part of the T-x phase diagram.

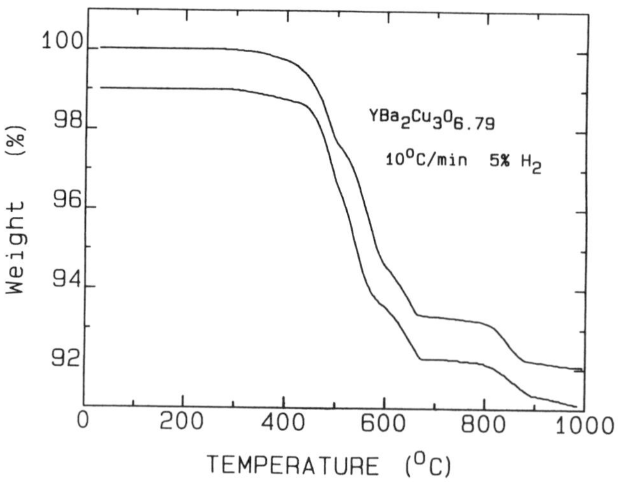

Fig. 2. TGA data for samples with x initially 0.10 (upper) and x = 1.00 (lower - offset by -1% along the weight axis to permit comparison) after separate equilibration at 345°C in the two arms of a sealed H tube.

CHARACTERIZATION

Thermogravimetric analysis

Thermogravimetric analyses were performed using a Dupont 1090/951 thermal analyzer system, using flowing O_2, 5%O_2, or 5%H_2/95%He and with heating rates of 2°C or 10°C min^{-1}. More extended scans were made of selected compositions in order to study weight uptake or loss under kinetically more limited conditions, or to examine the character of phase transitions at higher temperature.

The thermogravimetric analyses have performed an important role in characterizing these materials. First, they provide a means of determining oxygen contents. By monitoring weight losses of samples heated under reducing conditions to temperatures at which final phase identities are known, the weight differences provide measures of oxygen content in the starting materials. Thus in $YBa_2Cu_3O_{7-x}$, for example, reduction in a 5%H_2/95%He atmosphere (see Fig. 2) to 1000°C effects complete conversion to barium and yttrium oxides and Cu metal. The estimated errors on the initial oxygen contents determined in this way are typically 0.03. The thermogravimetric method is rapid and convenient for measuring oxygen contents. Other analytical techniques such as iodometric titration to determine the average copper oxidation state have also been used by other groups (refs. 33,34).

The tga scans shown in Figs. 2 and 3, convey more information than simply the weight changes. The rate of weight change during a temperature ramp is, generally, a structural characteristic. Thus, even though the system is not necessarily at equilibrium, changes in slope are indicative of phase changes. The weight loss to the point of the discontinuity indicates the composition at which the phase change occurs. A steady weight loss implies no structural change. The most notable discontinuity in the He atmosphere data reproduced in Fig. 3 occurs just above 800°C and corresponds to a composition x = 1.00. This reproducible characteristic discontinuity in the tga scans defined the conditions required to synthesize such a material. The x = 1.00 material was investigated structurally (ref. 13; see below), and forms a precursor in our preparative route to intermediate compositions by sealed tube equilibrations, as discussed above (refs. 13,32). Notably, tga scans of $La(La,Ba)_2Cu_3O_{7-x}$ materials also show pronounced discontinuities corresponding to x = 1.00 (ref. 35). This implies that the stability of the x = 1.00 composition is structural rather than electronic, because the mean copper valence state at x = 1.00 is different for differing Ba/La ratios.

Further discontinuities occur at other temperatures, depending on the atmosphere used (Figs. 2,3). However, there is no discontinuity corresponding to x = 0.5 for any atmosphere. Initially, this was surprising, as this composition would presumably correspond to all copper ions being in the +2 valence state. As is discussed below, the phase diagram proves to be somewhat more complex than would be expected on the basis solely of isolated-ion electronic considerations, and the composition x = 0.5, as indicated by the initial tga results, is apparently not a stable composition at lower temperatures.

This discussion of the changes in slope in the tga scans reflects that for regions where equilibrium with respect to oxygen content is attained, the tga scans represent isobaric lines in the T - x phase diagram. The high temperature (T > 350°C) phase diagram can therefore be explored in detail by using differing partial oxygen pressures for the tga scans. Such data have now been reported by several independent groups and the various results are generally in good agreement (refs. 36-40). These representations of the phase diagram have also been substantiated in detail by extensive structural analyses based on high-temperature powder neutron diffraction (ref. 30,31). Equilibrium data for temperatures T < 350°C are rather more difficult to obtain.

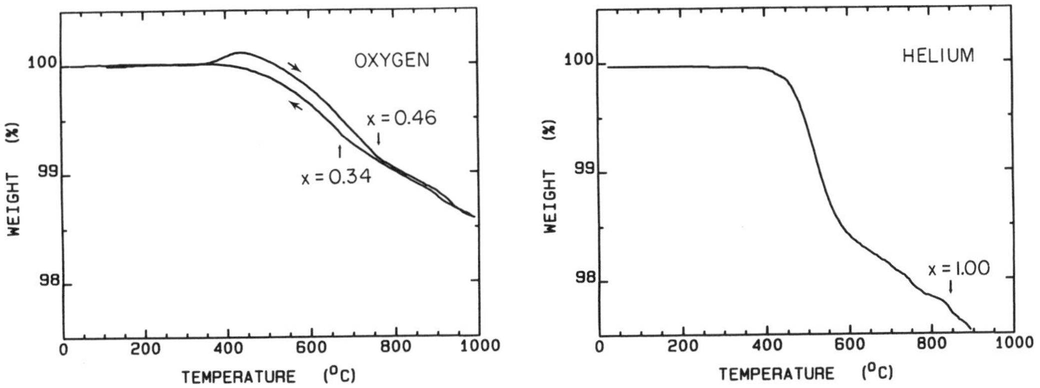

Fig. 3. TGA data for scans $YBa_2Cu_3O_{6.96}$ 10°C min-1 in O_2 (left) and helium (right). The long arrows indicate the direction of the heating cycle (left). The short vertical arrows point to discontinuities in the rate of weight loss with temperature, at the corresponding indicated compositions (x).

The tga data provide, indirectly, some information on the oxygen mobility in these systems. The substantial reversibility of the tga scans in O_2 (the heating and cooling curves shown in Fig. 3 nearly overlay each other) demonstrates high oxygen mobility for T > 500°C. For T ≃ 350°C oxygen mobility becomes more restricted. The weight gain reflecting formation of the stoichiometric x = 0.00 material at 350°C is, for example, not observed on cooling, although the starting composition is identically reproduced. This implies that an activation barrier is associated with replenishment of the oxygen deficiency corresponding to x ≃ 0.04. Such has been suggested earlier based on preliminary DSC scans under 500 psi oxygen pressure (ref. 13). The results of the equilibration experiments discussed above also demonstrate that a temperature of close to 350°C is the minimum required to equilibrate powders with respect to oxygen content in a reasonable time frame.

Finally, we note that the tga scans can be used to indicate phase integrity. Thus, atmospheric attack on $YBa_2Cu_3O_{7-x}$ is, as above, reflected by the gradual appearance of a weight loss feature at relatively low temperature, centered on 250°C.

Powder X-ray diffraction

Powder X-ray diffraction (PXD) data were measured on an automated Siemens D500 diffractometer using Ni-filtered Cu Kα radiation (fine focus sealed tube source operating at 1200W). Samples were loaded into 2mm deep aluminum sample trays or, where sample quantity was limited, spread over a thin layer of silicone grease on a single quartz crystal mounted within an aluminum plate. Diffraction profiles were scanned in $\omega - 2\theta$ mode, typically over the range $2.0 \geq \theta \geq 92.0°$ in steps of 0.02°, with count times of 2s at each point. Higher precision scans using steps of 0.01° and 20s count times were also made in many cases. Incident beam divergence slits of 1° (and symmetrical diffracted beam slits prior to the graphite 002 monochromator crystal) were used for this full data range. No correction for the spreading of the incident beam beyond the sample at low Bragg angles was applied because of the absence of intense peaks at low angles. No particular precautions were taken to minimize preferred orientation effects. Such effects, where noticeable, could generally be improved by regrinding the sample and taking care during the loading of the sample tray. In other cases, samples were sprinkled onto a layer of silicone grease on the quartz crystal mount.

An essential feature of the PXD experiments was the treatment of the full PXD profiles by Rietveld analysis (ref. 41). Such refinements enable accurate (zeropoint errors are well determined) and precise lattice constants to be determined and also prove sensitive to multiphasic behavior. The refinements provide a quantitative measure of the relative amounts of two (or more) phases when present, information which is particularly valuable in mapping out phase diagrams. One region of the PXD profile that is sensitive to biphasic behavior is the region between $2\theta = 44°$ and 51°. The data in Fig. 4 show this range for pure orthorhombic (x = 0.15) and tetragonal (x = 1.00) phases and for a two phase mixture with an average composition corresponding to x = 0.67 and with an orthorhombic to tetragonal ratio of 0.93/1.00. Rietveld refinements might generally be considered as rather complicated and time-consuming for routine application but because of the advantages and additional information they provide, such analyses were performed on all of our PXD data.

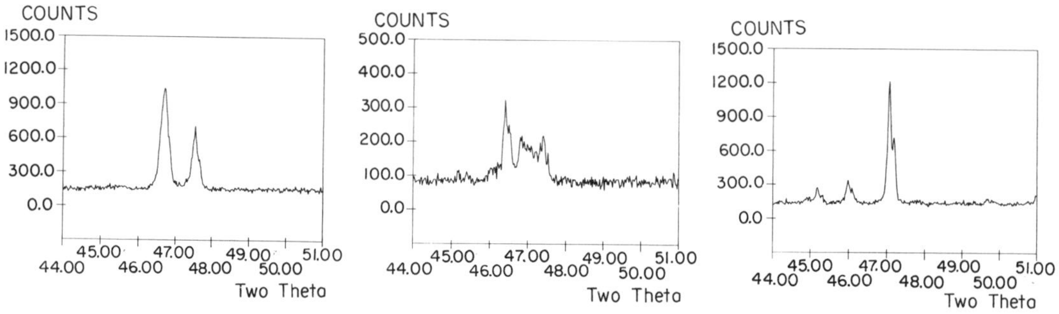

Fig. 4. The $44.0 \le 2\theta \le 51.0°$ regions of the PXD profiles of samples of $YBa_2Cu_3O_{6.85}$ (orthorhombic - left), $YBa_2Cu_3O_{6.33}$ (two phase mixture-center) and $YBa_2Cu_3O_{6.0}$ (tetragonal - right).

The full matrix least squares Rietveld refinements (ref. 41) were achieved using a locally modified version of the DBW3.2 code of Wiles and Young (ref. 42) as implemented on a VAX 11/750 computer system. The peak shapes were described by the pseudo-Voigt function (a simple sum of lorentzian and gaussian terms; the fractional contribution of the lonentzian term, γ, was generally fixed at 0.6 in the refinements) with an additional asymmetry correction applied at lower angles. The background was generated by linear interpolation from a set of points that were updated during the refinements. Scattering factors for the species Y^{3+}, La^{3+}, Ba^{2+}, Cu^{2+}, and O^{-1} were used. Space groups Pmmm (No. 47 (ref. 43)) or P4/mmm (No. 123 (ref. 43)) were assumed for the orthorhombic and tetragonal phases respectively.

Rietveld refinements on all data sets involved optimization of the scale factor(s), cell constants and zeropoint error. In certain cases additional variables including the peak half-width parameters, γ, and various atomic parameters were refined. Despite the generally routine methods used for data acquisition in most cases (necessitated by the need to evaluate large numbers of samples), reasonable fits between the observed and calculated diffraction profiles were usually achieved. Typical plots are shown in Fig. 5. It should be emphasized that the usefulness of PXD for full structural refinement is limited for the present systems. PXD provides only limited sensitivity to the oxygen atom parameters and cannot, for example, distinguish between La and Ba, nor between O and F.

<u>Powder neutron diffraction</u>

Powder neutron diffraction (PND) data were collected primarily on the powder diffractometer of the 8MW Missouri University Research Reactor (MURR) (ref. 44). Generally, approximately 5g of sample was contained in a 6.35mm outside diameter, 0.002" walled vanadium can. A wavelength of 1.2891Å was selected from the (220) planes of a Cu monochromator at a take-off-angle of 60.6°. Data from four 25° spans or five 20° spans of the linear position sensitive detector were each accumulated over, typically, some 4-6hrs and combined to yield the diffraction profile $5 \le 2\theta \le 105°$, rebinned in 0.1° steps. Data were generally collected at ambient temperature (298(5)K), although in some cases low temperature data were also accumulated, using a Displex refrigerator.

The PND data were analyzed by full matrix Rietveld refinement (ref. 41) using the DBW3.2 code (ref. 42) in a similar way to the PXD data. Scattering lengths of 7.750, 5.250, 7.718 and 5.805 fm (x 10^{-15} m) for Y, Ba, Cu, and O respectively were taken from the compilation of Koester and Yelon (ref. 45). The peak shapes were described by the pseudo-Voigt function, with, generally, a small correction for peak asymmetry for $2\theta \le 2\theta°$.

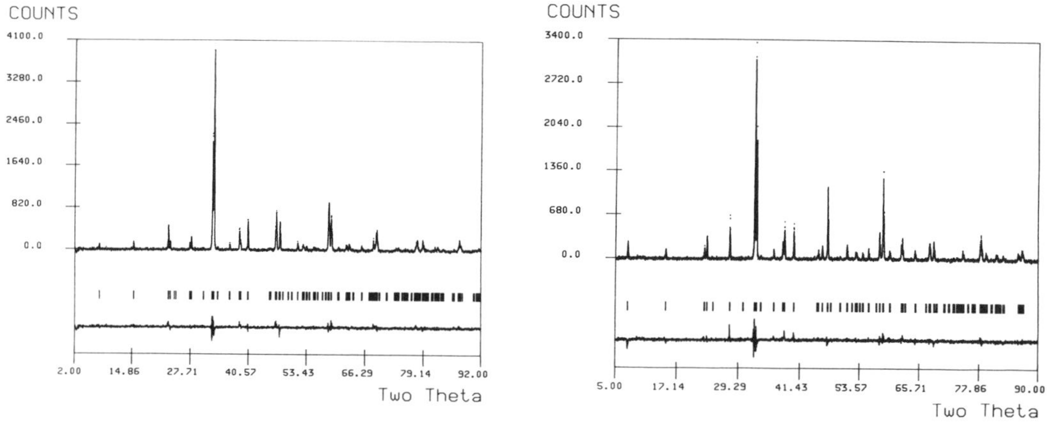

Fig. 5. Final observed (points), calculated (continuous line) and difference (lower - same scale) powder X-ray diffraction profiles for $YBa_2Cu_3O_{6.79}$ (orthorhombic - left) and for $YBa_2Cu_3O_{6.0}$ (tetragonal - right).

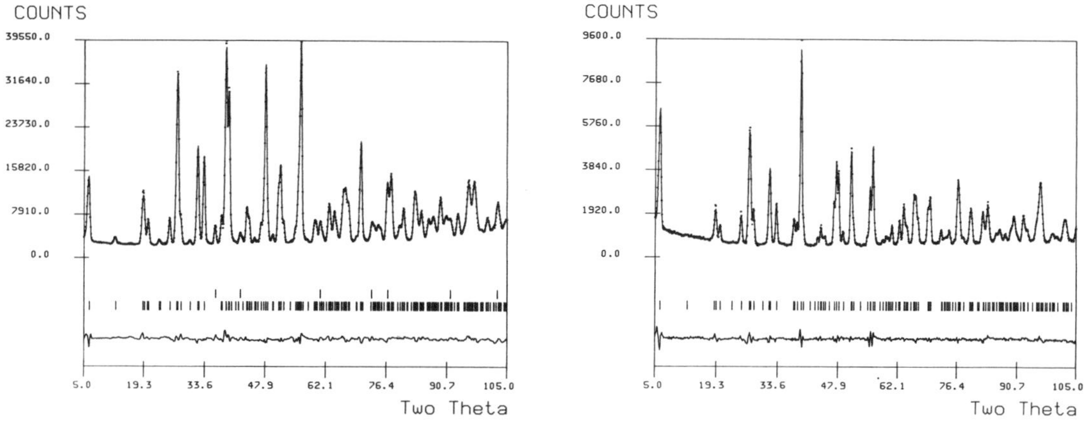

Fig. 6. Final observed (points), calculated (continuous line) and difference (lower - same scale) powder neutron diffraction profiles for $YBa_2Cu_3O_{6.96}$ (orthorhombic - left) and for $YBa_2Cu_3O_{6.0}$ (tetragonal - right).

Previous results for $YBa_2Cu_3O_{7-x}$ materials provided the starting points for structure refinements (ref. 13). The various overall (scale factor(s), detector zeropoint, half-width parameters, unit cell constants, asymmetry parameter, lorentzian-gaussian mixing term, γ) and atomic parameters (unrestricted coordinates, occupancies, isotropic B-factors) were introduced as variables at appropriate stages. The progress of the refinements was generally examined together with results of the Rietveld refinements of the corresponding PXD profile (preceding section). No evidence for symmetries other than Pmmm or P4/mmm was found for any of the materials studied. The moderate instrumental resolution of the MURR powder instrument (ref. 44) does not enable orthorhombic and tetragonal phases to be easily distinguished when the distortion is small ($b-a \leq \sim 0.005$Å) and, in common with other reactor powder instruments, provides only limited sensitivity to the presence of biphasic mixtures of orthorhombic and tetragonal components. Results for the compositions $x = 0.04$ (at 296K, 120K and 8K), 0.34 (296K), 0.60 (296K, 8K) and 1.00 (296K) were reported in some detail earlier (ref. 13). Final observed and calculated diffraction profiles for the $x = 0.04$ (296K) and $x = 1.00$ (296K) materials are shown in Fig. 6. Results for a range of compositions, predominantly close to $x = \sim 0.05$ and to $x = \sim 1.00$, have also been reported by other groups (see references cited in ref. 13). Results for materials close to these limiting compositions are generally in good agreement. Consistency is less good for intermediate compositions, as a result of the complex phase behavior of the $YBa_2Cu_3O_{7-x}$ system, coupled with the experimental difficulty of preparing homogeneous intermediate compositions. The structural results are discussed further below.

The widespread application of neutron diffraction to materials in the $YBa_2Cu_3O_{7-x}$ system (data from almost all neutron scattering centers are now represented in the literature) reflects the excellent sensistivity that neutrons provide to the parameters of lower atomic number (Z) atoms, such as oxygen, in the presence of atoms of much larger Z (Y, Ba, La etc.), and to scattering length differences between atoms of similar Z, such as between La and Ba, and La and Pr etc. The ability to make measurements in a simple fashion on sealed samples, and over a wide range of temperatures is an additional benefit. However, even though Rietveld analysis provides a relatively straightforward means of extracting precise structural data from the measured PND profiles, PND is experimentally inconvenient, requiring relatively large samples, and access to a suitable neutron scattering facility. The technique cannot, therefore, generally be applied routinely to a large number of samples but must rather be restricted to carefully selected compositions and conditions.

In addition to the nuclear structure, neutrons are, as a result of their magnetic moment, also susceptible to scattering by unpaired spin density. Magnetic interactions were early proposed (ref. 46) to play a significant role in the mechanism by which superconductivity occurs in these materials. Magnetic susceptibility measurements (see below) yield information about the behavior of local magnetic moments, but cannot provide information concerning magnetic structure. Following earlier susceptibility data, detailed PND studies have led to the establishment of the occurrence of antiferromagnetism first in $La_2CuO_{3.985}$ (ref. 47) and, more recently, in $YBa_2Cu_3O_{7-x}$ for $x > 0.5$ (ref. 48). The magnetic structure proposed for $YBa_2Cu_3O_{6.0}$ below the Neel point of $T_N \geq 500K$ is illustrated in Fig. 7. In this structure, the Cu1 atoms in the planes at $z = 0.0$ which are presumably in the $3d^{10}$ a Cu^{1+} state (see discussion below) are non-magnetic. The magnetic moments at the Cu2 sites (which are $S = 1/2$ $3d^9$ Cu^{2+}), are aligned approximately perpendicularly to the crystallographic c-direction, and, within any given plane (at $z \sim 0.36$ or $z \sim 0.64$), are coupled in an antiferromagnetic fashion. This structure resembles closely that observed in $La_2CuO_{3.985}$ (ref. 47). Successive Cu2 moments along the c-axis are also arranged in an antiferromagnetic fashion. Although it has been stated that the moments have no preferred direction within the Cu2 planes (ref. 48), the nature of the powder diffraction experiment is such that only the angle that the moment makes with the unique axis(c-direction) is accessible in the PND experiment.

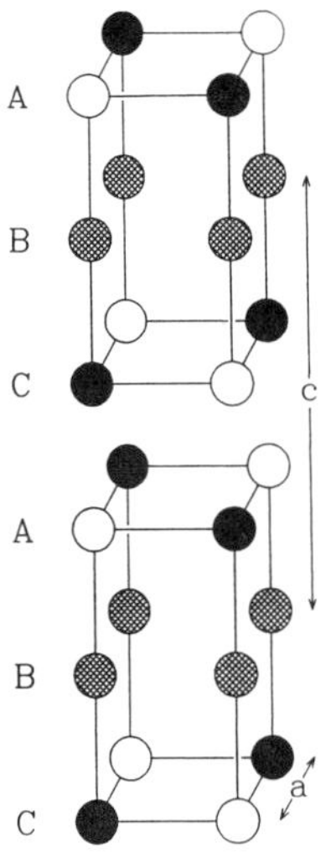

Fig. 7. Schematic diagram of the proposed magnetic structure for $YBa_2Cu_3O_{6.0}$ (ref. 48). The section of the structure corresponds to that shown in Fig. 13, but only copper atoms are shown. Cross-hatched, open and filled circles represent respectively non-magnetic Cu^+ ions, and antiparallel spins associated with Cu^{2+} at the Cu2 sites.

Magnetic reflections decline rapidly in intensity with increasing Bragg angle, because the scattering centers are the spatially distributed unpaired spin densities, rather than the effective point centers of the nuclei (as in the case of nuclear scattering). In addition, the net moments on the copper atoms in both $La_2CuO_{3.985}$ (ref. 47) and $YBa_2Cu_3O_{6.0}$ (ref. 48) were found to be anomalously low, typically ~0.4 - 0.5 BM, rather than about 1BM expected for a typical Cu^{2+} ion ($S=1/2$, $3d^9$). As a result of these factors, antiferromagnetic reflections, even at close to saturation, are rather weak, and their measurement was only possible with long counting times. The picture of the antiferromagnetic structure and the magnetic form factor (which yields information about delocalization of the spin density) in $La_2CuO_{3.985}$ (ref. 47) and the suggestion of magnetic fluctuations above the Neel point (ref. 49) determined by the Exxon group based on powder samples has since been confirmed by single crystal studies (refs. 50,51). Experimentally, the availability of large single crystals simplifies considerably the neutron scattering measurements and greatly extends the amount of information on the magnetic properties that can be extracted. Similarly sized homogeneous crystals of $YBa_2Cu_3O_{7-x}$ samples are not yet available, although several groups are working actively on this problem.

MAGNETIC, TRANSPORT, AND ELASTIC PROPERTIES OF $YBa_2Cu_3O_{7-x}$; $x \approx 0.0$

Magnetic properties

Meissner effect measurements (in which the magnetization M is measured upon cooling from above T_c in a small fixed magnetic field, H), were carried out using a Princeton Applied Research Vibrating Sample Magnetometer (VSM) with H=12G and/or with a George Associates Faraday Magnetometer with H=50G. The contribution of ferromagnetic impurities to M was determined by measuring M(H) isotherms at several temperatures for each sample. This contribution (which was then subtracted from the data) corresponded to that of only $\leq$ 20ppm of ferromagnetic iron impurities. All of the samples show, in addition, an upturn in the susceptibility, χ, at low temperatures, attributed to S=1/2, $3d^9$ Cu^{2+} defects and to trace amounts of $BaCuO_2$ and/or Y_2BaCuO_5 impurities. A correction for this defect contribution, χ_d = C/T was made by assuming that $\chi = \chi_o + C/T$ over a small temperature range near 100K.

In this section we concentrate on the properties of materials with $x \approx 0$. The variation of these properties with x is then discussed in the following section. A plot of χ-C/T versus T for a sample with x=0.04 (Fig. 8) shows a significant downturn in χ(T) below 140K. The existence of the downturn significantly above T_c is not a spurious result of the procedure for subtracting out the Curie tail, since even without this correction $d\chi/dT$ becomes negative for temperatures less than 105K. Also, even allowing a Curie-Weiss behavior for χ_d, $\chi_d = \chi/(T-\theta)$, yields $\theta \approx$-34K and an onset for the downturn in χ-C/(T-θ) of about 130 K. We believe that this high downturn temperature has not previously been observed in the susceptibility data on these compounds because very high precision is required in the measurements. The origin of this downturn may be due to superconducting fluctuations but is more likely due to the onset of granular superconductivity as discussed below.

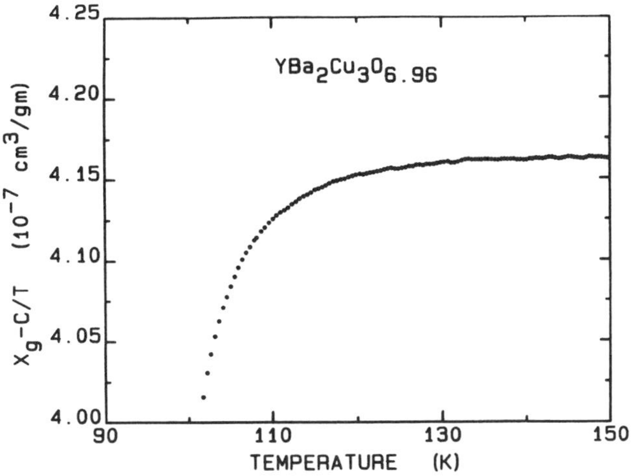

Fig. 8. Magnetic susceptibility versus temperature for $YBa_2Cu_3O_{6.96}$ after correction for the defect contribution.

Resistivity masurements

Four probe resistance measurements were made on ceramic samples. The powders were first pressed into 5mm diameter discs, approximately 2mm thick (Carver press, $\leq$ 8000psi) and then sintered at 940°C in air for 24h. After this sintering procedure, the pellets were annealed in 1 atm. O_2 for a range of times and temperatures.

A typical resistance versus temperature profile for $YBa_2Cu_3O_{7-x}$, x $\simeq$ 0.0 is shown in Fig. 9. These results are similar to other published data, showing a change in resistance that is linear below room temperature. However, there are two significant anomalies. First, above 300K the slope of the resistance change with temperature increases abruptly. This anomaly is discussed further below. Secondly, there is a pronounced downturn in the resistance well above T_c.

Due to the two-dimensional character of the $YBa_2Cu_3O_{7-x}$ structure, it has been suggested that the downturn in the resistance above T_c can be attributed to fluctuations into the superconducting state. Fitting the linear region of the resistance (typically from ~150 to ~250K) by least squares and subtracting this component from the total resistance curve yields a measure of the excess conductance (Fig. 10). In this plot any deviation from zero represents a deviation from the linear behavior. The excess conductance above T_c becomes obvious as a negative deviation from zero which commences at ~130K (labelled T_s) and continues to grow until superconductivity sets in at 91K. Other workers (ref. 52), in attributing this excess conductivity to fluctuations, have attempted power law fits in order to extract critical exponents. Such fits, however, have had limited success and dimensional crossover arguments have been invoked to explain the resulting non-linear behavior. Fig. 10 also shows a linear plot of the excess conductance versus log($T-T_c$) which demonstrates that it has a logarithmic divergence. Recent results on single crystals also show this logarithmic divergence, which was attributed to a two-dimensional pair breaking mechanism which suppresses T_c (ref. 53). We alternatively propose that the excess conductivity may mark the onset of granular superconductivity (ref. 54) at T_s. The excess conductance between T_c and T_s may reflect Josephson coupling of superconducting grains separated by semiconducting grain boundaries until long range coherence sets in at T_c.

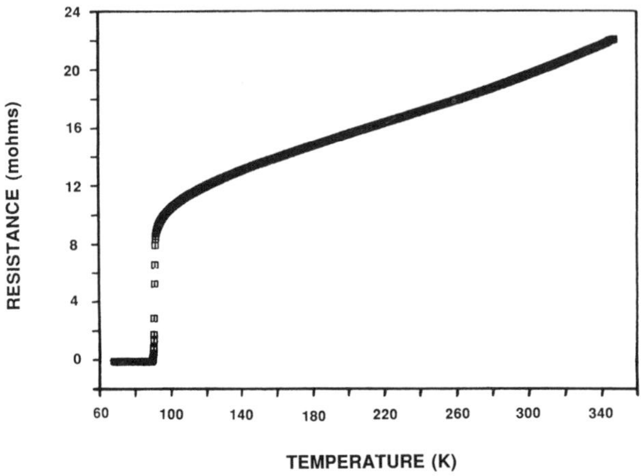

Fig. 9. Resistivity versus temperature for $YBa_2Cu_3O_{7-x}$, x ≃ 0.0.

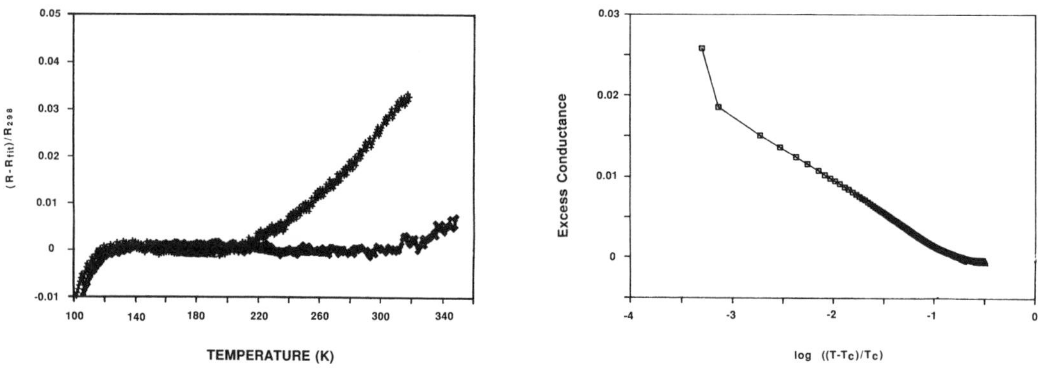

Fig. 10. Deviations from linearity in resistance versus temperature for $YBa_2Cu_3O_{7-x}$, x ≃ 0.0 and x ≃ 0.1 (left - see text for details) and excess conductivity versus log (($T-T_c$)/T_c for x ≃ 0.0 (right)

The second anomaly in the resistance, at ≃ 300K (Fig. 9), shows in Fig. 10 as a positive deviation from zero. Although we have not yet derived a convincing explanation for this anomaly, the temperature at which it occurs is very sensitive to the sample annealing conditions. Fig. 10 shows resistivity difference data for samples annealed at 440°C and at 940°C respectively. The anomalies appear at 310K and at 240K for the former and latter materials respectively. We estimate that the oxygen contents of these two materials are approximately x ≃ 0.0, and x ≃ 0.10. The strong dependence of the high temperature anomaly on annealing conditions is probably coupled to slight changes in the number and/or mode of ordering of oxygen vacancies, or, alternatively, to slight changes in the character of the grain boundaries.

Ultrasonic attennuation

Propagation of ultrasound is among the standard probes of superconductivity. In conventional superconductors, an exponential decrease in the sound attenuation below T_c signifies the opening of an energy gap. A step discontinuity in the velocity marks the mean field specific heat jump at T_c. Below T_c, the velocity has little temperature dependence. In the $YBa_2Cu_3O_7$ system and in $La_{1.8}Sr_{0.2}CuO_{4-\delta}$ the sound propagation characteristics are highly anomalous (ref. 55,56 and references cited). Ultrasound propagation experiments were performed on $YBa_2Cu_3O_{7-x}$ samples by using standard pulse-echo techniques. The velocity was measured by McSkimin interferometry to about the 50ppm level. The damping was measured from the decay of the echo train to the few percent level. Samples consisted of pressed ceramic pellets, typically 1cm in length and 1.2 cm in diameter. Equilibrated samples of the desired oxygen content (see synthesis section above) were formed into pellets using a Carver press at pressures ≤ 8000psi. Samples were subsequently sintered either in sealed tubes, or, in the case of the parent material with x=0.04 for which results are discussed below, at 950°C in air and then in O_2 at 450°C for 5.5d. The ends of the pellets were then repeatedly ground and polished. Several materials were examined for bonding the transducers (of quartz or lithium niobate) to the polished pellets. Armstrong C-1 epoxy was found to be adequate, able to withstand many slow thermal cycles between room and liquid-He temperature.

For a ceramic sample the longitudinal and transverse sound velocities are given by

$$\rho v_L^2 = B + 4/3G, \quad \rho v_S^2 = G \qquad (1)$$

where B and G are the macrosocopic bulk and shear moduli respectively. Both longitudinal and transverse velocities (Fig. 11) show anomalies at 91K (T_c) and at ≈125K (labelled T_s). The anomaly at T_c is dominated by the shear modulus and not the bulk modulus. A discontinuity in the temperature derivative of G occurs at T_c in agreement with the thermodynamic relation

$$\Delta G/G = (T.\Delta S)G \left[\frac{1}{T_c} \frac{d^2T_c}{d\sigma^2}\right] \qquad (2)$$

where σ is the shear stress and ΔS is the entropy change. For all materials studied, the temperature of this lower temperature discontinuity agrees within 1K with the transition temperature (zero resistance point) as measured by resistance. These results suggest that the superconducting order parameter has a large coupling with shear distortions. The same behavior is also found in the $La_{1.8}Sr_{0.2}CuO_{4-\delta}$ system and appears to be generic to this class of compounds (ref. 55,56).

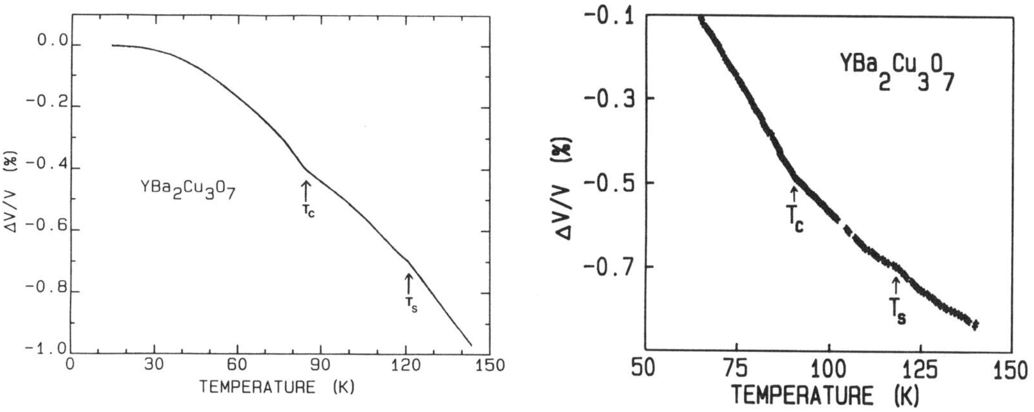

Fig. 11. Longitudinal (left) and transverse (right) sound velocity in $YBa_2Cu_3O_{7-x}$ (x ≈ 0), ΔV/V is [V(T)-V(4.2K)]/V(4.2K).

In typical superconductors the shear modulus softens below T_c, presumably due to enhanced screening of the lattice distortion by the condensate. In the present system, the modulus stiffens significantly. It is possible that this implies a co-condensation of excitations that are strongly coupled to transverse acoustic phonons in the normal state.

The damping of longitudinal ultrasound is also extremely anomalous (Fig. 12). In typical superconductors, the damping decreases exponentially below T_c (and the exponent provides, as above, a measure of the superconducting gap). In $YBa_2Cu_3O_{7-x}$ with x≈0.0, however, the damping increases below T_c, shows a pronounced peak and decreases approximately as a power law at lower temperatures. Similar behavior is also observed in $La_{1.8}Sr_{0.2}CuO_{4-\delta}$.

Comparison with elastic, structural, magnetic and transport data suggests that the second anomaly at T_S marks a phase transition of as yet unknown origin affecting primarily the (a x b) planes. As discussed above, both the resistance and the magnetization show a downturn beginning at a temperature corresponding to T_S. While others have suggested that the downturn in the resistance data may be due to fluctuations, the suggestion from the velocity measurements that T_S is a phase transition may rule out this explanation. For low-dimensional materials, fluctuations occur between the mean field transition temperature T_c^{MF} and the true three dimensional ordering temperature T_c. T_c^{MF} is not a true thermodynamic transition and should not show up in any measurements as a sharp anomaly.

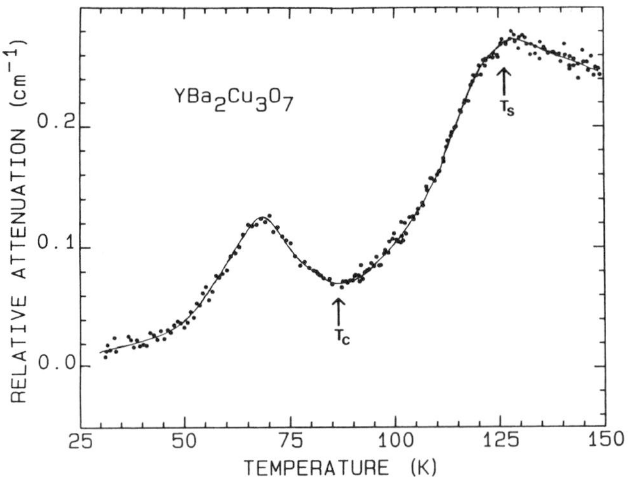

Fig. 12. Ultrasonic attenuation in $YBa_2Cu_3O_{7-x}$ ($x \simeq 0$)

Instead, we argue here that T_s may be the onset of granular superconductivity. At T_s some portion of an individual grain becomes superconducting while the rest of the grain remains conducting or semi-conducting. Since this is a true thermodynamic transition it would show up as a sharp anomaly in measurements sensitive to such a transition such as sound velocity. For resistance and magnetization (because only a fraction of the sample becomes superconducting) it will appear as a small effect which grows with decreasing temperature as Josephson coupling between grains increases the amount of material which superconducts. The bulk transition at T_c occurs when long range coherence sets in. While this is a plausible explanation for all of the above results, in the absence of further data we cannot definitively associate the transition at T_s seen in the ultrasonic measurements with the rather weak anomalies at the same temperature in the resistance and magnetization.

EVOLUTION OF THE PROPERTIES OF $YBa_2Cu_3O_{7-x}$ WITH OXYGEN CONTENT

In common with the $La_2CuO_{4-\delta}$ system, it was early recognized that the properties of $YBa_2Cu_3O_{7-x}$ depend sensitively on oxygen content. This sensitivity reflects three factors. First, the oxygen content implicitly determines the copper valence and therefore provides a means of adjusting the mean Cu oxidation state from +2.33 at x=0.0, through +2.0 at x=0.5 to +1.67 at x=1.0. Secondly, oxygen - vacancy ordering is expected at particular vacancy concentrations. As is discussed below, the details of such orderings are not yet fully understood, but it is likely that they play a significant role in determining the structural chemistry of this system. Thirdly, although a definitive theoretical treatment of the high temperature superconductivity in these systems has not yet emerged, the electron or hole density on the oxygen atoms that interleave between the copper atoms is generally believed to play an important role (refs. 57,58).

Structural evolution with oxygen content

Many detailed X-ray and neutron diffraction studies of $YBa_2Cu_3O_{7-x}$ materials have appeared. These studies are concentrated towards the more oxygen-rich side of the phase field, $x \simeq$ 0.0 - 0.1, although other compositions have also been studied. Aspects of our own studies of compositions x = 0.04, 0.34, 0.60 and 1.00 were discussed earlier (ref. 13).

For $x \simeq 0.0$, the system is orthorhombic, Pmmm (No. 47 (ref. 43)). The structure of the x = 0.04 material at room temperature (ref. 13) is illustrated in Fig. 13. Compared to the ideal perovskite lattice illustrated in Fig. 1, the unit cell length along the c-direction is tripled to accomodate the ordered YBa_2 content of the A-sites. The full oxygen complement of this tripled cell is O_9, but in $YBa_2Cu_3O_{7-x}$ at x=0.0, two oxygen sites are vacant. These are O6, at 0 0 1/2 in the yttrium atom plane, and O5, at 1/2 0 0 in the Cu1, O1 plane (Fig. 13). Site O6 is apparently always vacant in these materials, although might conceivably be occupied in monovalent anion substituted materials. Site O5, however, plays an important role in these oxides at other temperatures and compositions.

The two crystallographically distinct copper sites have different coordinations by oxygen and are affected differently by changes in composition. The relative roles of these two types of Cu site in the superconducting process remains largely an unresolved question (see below). The Cu2 planes (centred on $z \simeq 0.36$ and $\simeq 0.64$) have square pyramidal coordination

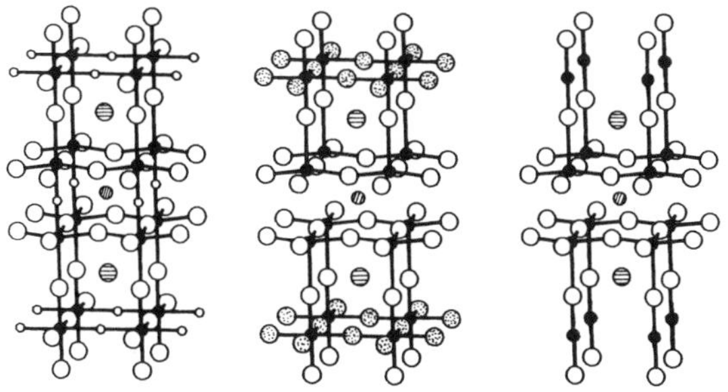

Fig. 13. Ball and stick representations of the room temperature structures of $YBa_2Cu_3O_{6.96}$ (left), $YBa_2Cu_3O_{6.4}$ (center) and $YBa_2Cu_3O_{6.0}$ (right). (● = Cu, ⊖ = Ba, ⊘ = Y, ○ = O and ◌ is the partially occupied oxygen position).

by O2 (x2), O3(x2), and O4 (apical), although the apical Cu2-O4 distance of 2.31Å is much longer that the basal distances (Cu2-O2 = 1.93Å, Cu2-O3 = 1.96Å). The Cu2 atoms are required by the accepted space group to be coplanar, but the basal oxygen atoms, O2 and O3, lie off the Cu2-Cu2 vectors, away from the O4 side. This slight distortion gives rise to the description 'dimpled sheet' for these planes in the structure. Both O2 and O3 are, however, displaced in the same sense, away from O4, so that their locations cannot be described in terms of effective tilting of $Cu2\text{-}O_4V_2$ octahedra (V = oxygen vacancy), similar to descriptions for the analogous planes in the $La_2CuO_{4-\delta}$ system. In the Cu1 planes, the ordered arrangement of oxygen atoms has the O1 sites completely filled (occupancy 0.96(1) (ref. 13)), and the O5 sites vacant. The Cu1-Cu1 separation is increased along the b-axis to accomodate the O1 species at 0 1/2 0, such that the b-axis (3.888Å) is longer than the a-axis (3.824Å). In the absence of communication between the parallel Cu1-O1-Cu1... chains, connectivity in the Cu1, z=0 planes across domains is disrupted at relatively low oxygen vacancy levels. Oxygen non-stoichiometry at intermediate compositions is accommodated by progressive depopulation of the O1 site until, at x = 1.0, the site is completely empty and the structure is accurately tetragonal (P4/mmm, Fig. 13). At the composition x = 1.00, the Cu1 atom has a bond distance to O4 equal to 1.81Å and is two coordinate consistent with a formal oxidation state of Cu^+. The geometry of the Cu2-O2,O3 planes is remarkably invariant across the $0 \leq x \leq 1.00$ composition field. The Cu2-O2 and Cu2-O3 distances vary, at most, from 1.93 to 1.94Å, and from 1.96 to 1.94Å respectively. This composition range includes both the orthorhombic and tetragonal phases. The Cu2-O2-Cu2 and Cu2-O3-Cu2 angles vary by 2.9° and 1.8° respectively across this same composition range. The Cu2-O2-Cu2 and Cu2-O3-Cu2 angles of 164.0° and 165.1° observed for the x = 0.04 composition compare with the Cu-O1-Cu angle of 174.2° in $La_2CuO_{3.985}$ (ref. 47). The absence of major change in the Cu2 coordination suggests that the Cu2 valence state varies little for $0 \leq x \leq 1.00$. The average Y-O distance is, however, very close to that predicted by a bond valence estimate and, consequently, maintaining optimal yttrium coordination may be an important factor in maintaining the geometry of the Cu2-O planes.

In addition to the results for the end members, we have also obtained structural data at two intermediate compositions, x = 0.34 and x = 0.60. The average structure at x = 0.34 shows no major differences from the structure with x = 0.04. The unit cell is orthorhombic and displays average depopulation of the O1 site consistent with the known oxygen content. It is clear from these and other structural data (ref. 13) that the major change in the structure throughout the composition range, $0.0 < x < 0.33$ is the introduction of oxygen vacancies onto the O1 site. The structural and superconducting properties (see below) are both consistent with random vacancies and a homogeneous non-stoichiometric phase in this composition range. However, electron diffraction data, at a composition of x = 0.125, has shown evidence for at least partial ordering of oxygen vacancies (ref. 59). These different techniques have different sensitivities to the presence of partial oxygen vacancy order and to the length scale over which it persists. More detailed characterization of specific compositions which can give ordered vacancy structures such as x = 0.33 is required to resolve the details of the low temperature phase behavior.

For still higher oxygen vacancy concentrations, $x > 0.33$, communication along the Cu1-O1-Cu1... chains is disrupted further, facilitating disorder of the oxygen atoms over the O1 and O5 sites. (We have, notably, seen no evidence for oxygen atom disorder involving any of the other oxygen atom sites.) The orthorhombic-tetragonal transition temperature is therefore progressively lowered as x tends towards 1.00. The preparation of tetragonal phases at room temperature by quenching is thus facilitated, and such materials do indeed show oxygen disorder over the O1 and O5 sites (now, of course, equivalent). The structure of a tetragonal phase with x = 0.6 is shown in Fig. 13. However, in this composition range, $0.33 \leq x \leq 1.00$, biphasic behavior (over relatively large length scales) can occur under suitable conditions. For example, Fig. 4 illustrates the result of a sealed tube equilibration of a mixture of phases with x = 0.1 (T_c = 91K) and x = 1.00 (semiconducting) appropriate for synthesis of an x = 0.70 material. The product is a mixture of approximately equal amounts of a tetragonal phase with $x \simeq 1.0$ and an orthorhombic phase with $x \simeq 0.3$ (T_c = 60K).

The intermediate phase behavior of typical samples prepared by the synthesis methods described above is reflected in the variation of the unit cell constants with composition shown in Fig. 14. The data can be approximately divided into three composition ranges: (i) $0.00 < x < 0.33$, where the samples are apparently single phase orthorhombic, (ii) a single phase tetragonal region for $0.60 < x < 1.00$, and (iii) $0.33 < x < 0.60$ where the data in Fig. 14 have been treated as orthorhombic but where Rietveld analyses of the PXD data indicate biphasic behavior in some samples depending on the annealing conditions. In such cases, significantly improved fits of the PXD data were obtained with mixtures of orthorhombic and tetragonal phases.

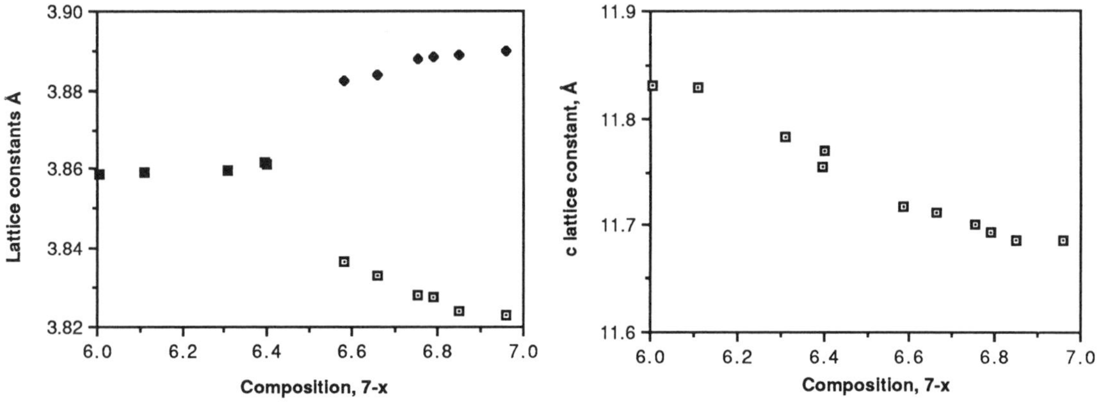

Fig. 14. Variation in the lattice constants with oxygen content (x) for $YBa_2Cu_3O_{7-x}$ samples prepared at T > 600°C.

Overall, there is a steady increase in the length of the c axis as x increases. This change does not reflect directly the reduction of the oxygen content but rather the change in the mean copper oxidation state. The M-O bond lengths reflect the metal oxidation state and its coordination number. The c-axis length in $YBa_2Cu_3O_{7-x}$ is the sum of three different separations, Cu2-O4(x2), Cu1-O4(x2) and Cu2-Cu2. A general decrease in the latter two distances with x is more than compensated by the increased Cu2-O4 separation. Interestingly, the ambient temperature structural data for x=0.04, 0.34, 0.60 and 1.00 provide respective Cu1-O4 bond lengths of 1.849(3)Å, 1.841(4)Å, 1.797(5)Å and 1.807(4)Å (ref. 13) suggestive of bimodal behavior.

Evolution in the magnetic properties with oxygen content

Normal state magnetic susceptibility data in a field of 6.3kG on $YBa_2Cu_3O_{7-x}$ samples with $0.04 \leq x \leq 1.00$ were obtained using the Faraday technique from 4K to ≃ 800K (ref. 13). The samples with x = 0.1 to 0.6 were prepared in situ as described above. The x = 1.0 sample was prepared separately in helium at 840°C. The data were corrected for ferromagnetic and defect impurities as described above. These corrections are very important for materials with low susceptibilities. Otherwise, merely assuming $\chi = M/H$ can give large errors in χ. Such factors may be responsible for discrepancies between different results for the same compound reported by different groups.

The corrected $\chi(T)$ data for a selected set of samples are shown in Fig. 15 (uncorrected data have been shown previously (ref. 13)). There is a smooth evolution in the shape and magnitude of $\chi(T)$ with x. The broad maxima near 800K are attributed to (dynamic) short range antiferromagnetic (AF) ordering within the Cu-O layers. Indeed, as described above the x = 1.00 sample exhibits static long range antiferromagnetic order below ≃ 500K (ref. 46). This latter ordering cannot be seen on the scale of Fig. 15, but is manifested as a small slope discontinuity in $\chi(T)$ upon closer inspection (ref. 60).

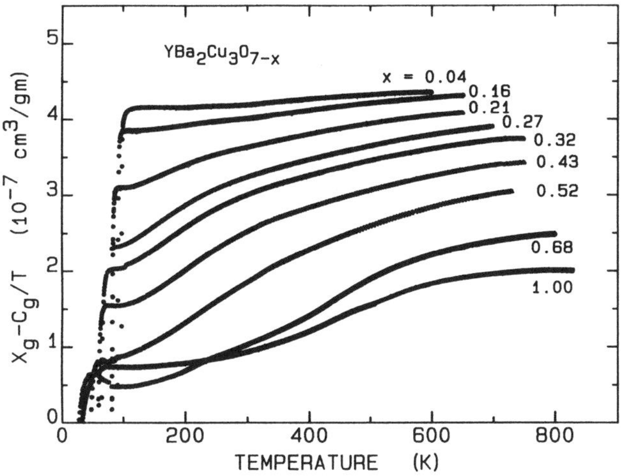

Fig. 15. Corrected magnetic susceptibility data for $YBa_2Cu_3O_{7-x}$ versus x.

<u>Evolution in the superconducting properties with oxygen content</u>

The Meissner Effect data for various samples prepared by the sealed tube equilibration route at 650-670°C (ref. 32), and the data for materials prepared 'in situ' in the Faraday balance (ref. 13) are similar (Fig. 16). Between x = 0.04 and x = 0.40, the computed superconducting volume fractions are 38 - 55%, reasonably typical of this class of materials as prepared by the ceramic route. The volume fractions, however, decrease rapidly for x > 0.4. The midpoint temperature of the superconducting transition for these various samples is shown as a function of x in Fig. 17. The 'in situ' and sealed tube equilibration data are consistent and show the presence of two plateau regions (at ≃ 90K for $0.0 \leq x \leq 0.1$ and at ≃ 50K for $0.30 \leq x \leq 0.45$) connected by an S-shape curve. Similar $T_c(x)$ behavior has also been seen recently in the $HoBa_2Cu_3O_{7-x}$ system (ref. 61). Several explanations for this behavior might be proposed (refs. 13,32,61).

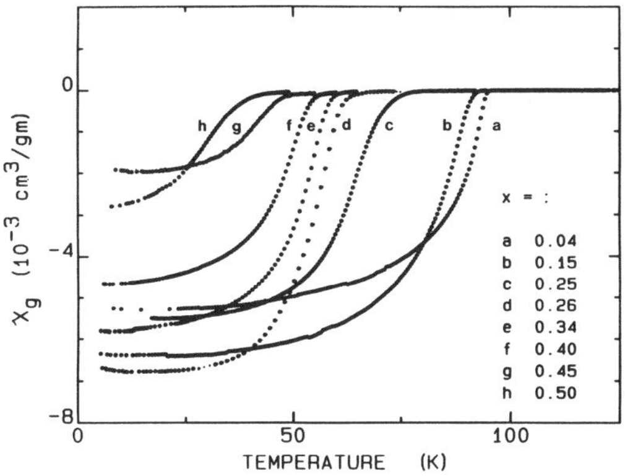

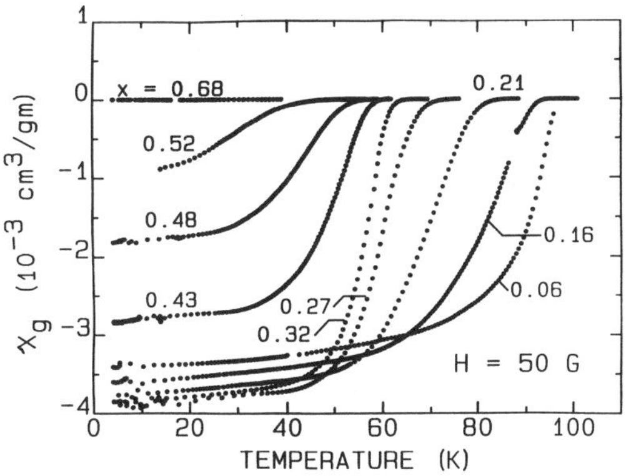

Fig. 16. Meissner effect data for samples prepared by sealed tube equilibration (left) and for samples prepared in situ in the Faraday balance (right).

Different superconducting transition temperatures could be associated with the two crystallographically distinct types of copper atom that are present in the structure. The T_c = 90K transition would then reflect a superconducting path in the Cu1 planes, which persists in the composition field only as long as adequate interconnecting oxygen content is maintained. The T_c = 50K transition would reflect superconductivity supported by the Cu2 planes. The structural character of the Cu2 planes is similar in $YBa_2Cu_3O_{7-x}$ and in $La_2CuO_{4-\delta}$ as noted above. Other similarities between the 50K superconducting state of $YBa_2Cu_3O_{7-x}$ and $La_{2-y}(Ba,Sr)_yCuO_{4-\delta}$ are evident. The superconducting transition temperature is similar. The mean Cu valence state (albeit averaged over both Cu1 and Cu2 planes - it proves difficult to ascertain separately the state of the Cu1 and Cu2 planes) at x = 0.33 is 2.11, similar to that of ≃ 2.15 observed for optimal superconducting properties in $La_{2-y}(Ba,Sr)_yCuO_{4-\delta}$. The magnitudes and temperature dependences of the magnetic susceptibilities are also comparable (compare Fig. 15 above with data given in ref. 62).

Although the Cu1 and Cu2 planes are crystallographically distinct, they are strongly interconnected and their mutual interaction needs also to be considered. We have suggested earlier (ref. 13) that the T_c = 50K transition is, as above, associated with the Cu2-O

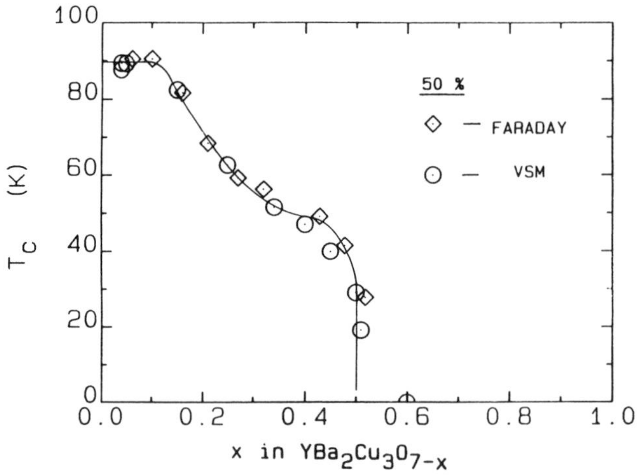

Fig. 17. The superconducting transition temperature, T_c, versus oxygen content, x, as determined by Meissner effect measurements. Data from samples produced both by sealed tube equilibrations and by 'in situ' control of x in the Faraday balance are plotted.

layers, but that the enhancement of T_c to ≃ 90K for $0.0 \leq x \leq 0.1$ is due to increased electronic coupling of the Cu2-O layers by the interleaving O4-Cu1-O4 linkages. The increased coupling would reduce superconducting fluctuation effects, allowing the observed T_c to more nearly approach the mean field T_c (i.e. the T_c in the absence of fluctuations). Evidence that superconducting fluctuations may indeed be present significantly above T_c may be found from susceptibility data in moderate fields(H = 6.3kG, see below).

A detailed theoretical treatment of the variation of T_c with x according to any of the possible mechanisms is lacking. In the absence of such a treatment we, and other groups (refs. 13,61,63), had initially assumed that the S-shape of the T_c versus x curve (rather than a square, stepped form, ref. 55) reflected in some undefined way the effect of phase separation into x = 0.0 and x = 0.33 materials at intervening compositions. The S-curve, however, is followed by samples produced both 'in situ' and by sealed tube equilibrations at temperatures as low as 345°C, the lowest temperature at which we have achieved equilibration with respect to oxygen content. This implies that the S-shape curve behavior is intrinsic, or at least that the length scale of the phase separation is very limited, since only one T_c was seen for each sample.

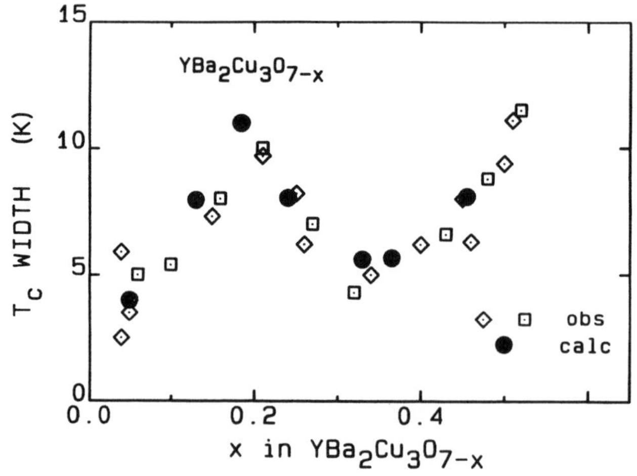

Fig. 18. A measure of the transition width versus oxygen content, x, for various samples, including both 'in situ' Faraday data and VMS data. The solid circles are calculated values (see text).

A measure of the transition width (defined as the difference in temperature between the points of achieving 10% and 50% of the maximum diamagnetic susceptibilites) versus oxygen content, x, (ref. 32) is shown in Fig. 18. If the form of the T_c versus x plot is intrinsic and there is a small range of oxygen content in the samples, the greatest transition width would be expected to occur at the points where T_c is varying most rapidly. Points calculated on this basis (using values for dT_c/dx read from Fig. 17 and with fitted values for the minimum width of ΔT_c = 4K, and Δx = 0.027, (ref. 32)) are included in Fig. 18. This approach reproduces quite well the observed variation in the transition width. It is noteworthy that the ME data for other groups obtained from quenched samples are also similar to that shown in Fig. 16 and that the observed behavior for $0.0 \leq x \leq 0.33$ is thus apparently independent of synthesis temperature for T > 345°C. If significant phase segregation were to occur over length scales larger than twice the coherence length of roughly 30Å (ref. 54), evidence for two distinct superconducting transitions would be expected, and the variation of the width with x would probably not be describable in this simple fashion. The superconducting properties in the $0.0 \leq x \leq 0.33$ regime can thus be interpreted in terms of a state which is homogeneous on a length scale of greater than about 60Å, but which has an oxygen inhomogeneity, Δx, of about 0.03. This distance represents some 10-20 unit cells, and we do not yet have a complete description of the homogeneity within this length scale.

The superconducting properties for $0.33 \leq x \leq 1.00$, are, in one sense, simpler. Superconductivity is not observed in the tetragonal x = 1.00 phase. Single phase tetragonal materials with $x \geq 0.5$ can be prepared quite readily by quenching from $T \geq$ 450°C, and such materials are also semiconducting, at least down to 4.2K. As mentioned above, however, biphasic behavior for some compositions in the range $0.33 \leq x \leq 1.00$ has been observed and, phase segregation into orthorhombic $T_c \simeq$ 60K, and tetragonal semiconducting materials may therefore occur under suitable conditions. Such phase separated materials do show superconductivity, with T_c ~60K, but with reduced superconducting volume fractions that reflect the relative proportions of the orthorhombic and tetragonal components. Superconductivity has not been observed in any material with tetragonal symmetry in the $YBa_2Cu_3O_{7-x}$ or related systems, although tetragonal $La_{2-y}(Sr,Ba)_yCuO_{4-\delta}$ is apparently superconducting (ref. 66).

CONCLUSION AND OUTLOOK

In conclusion, measurements of the properties of $YBa_2Cu_3O_{7-x}$ as a function of oxygen content have provided some useful insights into the relative importance of the different structural units (Cu-O layers and chains) and the influence of the average copper oxidation state. Many structural aspects of the system are now well understood but several important issues remain, particularly concerning the low temperature (<350°C) part of the phase diagram. For example, the details of phase separation and oxygen vacancy ordering, the T-x dependence of the orthorhombic-tetragonal phase transition and the origin of the shape of the $T_c(x)$ data are all areas which require further study. A comparison of the $YBa_2Cu_3O_{7-x}$ system with other chemical modifications promises to be particularly helpful in separating the effects of structure, symmetry, vacancy concentration and mean copper oxidation state on the above properties and on other aspects of the superconductivity in these materials. In this respect, investigations of the composition dependence of the properties of $La(La,Ba)_2Cu_3O_{7-x}$ (refs. 35,64) and $(Y,Pr)Ba_2Cu_3O_{7-x}$ materials (ref. 65) are currently of particular interest.

ACKNOWLEDGEMENTS

We thank many colleagues at Exxon for helpful discussions (particularly A. N. Bloch, M. H. Cohen and S. K. Sinha), W.B. Yelon and D. Xie (MURR) for a large number of PND measurements, and authors for kindly sending preprints. We also thank M. Hillpot for typing the manuscript.

REFERENCES

1. J.G. Bednorz and K.A. Müller, Zeit. Phys., B64, 189-193 (1986).
2. J.G. Bednorz, K.A. Müller and M. Takashige, Science, 236, 73, (1987)
3. H. Takagi, S. Achida, K. Kitazawa and S. Tanaka, Jpn. J. Appl. Phys., 26, L123-L124 (1987).
4. M.K. Wu, J.R. Ashburn, C.J. Torng, P.H. Hor, R.L. Meng, L. Gao, Z.J. Huang, Y.Q. Wang, C.W. Chu., Phys. Rev. Lett., 58, 908-910 (1987)
5. R.J. Cava, B. Batlogg, R.B. Van Dover, D.W. Murphy, S. Sunshine, T. Siegrist, J.P. Remeika, E.A. Rietman, A. Zahurak and G.P. Espinosa, Phys. Rev. Lett., 58, 1676-1679, 1987.
6. C.N.R. Rao, P. Ganguly, A.K. Raychaudhuri, K. Shreedhar and R.A. Mohan Ram, Nature, 326, 856-857 (1987).
7. Advanced Ceramic Materials, 2(3B), (1987) Special Supplementary Issue on "Ceramic Superconductors"
8. Jpn. J. Appl. Phys., Part 2, 26 (4 and 5), (1987).
9. Chemistry of High Temperature Superconductors, ACS Symposium Series, 351, Eds D.L. Nelson, M.S. Whittingham and T.F. George, Sept. 1987.
10. Extended Abstracts, Materials Research Society Fall Meeting, November 1987.

11. R.M. Hazen, L.W. Finger, R.J. Angel, C.T. Prewitt, N.L. Ross, H.K. Mao, C.G. Hadidiacos, P.H. Hor, R.L. Meng and C.W. Chu, Phys. Rev., B35, 7238-7241 (1987).
12. M.A. Beno, L. Soderholm, D.W. Capone II, D.G. Hinks, J.D. Jorgensen, J.D. Grace, I.K. Schuller, C.U. Segre and K. Zhang, Appl. Phys. Lett., 51, 57-59 (1987).
13. D.C. Johnston, A.J. Jacobson, J.M. Newsam, J.T. Lewandowski, D.P. Goshorn, D. Xie and W.B. Yelon in "Chemistry of High-Temperature Superconductors" Eds. D.L. Nelson, M.S. Whittingham, T.F. George, ACS Symp. Ser., 351,136-151 (1987)
14. C.N.R. Rao and P. Ganguly, Jpn. J. Appl. Phy., 26, L882-L884 (1987).
15. C. Michel, F. Deslandes, J. Provost, P. Lejay, R. Tournier, M. Hervieu and B. Raveau, C.R. Acad. Sc. Paris, 304, 1059-1061 (1987).
16. J.B. Goodenough and J.M. Longo, Landolt-Bornstein Tabellen, New Series, GroupIII/Vol. 4a, Springer Verlag, Berlin, 126-314 (1970).
17. C.N.R. Rao, J. Gopalakrishnan and K. Vidyasagar, Indian Journal of Chemistry, 23A, 265-284 (1984).
18. P.G. Dickens and M.S. Whittingham. Quart. Rev., 22, 30-44 (1968).
19. K.R. Poeppelmeier, A.J. Jacobson and J.M. Longo, Mater. Res. Bull., 15, 339-345 (1980).
20. J.C. Grenier, M. Pouchard and P. Hagenmuller, Structure and Bonding, 47, 1-25 (1981).
21. K. Vidyasagar, J. Gopalakrishnan and C.N.R. Rao, Inorg. Chem., 23, 1206-1210 (1984).
22. K.R. Poeppelmeier, M.E. Leonowicz and J.M. Longo, J. Solid State Chem., 44, 89-98 (1982).
23. K.R. Poeppelmeier, M.E. Leonowicz J.C. Scanlon, J.M. Longo and W.B. Yelon, J. Solid State Chem., 45, 71-79 (1982).
24. R.A. Beyerlein, A.J. Jacobson and L.N. Yacullo, Mater. Res. Bull., 20, 877-886 (1985).
25. A. Reller, D.A. Jefferson, J.M. Thomas and M.K. Uppal, Proc Roy Soc. London A, 394, 223-241 (1984).
26. M.A. Alario-Franco, J.M. Gonzalez-Calbet, M. Vallet-Regi and J.C. Grenier, J. Solid State Chem., 49, 219-231 (1983).
27. J.M. Gonzalez-Calbet, M. Vallet-Regi, M.A. Alario-Franco and J.C. Grenier, Mater. Res Bull., 18, 285-292 (1983).
28. C.C. Toradi, E.M. McCarron, M.A. Subramanian, H.S. Horowitz, J.B. Michel, A.W. Sleight and D.E. Cox in "Chemistry of High Temperature Superconductors." Eds. D.L. Nelson, M.S. Whittingham and T.F. George, ACS Symposium Series, 351, 152-163 (1987).
29. S.W. Keller, K.J. Leary, T.A. Faltens, J.N. Michaels and A.M. Stacy in "Chemistry of High Temperature Superconductors". Eds. D.L. Nelson, M.S. Whittingham and T.F. George, ACS Symposium Series, 351, 114-120 (1987).
30. J.D. Jorgensen, M.A. Beno, D.G. Hinks, L. Soderholm, K.J. Vollin, R.L. Hitterman, J.D. Grace, I.K. Schuller, C.U. Sergre, K. Zhang and M.S. Kleefisch, Phys. Rev. B36, 3608-3616 (1987).
31. D.E. Cox et al., submitted for publication.
32. A.J. Jacobson, J.M. Newsam, D.C. Johnston, J.T. Lewandowski, D.P. Goshorn and M. S. Alvarez, submitted to Phys. Rev. B (1987)
33. D.C. Harris and T.A. Hewston, J. Solid State Chem., 69, 182-185 (1987).
34. E.H. Appelman, L.R. Morss, A.M. Kini, U. Geiser, A. Umezawa, G.W. Crabtree and K. D. Carlson, Inorg. Chem., 26, 3237-3239 (1987).
35. J.M. Newsam, A.J. Jacobson, D.B. Mitzi, D.P. Goshorn and J.T. Lewandowski, to be published.
36. P.K. Gallagher, H.M. O'Bryan, S.A. Sunshine and D.W. Murphy, Mater. Res. Bull, 22, 995-1006 (1987).
37. P.K. Gallagher, Adv. Ceram. Mater., 2, 632-639 (1987).
38. P.K. Gallagher and H.M. O'Bryan, Adv. Ceram. Mater., 2, 640-648 (1987).
39. H. Oesterreicher and M. Smith, Mater. Res. Bull., 22, 1709-1714 (1987).
40. P. Strobel, J.J. Capponi II, C. Chaillout, M. Marezio and J.L. Tholence, Nature, 327, 306-308 (1987).
41. H.M. Rietveld, J. Appl. Cryst., 2, 65-72 (1969)
42. D.B. Wiles and R.A. Young, J. Appl. Cryst., 14, 149-151 (1981)
43. International Tables for Crystallography, Volume A, D. Riedel, Dordrecht, Holland (1983) .
44. C.W. Tompson, D.F.R. Mildner, M. Mehregany, J. Sudol, R. Berliner and W.B. Yelon, J. Appl. Cryst., 17, 385-394 (1984).
45. L. Koester and W.B. Yelon, Summary of Low Energy Neutron Scattering Lengths and Cross Sections Netherlands Energy Research Foundation (1982).
46. P.W. Anderson, Science, 235, 1196-1198 (1987)
47. D. Vaknin, S.K. Sinha, D.E. Moncton, D.C. Johnston, J.M. Newsam, C.R. Safinya and H.E. King Jr., Phys. Rev. Lett., 58, 2802-2805 (1987)
48. J.M. Tranquada, D.E. Cox, W. Kunnmann, H. Moudden, G. Shirane, M. Suenaga, P. Zolliker, D. Vaknin, S.K. Sinha, M.S. Alvarez, A.J. Jacobson and D.C. Johnston, Beano, submitted (1987)
49. S.K. Sinha, C.S. Stassis and G. Shirane unpublished (1987).
50. S. Mitsuda, G. Shirane, T. Frettoft, J.P. Remeika et al., to be published.
51. G. Shirane, Y. Endoh, R.J. Birgeneau, M.A. Kastner, Y. Hidaka, M. Oda, M. Suzuki and T. Murakami, Phys. Rev. Lett., 59, in press (1987)
52. P.P. Fretias, C.C. Tsuei and T.S. Plaskett, Phys. Rev., B36, 833-835 (1987)

53. A.T. Fiory, A.F. Hebard, L.F. Schneemeyer and J.V. Waszczak, In proc. MRS Fall 1987 Mtg. (MRS Symp. Proc., Washington, DC) in press.
54. X. Cai, R. Joynt and D.C. Larbalestier Phys. Rev. Lett., 58, 2798-2801 (1987).
55. S. Bhattacharya, M.J. Higgins, D.C. Johnston, A.J. Jacobson, J.P. Stokes, J.T. Lewandowski and D.P. Goshorn, Phys. Rev. B, submitted (1987).
56. S. Bhattacharya, M.J. Higgins, D.C. Johnston, A.J. Jacobson, J.P. Stokes, D.P. Goshorn and J.T. Lewandowski, Phys. Rev. Lett., submitted (1988).
57. J. Yu, A.J. Freeman and J.-H. Xu, Phys. Rev. Lett., 58, 1035-1038 (1987).
58. A. Manthiram and J.B. Goodenough, Nature, 329, 701-703 (1987).
59. M.A. Alario-Franco, C. Chaillout, J.J. Caponi and J. Chenavas, Mater. Res. Bull., 22, 1685-1693 (1987).
60. D.C. Johnston to be published (1988).
61. B.C. Sales et al. to be published (1988).
62. L.F. Schneemeyer, J.V. Waszczak, E.A. Rietman and R.J. Cava, Phys. Rev., B35, 8421-8424 (1987); D.C. Johnston, to be published.
63. R.J. Cava, B. Batlogg, C.H. Chen, E.A. Rietman, S.M. Zahurak and D. Werder, Nature, 329, 423-425 (1987)
64. D.B. Mitzi, A.F. Marshall, J.Z. Sun, D.J. Webb, M.R. Beasley, T.H. Geballe and A. Kapitulnik, Appl. Phys. Lett. (in press).
65. L. Soderholm, K. Zhang, D.G. Hinks, M.A. Beno, J.D. Jorgensen, C.U. Segre and I.K. Schuller, Nature, 328, 604-605 (1987).
66. R.M. Fleming, B. Batlogg, R.J. Cava and E.A. Rietman, Phys. Rev., B35, 7191-7194 (1987).

High T_c superconductivity in $LnBa_2Cu_3O_7$ (Ln = La–Yb, Y) compounds

G.V. Subba Rao
Materials Science Research Centre, Indian Institute of Technology, Madras - 600 036, India

Abstract - The discovery of superconductivity above the liquid nitrogen temperatures (77K) in March, 1987 in the mixed copper oxide system, Y-Ba-Cu-O, aroused worldwide interest in the study and exploitation of these oxide materials. The 90K superconductor, $YBa_2Cu_3O_7$, is amenable to chemical substitution at the Y, Ba and Cu sites. The structure and superconducting behavior of $LnBa_2Cu_3O_7$ (Ln=rare earth) phases are described and discussed with emphasis on the work done in the author's laboratory. Mention has also been made on the effects of chemical substitution at the Ba- and Cu- and O- sites on the T_c of $YBa_2Cu_3O_7$ and possible mechanism of high T_c superconductivity.

INTRODUCTION

After the identification of $YBa_2Cu_3O_7$ as the single phase material responsible for the high T_c (91K) superconductivity in the Y-Ba-Cu-O system (ref. 1-3), several research groups reported preliminary results of the effects of chemical substitution at the Y, Ba and Cu-sites on the T_c behaviour (ref. 4-9). One interesting finding is that the yttrium can be substituted partly or fully by other trivalent rare earth (Ln^{3+}) ions and that the phases, $LnBa_2Cu_3O_7$ also exhibit high T_c's including the Ln-ions which possess localized magnetic moments. This is a significant observation and is contrary to the expectations, since it is known that the localized electron-conduction electron (Cooper pair-breaking) interaction reduces drastically the transition temperature (T_c) of a superconductor. However, a behaviour similar to $LnBa_2Cu_3O_7$ is encountered in 'low temperature' ternary superconductors (e.g., Chevrel phases, AMo_6Ch_8, A = metal; Ch = S,Se and rare earth rhodium borides, $LnRh_4B_4$) (ref.10). This definitely indicates that the superconducting electrons arise from Cu in $LnBa_2Cu_3O_7$ and there is negligible interaction between the localized (f-) electrons and conduction electrons. High temperature magnetic susceptibility and Mossbauer studies on $LnBa_2Cu_3O_7$ (ref. 9, 11) have also shown that the Ln ions are present in 3+ oxidation state and exhibit almost spin- only magnetic moments ($4f^n$, n ≠ 0 or 14). Many of the f^n-$LnBa_2Cu_3O_7$ order antiferromagnetically below 2K retaining the high T_c superconductivity behaviour (ref. 8, 12). The low T_c values in these antiferromagnetic superconductors can be attributed to the typical magnetic-dipolar interactions. It may be mentioned that eventhough $LnBa_2Cu_3O_7$ compounds are isostructural to $YBa_2Cu_3O_7$ and exhibit high T_c behavior, subtle changes with respect to T_c, resistivity (ρ_{300K}) and ρ-T behavior, Seebeck coefficient and significant changes in H_{c2}, J_c and related properties are expected in $LnBa_2Cu_3O_7$. Many of the above properties are yet to be examined.

Structural Aspects

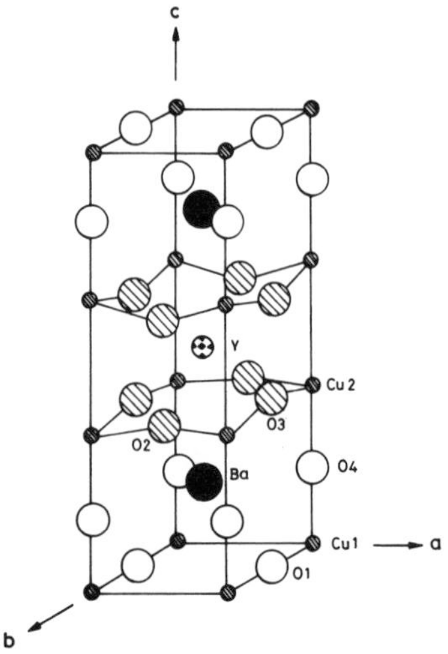

Fig.1 Structure of $YBa_2Cu_3O_7$

The compound $YBa_2Cu_3O_7$ is an oxygen-deficient, orthorhombically distorted perovskite with a tripled unit cell along the c-axis. The detailed crystal structure has been elucidated by single crystal X-ray and neutron powder diffraction studies (ref. 13-16). The lattice parameters are: a=3.823Å; b=3.886Å; c=11.681 (3x3.89Å); Z=1, space group, Pmmm. The yttrium, and two barium atoms occupy crystallographically distinct sites with 8-fold and 10-fold oxygen coordination respectively (Fig. 1). There exist two different kinds of Cu-atoms (Cu1(x1) and Cu2(x2)) and four kinds of oxygen atoms (O1 to O4) in the structure. Cu1 atoms have 4-fold planar oxygen coordination (in the b-c plane) but essentially there exist linked -O1-Cu1-O1-Cu1-linear (1D) chains along (parallel to) the crystallographic b-axis. The Cu2 atoms have a five-fold oxygen coordination. However, since the apical Cu2-O4 distance is large, these Cu2 atoms can be considered as existing as Cu2-O4 planar-like groups corner linked in the a-b plane (perpendicular to c-axis).

The disposition of the CuO_2 planes and Cu-O chains is such that there is no significant interaction between the planes and chains. Due to the oxygen deficiency in $YBa_2Cu_3O_7$ there exist mixed valency in the Cu and the nominal chemical formula can be written as $YBa_2Cu_2^{2+}Cu_1^{3+}O_7$. However, structural studies show that there are significant Y-Cu, Ba-Ba (ref. 16) and O-O interactions (O1-O4 = 2.68Å, less than the diameter of oxide ion) (ref. 15). This observation and other considerations (e.g. XPS data) led researchers (ref. 17-22) to propose the existence of only Cu^{1+}/Cu^{2+} and O_1^- in the structure. The metallic conduction (and superconductivity behavior) is attributed to the planes and/or chains within the $YBa_2Cu_3O_7$ structure.

Eventhough nominal composition is 7.0 for oxygen, careful studies have shown that high T_c superconductivity is encountered only for δ = 0.05 - 0.2 in $YBa_2Cu_3O_{7-\delta}$. Detailed studies by Uchida et al (ref. 23), Rao et al (ref. 22, 24) and others (ref. 25, 26) have shown that T_c is highly sensitive to δ ; for δ = 0.2 - 0.4, the T_c value drops to around 60K (retaining the orthorhombic structure) whereas for δ = 0.4-0.8, no superconductivity has been observed and the structure changes into a tetragonal lattice. It is established that this nonstoichiometry arises due to missing O1 oxygen atoms in the structure thereby disrupting Cu1-O1 linear chains. Hence, it is clear that to obtain high T_c superconductivity in $YBa_2Cu_3O_7$ and $LnBa_2Cu_3O_7$, oxygen annealing in the final stages of synthesis is one of the essential steps.

HIGH T_c SUPERCONDUCTIVITY IN $LnBa_2Cu_3O_7$

We have synthesized single phase $LnBa_2Cu_3O_7$ (Ln = La,Pr,Nd,Sm,Eu,Gd,Dy,Ho,Er, Tm and Yb) and $YBa_2Cu_3O_7$ by the high temperature solid state reaction from high purity starting materials. The synthesis procedure has been standardized w.r.t. time and temperature of heating in air and O_2 atmosphere to obtain well-defined phases exhibiting reproducible physical properties (ref. 27). We have established superconductivity by the four probe dc electrical resistivity, ac susceptibility and Seebeck coefficient studies as a function of temperature. The crystal and T_c data, shown in Table I, indicate that

Table I. X-ray and T_c data on high T_c $LnBa_2Cu_3O_7$ superconductors

Compound, $LnBa_2Cu_3O_7$ Ln	Lattice parameters, Å a (±0.01)	b (±0.01)	c (±0.03Å)	T_c^{onset}[a], K	T_c^{zero}[b], K	ΔT_c[c], (90–10%) K	ρ300K′ Ωcm(×10⁻³)	(1/ρ300K)(∂ρ/δT) × 10⁻³ (range 140–240K)
Y	3.83	3.89	11.68	104	92	4.5	2.75	2.04
La	3.91	3.91	11.73	—	—	—	2.75	—
Pr	3.99	3.99	11.98	—	—	—	31.70	—
Nd	3.87	3.92	11.74	91	85	4.0	1.90	2.26
Sm	3.85	3.90	11.70	120	90	13.0	1.09	2.61
Eu	3.85	3.90	11.70	113	93	3.0	1.24	1.81
Gd	3.82	3.88	11.63	115	91	6.0	1.40	1.62
Dy	3.82	3.88	11.66	98	91	1.5	1.80	2.11
Ho	3.82	3.89	11.66	97	92	2.0	0.90	2.92
Er	3.82	3.88	11.65	114	90.5	5.5	2.60	1.11
Tm	3.82	3.88	11.61	96	91	3.0	2.40	2.77
Yb	3.81	3.88	11.64	97	88	2.8	2.20	2.88

[a]Temperature at which resistivity dropped significantly from linearity.
[b]Temperature at which zero resistance is recorded.
[c]Difference in temperature corresponding to 90% and 10% drop in resistivity values.

$LnBa_2Cu_3O_7$ are isostructural to the orthorhombic $YBa_2Cu_3O_7$, except when Ln=La and Pr, which have a tetragonal structure. There is a decrease in the c lattice parameter and slight variations in the a and b axes from La → Yb in the $LnBa_2Cu_3O_7$ phases. However, within the limits of experimental error, we do not see any systematic variation in the lattice parameters. Several groups have reported systematics in the variation of a, b, c and (b-a)/a (ref. 28, 29). Perhaps more careful measurements, preferably on single crystals, may reveal the subtle differences.

Four probe dc electrical resistivity data (van der Pauw's method) on sintered (oxygen annealed) pellets in the range 77-300K show that all the $LnBa_2Cu_3O_7$ compounds are metallic with low room temperature resistivity (1-3 milli ohm cm) (Table I). The values of $(1/\rho_{300K})(d\rho/dT)$ (range 140-240K) are also typical of the metallic behavior (little departure from linearity). All the $LnBa_2Cu_3O_7$ with the exception of La and Pr showed high T_c (>77K) superconductivity behavior. The values of T_c^{onset}, T_c^{zero} and ΔT_c are listed in Table I. Very sharp superconducting transitions are seen with Ln=Eu, Dy,Ho,Tm and Yb. The highest T_c^{zero} has been noted in Ln=Eu and the lowest in Ln=Nd (Fig. 2). It is significant to note that all $LnBa_2Cu_3O_7$ with Ln=Gd Dy,Ho,Er and Tm (heavy rare earths) have T_c= 91 ± 1K. Similar behavior has been noted by other workers as well (ref. 4-9, 28-29). For Ln=Nd, our value of T_c^{zero} is higher than that noticed by Murakami et al (ref. 30) but is in good agreement with Yang et al (ref. 29) and other workers (ref. 5, 6, 31).

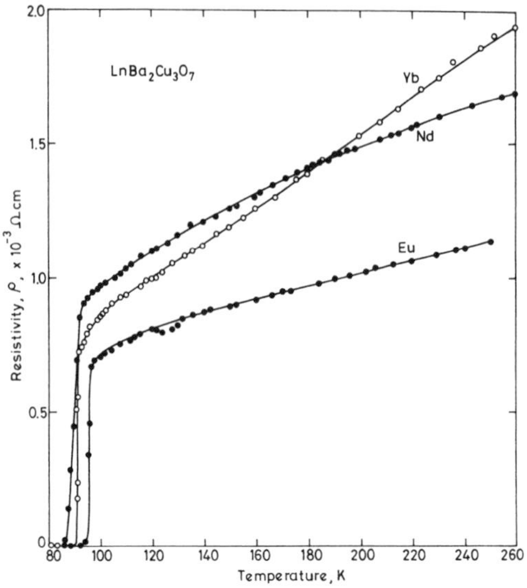

Fig.2 ρ-T plots of $LnBa_2Cu_3O_7$ (Ln= Nd,Eu and Yb) showing superconducting transitions

The superconducting behavior of $LnBa_2Cu_3O_7$ have also been established by ac susceptibility studies on a few select compounds. The transition temperatures coincide very well with T_c^{zero} values obtained by resistivity data. The magnitude of the Meissner signal is estimated as 60-70% of the Pb-metal signal (at 7K) indicating that the superconducting behavior in these compounds is a bulk property and not of the filamentary type.

Seebeck coefficient measurements were carried out on typical $LnBa_2Cu_3O_7$ compounds. As expected for a bulk superconductor, the Seebeck coefficient (α) value dropped to zero at the transition temperature. The temperature at which α reached zero is in agreement with the T_c^{zero} observed by resistivity studies. The sign of α is positive indicating holes as the majority charge carriers. The numerical value is +2 to +4 μV/K typical of a metallic material and increases with increase in temperature (150-300K). We observed an anamolous enhancement in Seebeck coefficient below 150K for $LnBa_2Cu_3O_7$ Ln=Y, Dy and a few solid solutions. The anamolous enhancement noticed by us (ref. 32) and also by others (ref. 17, 33) can not be explained by simple phonon-drag mechanism. Perhaps in these materials, multi-band conduction is involved in addition to other factors governing the transport properties.

As mentioned earlier, we have not found superconductivity above 77K in La and Pr compounds which have tetragonal structure (Table I). This is in agreement with observations by other groups (ref. 34). However, it is now known that $LaBa_2Cu_3O_7$, on annealing does undergo orthorhombic distortion and exhibits high T_c (> 77K) superconductivity (ref. 35, 36).

We have made efforts to synthesize $LnBa_2Cu_3O_7$ with Ln=Ce,Tb and Lu starting from CeO_2, Tb_4O_7 and Lu_2O_3. Single phase materials of the '123' type were not obtained and always $BaLnO_3$, Ln=Ce and Tb, was obtained as impurity along with other phases. With Lu_2O_3, the mixture was found to melt under the preparative conditions employed, and yielded a mixed-phase material with small admixture of '123' phase. Further studies have not been carried on the above. Efforts are now on to synthesize Ce and Tb phases using the presynthesized Ce_2O_3 and Tb_2O_3.

Effect of chemical substitution on T_c of $YBa_2Cu_3O_7$

We have prepared and studied solid solutions of the type $Y_{1-x}Ln_xBa_2Cu_3O_7$, Ln=La,Gd,Ho and Tm (ref. 37, 38). The results show that isostructural solid solutions are formed which also exhibit high T_c but Vegards law behaviour is not obeyed. Some compositions exhibit T_c^{zero} as high as 94K. In the case of $(Y_{1-x}La_x)Ba_2Cu_3O_7$, $T_c > 77K$ was observed only for $x \leq 0.5$ and the phases adopt an orthorhombic structure. For $0.5 < x \leq 1.0$, the phases are tetragonal and no superconductivity has been found above 77K.

Strontium can be substituted at the Ba site in $YBa_2Cu_3O_7$ and single phase materials can be obtained for $x = 0.0 \leq x \leq 1.25$ in $YBa_{2-x}Sr_xCu_3O_7$. The compounds are isostructural and high T_c ($> 77K$) is retained for $x = 1.0$ with T_c decreasing monotonically with increasing x. Similar results have been reported by others in the literature (ref. 39-41). Substitution at the Cu site by a transition metal (Fe,Co,Ni) or Zn in smaller amounts reduces the T_c drastically (ref. 8,42,43) which indicates that the integrity of the metal atoms at the Cu - site is crucial in determining the high T_c behaviour in $YBa_2Cu_3O_7$. Detailed studies are needed especially with those dopents, which can occupy specific crystallographic sites (Cu1 or Cu2 sites) to understand the variation of T_c with the dopent concentration.

Attempts have been made in the literature to substitute fluorine partly at the oxygen site in $YBa_2Cu_3O_7$ (ref. 44,45,46). T_c values as high as 155K have been reported (ref. 45,46) but the phases have not been found stable nor the results reproducible. We have attempted the synthesis of the following F-containing phases: Series I: $YBa_2Cu_3O_yF_x$ ($x = 0.1, 0.25, 0.50, 0.75$ and 1); Series II: $Y_{1-x}Ba_{2+x}Cu_3O_{7-x}F_x$ ($x = 0.25, 0.33, 0.50$ and 0.75); Series III: $YBa_2Cu_3O_yF_x$ (x=2,3,4) (Ovshinsky compositions (ref.45)). X-ray diffraction patterns of the above F-containing phases showed the presence of BaF_2, CuO, CuF_2 and $BaCuO_2$ impurities in addition to the '123' phase. There is no significant change in T_c and the observed T_c was attributed to the presence of $YBa_2Cu_3O_7$ phase. In addition, $YBa_2Cu_3O_yF_2$ was semiconducting and F_3 and F_4 analogues (series III) were found to show insulating behavior. Hence, it is concluded that under the preparative conditions employed (950°C, air and O_2 ; BaF_2 as the starting material), F does not get incorporated into the $YBa_2Cu_3O_7$ lattice. Perhaps, novel methods may have to be adopted to incorporate fluorine (part or full) in the lattice.

MECHANISM OF HIGH T_c IN $YBa_2Cu_3O_7$

There exists, at present, no suitable theory to explain the high T_c behavior of Y-and $LnBa_2Cu_3O_7$ compounds (ref. 8,22,47). The conventional BCS (electron-phonon) mechanism seems to be insufficient and other mechanisms (excitonic, resonating valence bond and antiferromagnetic correlations) are being explored. Further, though it is known that electrons from copper in $LaBa_2Cu_3O_7$ are responsible for Cooper-pair formation and high T_c superconductivity, it is not yet clear whether the latter occurs in the CuO_2-planes or Cu-O chains in the structure. Arguments have been put forward in favor of both. For example, the extreme sensivity of T_c to the variation in oxygen concentration, which mainly occurs in Cu-O1 chains (in annealed and unannealed samples) is taken as the evidence for superconductivity exclusively attributed to the chains within the structure. However, since it is known that 2D-CuO_2 planes are responsible for superconductivity in $La_{2-x}A_xCuO_4$ (A=Ca, Sr,Ba) which are the 40K superconductors, it is likely that superconductivity occurs in 2D-CuO_2 planes also in $YBa_2Cu_3O_7$. Further, if the Cooper-pair formation interaction is very strong to give rise to T_c's around 90K, the localized spin (Ln-4f)-conduction electron (Cu-3d) interactions (pair-breaking) can only have a secondary influence and need not significantly affect the T_c. Perhaps, the high T_c superconductivity in $YBa_2Cu_3O_7$ occurs mainly due to the planes mediated by exciton interactions from the chains through -Cu2-O4-Cu1-O4- type superexchange interactions. Detailed studies on single crystals with respect to 2Δ (energy gap), ρ and H_{c2}-anisotropy, J_c and related properties will throw more light on this aspect.

Acknowledgements

Thanks are due to Dr.U.V. Varadaraju and students at MSRC, Prof.R. Srinivasan and his team of researchers at the Low Temperature Laboratory, IIT Madras for excellent collaborative efforts. Thanks are due to the Department of Science and Technology, Government of India, New Delhi for the award of a research grant.

References

1. R.J. Cava et al., Phys. Rev. Lett. 58, 908 (1987)
2. C.N.R. Rao et al, Nature 326, 856 (1987)
3. T. Siegrist et al, Phys. Rev. B35, 7137 (1987)
4. G. Xia et al, Solid State Commun. 63, 817 (1987)
5. P. Hor et al, Phys. Rev. Lett. 58, 1891 (1987)
6. E.M. Engler et al, J. Amer. Chem. Soc. 109, 2848 (1987)
7. C.N.R. Rao et al, J. Solid State Chem. 69, 182 (1987); Phil. Mag. Lett. 56, 29 (1987)
8. B.D. Dunlop et al, J. Magn. Magn. Mater. (in press)
9. P.A.J. de Groot et al, J. Phys F. 17, L185 (1987)
10. O. Fisher and M.B. Maple (Eds.), Supercond. in Ternary Compds., I and II, Springer Verlag, NY, 1982; G.V. Subba Rao and G. Balakrishnan, Bull.

Mater. Sci. (India) 6, 283 (1984); G.V. Subba Rao, Proc. INSA, 52A, 292 (1986)

11. R. Nagarajan et al, Hyperfine Interactions (1987) (in press)
12. J.R. Thompson et al, Phys. Rev. B 36, 718 (1987)
13. M.A. Beno et al Appl. Phys. Lett. 51, 57 (1987); F. Beech et al, Phys. Rev. B 35, 8778 (1987)
14. J.E. Greeden et al, Phys. Rev. B 35, 8770 (1987); W.I.F. David et al, Nature 327, 310 (1987)
15. M. Francois et al, Solid State Commun. 63, 1149 (1987)
16. I. Nakai et al, Jap. J. Appl. Phys. 26, L788 (1987)
17. C.F. Van Bruggen Phys. Status Solidi 142 537 (1987); C.F. Van Bruggen et al, Mat. Res. Bull (in press)
18. C.N.R. Rao et al, Mat. Res. Bull., 22, 1059 (1987)
19. H. Oyanagi et al, Jap. J. Appl. Phys. 26, (1987), L638
20. D.D. Sarma et al, Phys. Rev. B 36, 2371 (1987)
21. D.D. Sarma and C.N.R. Rao, J. Phys. C. Solid State, 20, L659 (1987)
22. C.N.R. Rao et al, J. Amer. Chem. Soc., (in press); C.N.R. Rao, IUPAC Symp. (in press).
23. S. Uchida et al, Proc.Adriatico Conf., Trieste), Int. J. Mod. Phys. (1987)
24. C.N.R. Rao et al, Mat. Res. Bull., preprint (to be published)
25. B. Batlogg et al, Yamada conf. on Supercond., Sendai, Japan, Aug, 1987 (Proc. to be published)
26. B. Raveau et al, Europhysics Lett. 4, 205, 211 (1987)
27. G.V. Subba Rao et al, Proc. 57th National Acad. of Sci. (India), Trichy, Oct., 1987 (to be published); Proc. of Indo-U.S.S.R. Symp. on Supercond., Novosibirsk, U.S.S.R., Oct., 1987 (to be published)
28. T. Yamada et al, Jap. J. Appl. Phys. 26 Suppl. 3, 1035 (1987) (Proc. 18th Intl. Conf. on Low Temp. Phys., Kyoto (Japan),(1987)
29. K.N. Yang et al, Jap. J. Appl. Phys. 26 (Suppl.3) 1037 (1987)
30. M. Murakami et al, Jap. J. Appl. Phys. 26 (Suppl.3) 1061 (1987)
31. R.J. De Angelis et al, Solid State Commun., 64, 1353 (1987)
32. R. Srinivasan et al, J. Phys. Pramana 29, L225 (1987)
33. A. Mawdslay et al, Nature 328 (1987)
34. C.U. Segre et al, Nature (Lond.), 329, 227 (1987)
35. K. Kitazawa, Univ. of Tokyo (Private commun.)
36. A. Maeda et al, Jap. J. Appl. phys. 26, L1366 (1987)
37. S. Natarajan et al, Abstr. DAE Symp., Bombay, Dec., 1987
38. S. Natarajan et al, Abstr. DAE Symp., Bombay, Dec., 1987
39. Y. Saito et al, Jpn. J. Appl. Phys. 26, 1081 (1987)
40. B.W. Veal et al, Appl. Phys. Lett. 51, 279 (1987)
41. J.M. Tarascon et al, ACS Symp. Series, 1987
42. G. Xiao et al, Phys. Rev. B 35, 8782 (1987)
43. T.A. Mary et al, Abstr. DAE Symp., Bombay, Dec., 1987
44. T.S.Radhakrishnan et al, in Ref.28; M.Paranthaman et al, Abstr. DAE Symp., Bombay, Dec., 1987
45. S.R. Ovshinsky et al, Phys. Rev. Lett., 58 2579 (1987)
46. M. Xian-Ren. et al, Solid State Commun. 64, 325 (1987)
47. A.J. Freemamn et al, Jap. J. Appl. Phys. 26, 1153 (1987)

Chemistry–T_c relations in oxide superconductors

A.S. Bhalla, Rustum Roy and L.E. Cross

Materials Research Laboratory, The Pennsylvania State University, University Park, PA 16802 USA.

Abstract - This paper presents an overview of a large number of systematic studies of ionic substitutions at each site in the $YBa_2Cu_3O_{7-\delta}$ phase. Superconducting properties (T_c, Meissner effect, Jc) have been measured and correlated with phases formed, the kinetics of reaction, the orientation of the ceramic product, etc.

INTRODUCTION

Superconductivity, the property early thought to be associated with metals and alloys, was demonstrated for the first time to exist in oxide materials like $SrTiO_3$ (in 1964) by Hein, et al (ref. 1). In the successive years the phenomenon of superconductivity was discovered in a variety of oxide and sulfide compounds (ref. 1-10) in the spinel, tungsten bronze, K_2NiF_4 and perovskite crystal structure families (table 1) with the highest Tc attained being in the 13K range. After a ten year quiet period, superconductivity was shown to be present in oxides of $(LaBa)_2CuO_{4-x}$ system at 40K higher than in any metal or oxide material. In less than one year the high superconductivity was discovered at 90K in a closely related family, $RBa_2Cu_3O_{7-x}$ (where R = Y or rare earth ion except Ce, Pr and Tb), belonging to the defect perovskite crystal structure. The chronology of the development (refs. 11-36) in oxide superconductivity may be summarized as shown in Figure 1.

Although the new class of superconductor oxides, $R_1Ba_2Cu_3O_{7-\delta}$ (henceforth referred to as 123 or YBC) have very attractive Tc, these materials also have serious shortcomings from a practical material processing and device fabrication point of view. The 123 ceramics generally sinter with low density and the superconducting properties are sensitive to the ambients. Poor chemical stability and low critical current are the key intrinsic problems associated with these materials.

This paper presents a summary of the results of many studies in this laboratory. Many of these were aimed i) to improve the processibility of the high-T_c supeconductors through various different approaches; ii) to enhance the current carrying capabilities of the materials through the improvement in the homogeneity of the oxides and then synthesizing the high density materials with superior mechanical properties; iii) to develop new materials by designing composites of polymers with the oxide superconductors. This achieves many goals at once. It reduces their sensitiveness to ambient

TABLE 1. Superconducting Oxides and Sulfides (refs. 1-10).

Compound	Crystal Structure	T_c(K)	Refs.
$SrTiO_3$	Perovskite	0.7	Hein, et al. (1964)
A_xWO_3	Tungsten Bronze	6.0	Raub, et al. (1965)
A_xMO_3	"	4.0	Sleight, et al. (1966)
A_xReO_3	"	4.0	Sleight, et al. (1969)
$Li_xTi_2O_4$	Spinel	13.7	Johnston, et al. (1973)
$Ba(PbBi)O_3$	Perovskite	13.0	Sleight, et al. (1975)
TiO	NaCl	2.3	Doyle, et al. (1968)
$Li_xTi_{1.1}S_2$		10-15	Barz, et al. (1972)
$Mo_{6-x}A_xS_6$ (A - Ag, Cu, ..)		2.5-13	Matthias, et al. (1972)
$CuRh_2S_4$	Spinal	4.8	Van Maaren, et al. (1967)
$CuRh_2Se_4$	Spinel	3.5	Van Maaren, et al. (1967)

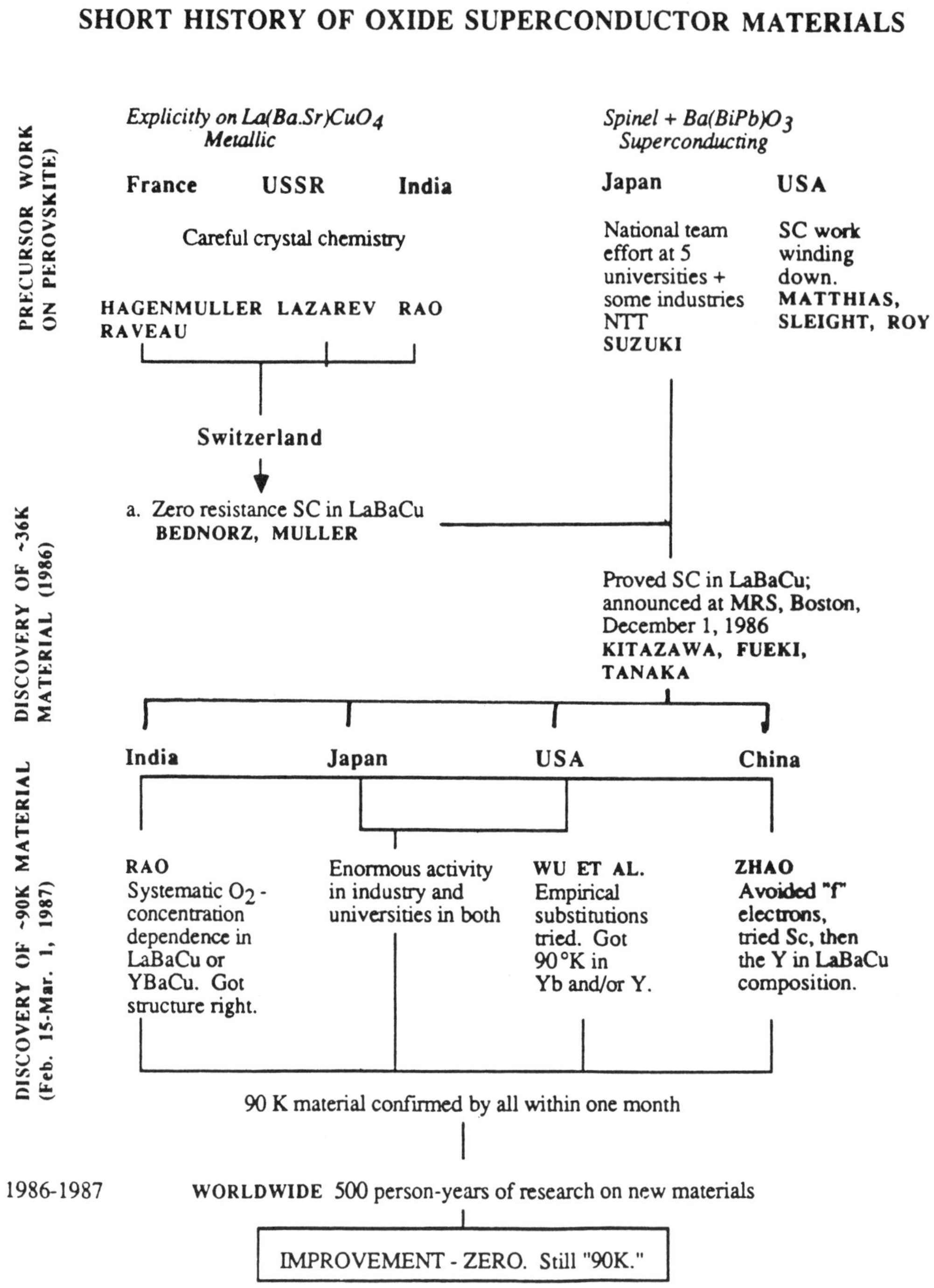

Fig. 1. Chronological developments in the area of oxide superconductors (refs. 11-36).

conditions; and iv) to investigate the effects of various impurities on the properties of 123 oxide superconductors and evaluate their processibility and chemical compatibility with other potential 'ferroic' materials. The BHK (Barsch, Horovitz and Krumhansl, refs. 37-39) recognition that the special twinning in the orthorhombic phase is responsible for the enhancement of T_c is perhaps the key in understanding the unique feature of these new (superconductor) 'ferroic' materials. This new theoretical model of domain wall enhanced superconductivity in the 123 ceramics also emphasizes the critical importance of the highly regular domain structure and related nanoscale organization of the material.

Because it is difficult to introduce the vast literature on the 123 materials in a coherent way in this section, a brief introduction pertinent to the subject matter of that particular section, is given in with each relevant part of the text.

SYNTHESIS AND PROCESSING OF YBC

BaO_2-Method

Several routes have been followed by various workers for preparing the 123 compounds, e.g., mixed oxide method, co-precipitation of oxalates, mixed nitrates and various chemical preparation methods. In practice most workers have prepared their materials by the mixed oxide method and used $BaCO_3$ as starting material for obtaining BaO, required in the preparation of 123 compounds. The complete decomposition of $BaCO_3$ to BaO occurs at higher temperatures and by using the 950°C temperature limit imposed on the synthesis of 123 compound, much longer firing times are required for the completion of the reaction. Also the slight unreacted material resulted in the multiphase powder.

By simply using BaO_2 instead of $BaCO_3$ in the 123 powder preparation the firing times were considerably reduced and pure 123 phase with improved superconducting properties was obtained directly. A few degrees increase in T_c and improved Meissner effects were also noticed on such samples (Fig. 2). Under the optimized firing time-temperature cycle, the post sintered annealing conditions were realized to be less critical. The reason for the value of BaO_2 may in part be connected to the presence of peroxo-bridges in the 123 phase itself. Many chemists are looking increasingly at the possibility of describing the structural chemistry of the 123 phase in these terms (ref. 41).

Preparation of Fine Particle Superconducting Oxides by the Sol-Gel Method

Sol-gel methods are used by many workers. In our case (ref. 42) Yttrium acetate was dissolved in methoxyethanol and the solution was refluxed. As a result a brown color solution of the corresponding ethoxide derivative of Yttrium was obtained. Barium isopropoxide was added to this solution and the mixture was heated at 70°C for a few hours. In a separate experiment, copper was dissolved in methoxymethanol and the solution was refluxed to obtain Cu-methoxymethanol derivative. Finally, Y, Ba and Cu alkoxide derivatives were mixed and the mixture was diluted with deionized water. The metal hydroxides were precipitated and separated out from the solution and dried for 10 hours at 110°C. The dried powders were calcined at 800°C for several hours in a flow of partial pressure of oxygen to produce fine particles of single phase superconducting oxide - $YBa_2Cu_3O_{7-\delta}$ (as revealed from the x-ray diffraction studies).

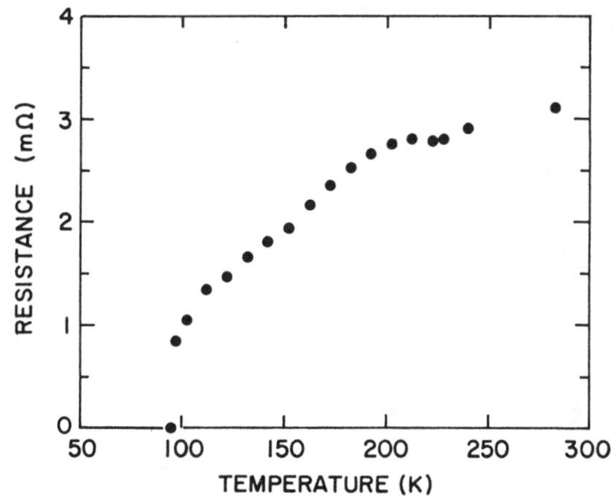

(a)

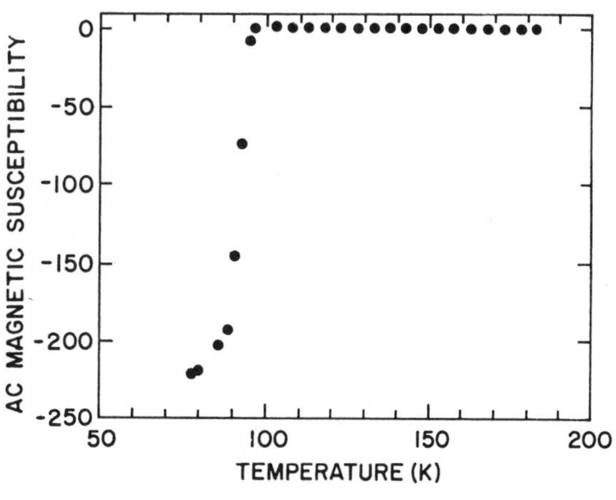

(b)

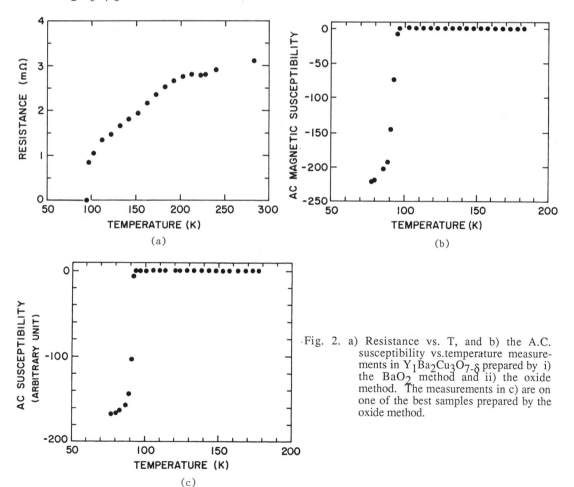

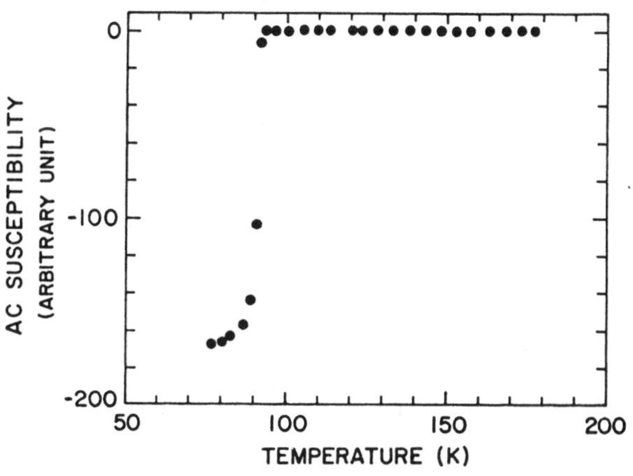

(c)

Fig. 2. a) Resistance vs. T, and b) the A.C. susceptibility vs.temperature measurements in $Y_1Ba_2Cu_3O_{7-\delta}$ prepared by i) the BaO_2 method and ii) the oxide method. The measurements in c) are on one of the best samples prepared by the oxide method.

The 123 powder was pressed into discs and then sintered at 900°C. The sintered pellets densified to 95% the theoretical density. The superconducting Tc measured on these samples was ~ 92K and the Meissner effect was stronger by a factor of two as compared to the 123 materials prepared by the mixed oxide method.

Preferred Orientation of the Grains in the Ceramics of High T_c-Superconductors:
The crystalline structures of the $(LaBa)CuO_{4-x}$ and $LnBa_2Cu_3O_{7-\partial}$ are strongly anisotropic. It is reported that the electrical, magnetic, and superconducting properties of these compounds are also highly anisotropic. In the case of $(LaSr)CuO_{4-x}$ both the electrical resistivities and H_{c2} along the a and c axes differ by at least a factor of five. Similar observations in the case of YBC or 123 compounds have been made by several workers. YBC and other related 123 high T_c superconductors are orthorhombic with the point group mmm at room temperature. At higher temperature the orthorhombic to tetragonal phase transition occurs at ~ 600°C for various 123 compounds.

The reported superconducting properties, e.g., J_c, H_c, etc. on single crystal YBC samples are much superior to those observed on the bulk ceramic bodies of YBC. Current densities of ~ 10^5 A/cm^2 have been measured on single crystal thin films deposited on $SrTiO_3$ substrate whereas the values on the bulk samples range from 10 A/cm^2 to 10^3/cm^2. The large differences in J_c on the bulk compared to the single crystal samples as well as for the ceramics synthesized by various processing techniques are attributed to the random grain orientation and the density of the ceramics. Also the transition temperature spread, ΔT_c, is much smaller, ~ 1K, in the single crystal samples compared to the $\Delta T_c \geq$ 1K - 10K measured in the bulk samples. However, it is not clear whether the single crystal samples are chemically and phase pure.

Thus, in order to establish the usefulness of the bulk ceramics, grain oriented samples are desirable, even though other secondary factors such as grain-grain boundary, low density, multichemical phases could also be responsible for the low J_c.

There are several reports on the ceramic preparation by hot pressing (ref. 43) and hot forging techniques and the enhancement of the preferred orientation of the grains in the resultant ceramic bodies has been noticed. Grain orientation has also been observed after the particles are suspended in the weakly diamagnetic epoxy and then the mixture set under a magnetic field (ref. 44). The weak torque produced by the epoxy on the anisotropic particles of YBC under the magnetic field, produced the composite with a preferred b-axis after the mixture was cured.

For the grain orientation studies, we prepared YBC ceramics by the mixed oxide method (ref. 45). The compound CuO (98%), Y_2O_3 (99.9%) and $BaCO_3$ (anhydrous) were mixed in the proper ratio and milled for 6 hours and calcined twice at 920°C and 960°C in air and oxygen respectively. The resulting large grain powder was mixed with acetone in a container and settled down under gentle ultrasonic vibrations. The powder was then pressed into pellets. The pellets were sintered at 960°C for 12 hours in partial pressure of oxygen, then cooled down and held at 500°C for 4 hours and finally brought to room temperature while cooling at 3°C/min.

SEM studies showed large grains (10-50μ) of platy morphology on the surface of the ceramic samples. X-ray diffraction on such samples revealed the highly preferred orientation of the grains with their crystallographic c-axis perpendicular to the surface (Fig. 3a). By successive polishing and the corresponding x-ray diffraction studies the depth of the layer on each side of the pellet was estimated to be ~ 100μm. It is clear from the intensities of the 001 diffraction lines that the c-plates are predominantly oriented in the surface and after 100μ depth along the c-direction in the plates, the x-ray diffraction pattern resembles to that of a typical YBC powder with random orientations of the grains (Fig. 3b).

The T_c measured on these samples was 92K (zero resistance) and the transition width ΔT_c~ 1-2K.

Similar studies on $GdBa_2Cu_3O_7$ revealed the same strong orientation effects in the grains. SEM studies and the x-ray diffraction of as-prepared powder clearly showed that the GBC grains have large growth anisotropy with a predominantly platy growth habit and having their c-axis perpendicular to the major faces. This growth feature of the grains resembles the typical growth habit of the single crystal samples of YBC or GBC grown in this laboratory or elsewhere. Though the material is prepared at 960°C, a temperature at which the crystallites have the tetragonal crystal structure, the morphology is governed by the growth habit, reaction kinetics and the nature of the impurities. In the YBC family the tetragonal crystals have the same platy morphology with the c-axis perpendicular to the major surfaces of the single crystal plates. Thus, it is logical to think that the morphology of the particles in the 123 family members may be platy but the degree of anisotropic growth behavior may be different from one member to another (i.e., for different ions of the lanthanide series). Also certain substitutions in 123 compounds may produce an even higher degree of platy growth of the grains and make these materials more suitable for the tape casting. Such impurities may also change the reaction kinetics and favor the growth of large b-surfaces. Studies in these directions are underway and details along with the current carrying capabilities of such oriented ceramics will be published in successive specialized papers.

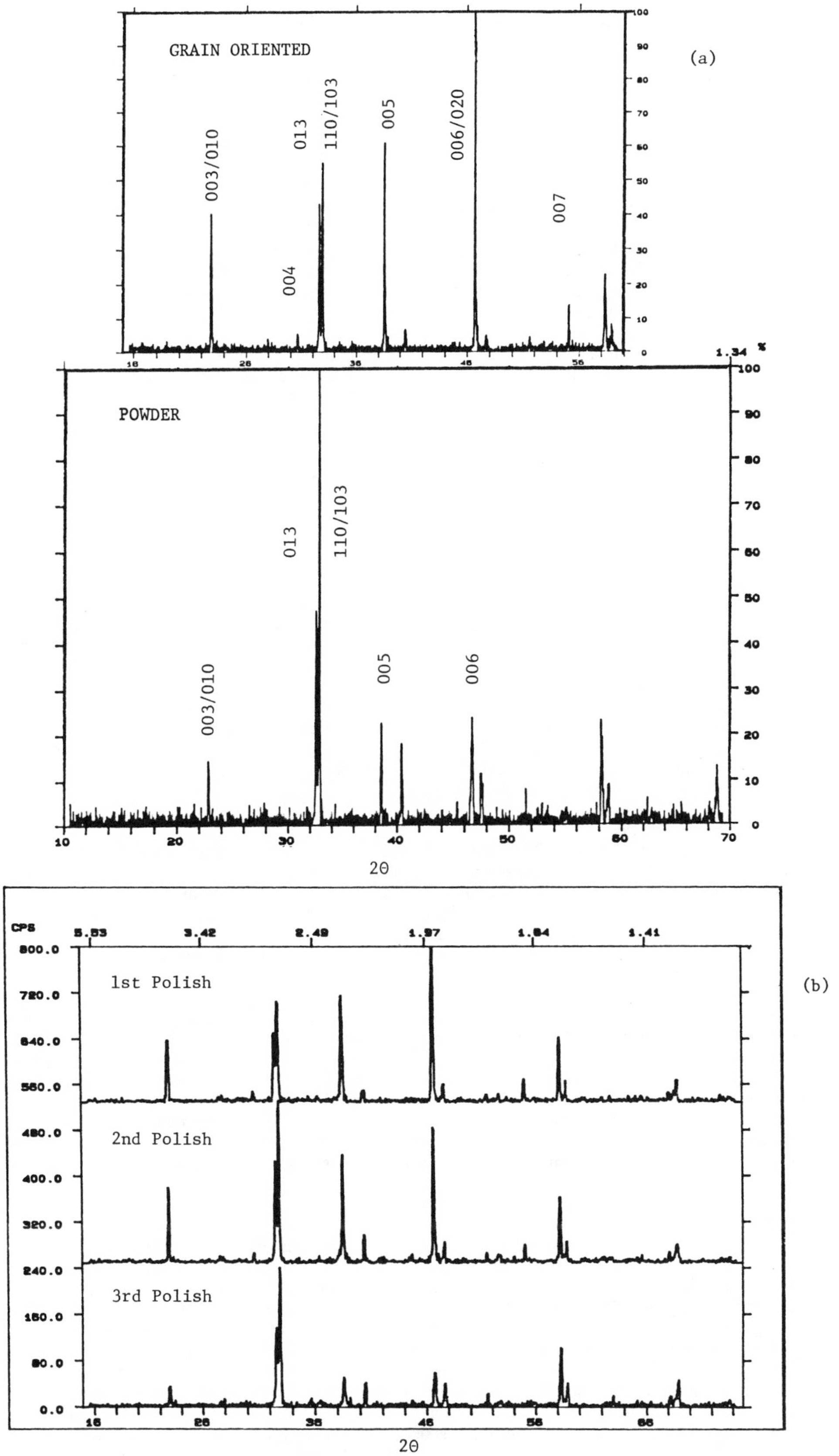

Fig. 3. a) X-ray diffraction patterns of a grain oriented 123 ceramic and the powder.
b) X-ray diffraction pattern of a ceramic surface after first, second and third polishing of the surface

ADDITION OF IMPURITIES IN 123 OXIDE COMPOSITIONS

Addition of other ions as a substitution for one or more ions in the 123 compositions or to form a second phase has had several important outcomes and the studies could provide valuable directions for the processibility of the ceramics, improvements in the superconducting properties, and important clues towards the compatibility of the 123 materials with the other electronic materials to be incorporated in a variety of devices.

Based on extensive studies of the additives to the 123 compositions, it is clear that a very large number of ions can be substituted for Y, Ba, Cu, or O. Remarkably, no major improvement in the superconducting T_c was noticed in any single material. This was identical to our experience 10 years ago in $Ba(BiPb)O_3$ (ref. 7). Our studies on the additives was done with the specific aim to look into a) the possibilities of using these materials with other important device materials, and b) to look for new coupled phenomena with semiconductors or ferroics. Specific ions have been tried for substitutions to meet the following goals:

a) reveal the acceptability of the impurity and its level at the specific position in the structure and the crystal chemical consequences, e.g., single or multiphase system,
b) reveal the effect on the superconducting properties of new single phase or multiphase material,
c) reveal other important effects such as
   i) the chemical stability in the ambient
   ii) physical properties, e.g., density, mechanical strength
   iii) reliability, e.g., electrical and magnetic properties with time, i.e., aging effects

Substitution of Ba by Alkali Ions
In general long procedures and times are required for processing of 123 superconductors. Several time consuming stages of materials preparation such as mixing, multiple calcining, sintering, and post-sintering treatments are necessary in order to achieve the good superconducting properties in a sample. In our present studies (ref. 46), we introduced several alkali metal oxides of Na, K, Rb, Cs (up to 10 at %) for Ba ions. There were significant changes observed in the calcining and sintering process. Apparently the reaction kinetics improved and as a result only single calcination produced single phase material and shorter sintering times (at 900°C) were enough for obtaining the final superconducting materials. Also there were no significant differences in the superconducting properties of the as-prepared and post-sintered oxygen annealed samples (Fig. 4). Thus, by the addition of K, Rb, Cs the severe and tight conditions of processing were relaxed. The T_c in each case remained ineffective before and after the post sintering oxidation and the ceramics could be densified beyond 85% of the theoretical density.

Substitution of Ba by Pb
On substituting Ba by Pb the 123 compounds showed good superconducting properties. The density of the 123 compounds increased to › 88% and T_c was unchanged up to the 10 at % doping levels. Such studies showed the potentialities of processing 123 superconductors in juxtaposition with Pb-based compounds. The studies done in this laboratory established that the interface produced between the Pb compounds and 123 materials has no deleterious effects on either the superconducting properties of 123 compound or the characteristics of Pb based materials (Fig. 4d).

Substitution of Ba and Y by Al
The tolerance of the two high T_c-superconductor compositions $(LaBa)_2CuO_{4-\partial}$ and $YBa_2Cu_3O_7$ for the Al impurities is very different. The 123 compounds could accept the substitution of Y by Al up to 0.5 atoms and sowed the T_c above 77K. In contrast, only very small amounts of Y = 0.05 of Al in $(La_{1-x}Ba_{x-y}Al_y)_2CuO_{4-\partial}$ drastically degraded the superconducting properties of the compound. These results established the potential use of 123 compounds for the interconnects on the alumina substrates and also set the safer limit of Al that can be incorporated in the interface without affecting the usefulness of the superconductor strip lines on alumina for applications at or above liq. N_2 temperatures. Similar results with Al impurities are also shown by other workers (refs. 47-48).

Substitutions at the Cu Sites
Several workers have tried, without much success, the replacement of Cu by various transition element ions. In all cases the superconducting properties of the 123 materials degraded even by 0.05 at % replacement of Cu by Ni, Co, Fe. A small impurity of silicon degrades drastically the superconducting properties of the 123 compounds. While we have noticed in a very large number of cases a wide variety of apparently superconducting transitions well above 100K, we have not regarded these as true bulk superconductivity. We have no satisfactory explanation for this behavior but it was not surprising as it was also being reported by several workers. We noted that Cu electrodes as interconnects on silicon chips generally produce unsatisfactory performance of the devices and in several instances serious degradation of the electrical properties of silicon. This may also explain why the 123 superconductor films deposited on the silicon substrates invariably are of poor quality.

It is found that titanium and cadmium are some of the best substituents in the Cu sites. In both the cases, the density of the 123 compounds was 90-95% of the theoretical density. Both the

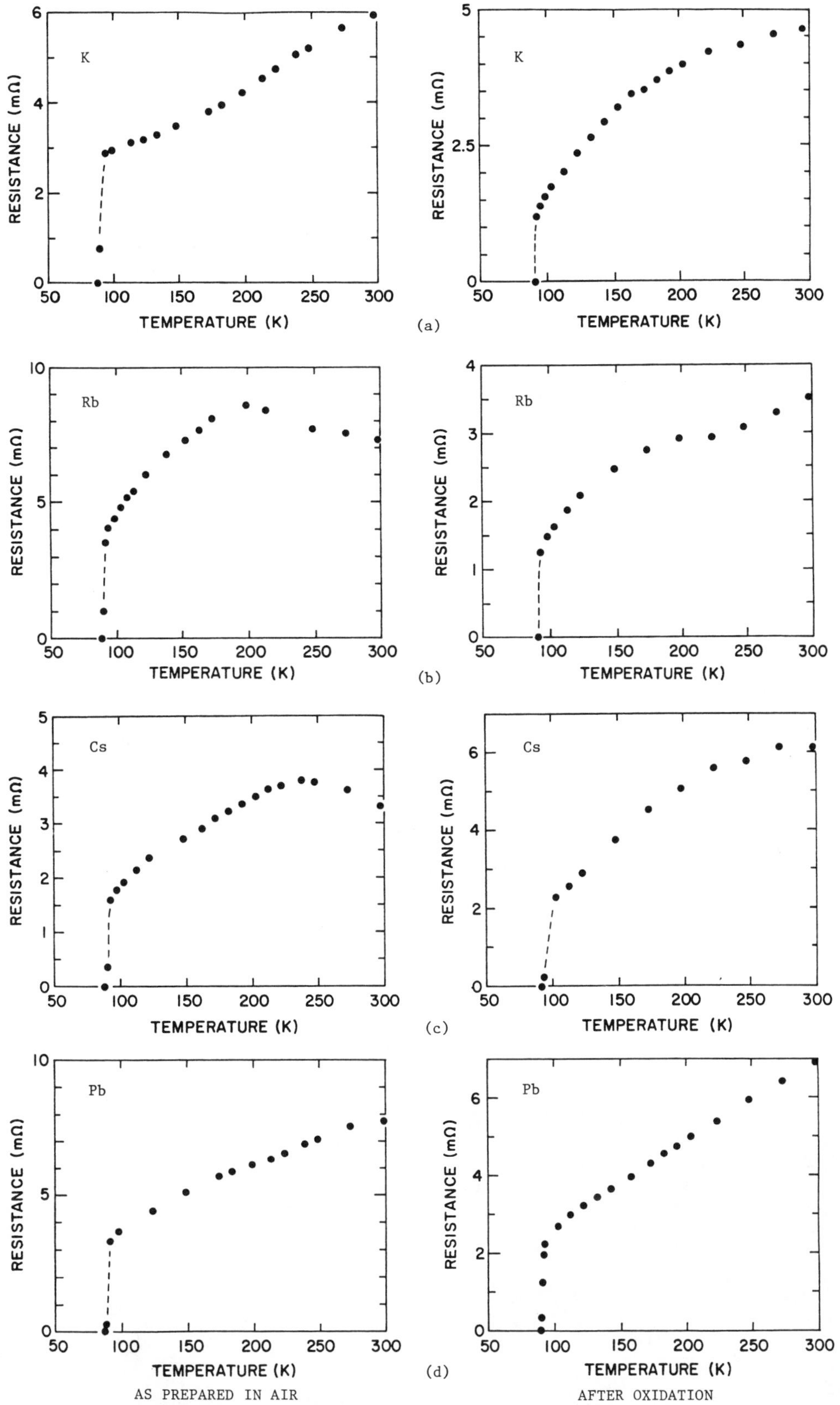

Fig. 4. Resistance vs. temperature characteristics of 123 compounds when Ba is substituted by a) K, b) Rb, c) Cs, and d) Pb. Measurements on as prepared (in air) and oxygen annealed samples showed no significant differences in the T_c's.

compositions showed good superconductor properties and a narrower transition width, $\Delta T_c \sim 1K$. We have tried several other dopants for possible replacement of Cu. It is rather intriguing that substitution of Cu by Nb, up to 15% has not shown any degradation of the superconducting properties in the doped 123 compounds.

These results suggest the compatibility of 123 superconductors with the important perovskites, e.g., $BaTiO_3$, $SrTiO_3$, and $PbTiO_3$ and the $Sr_{1-x}Ba_xNb_2O_6$ tungsten bronze family (ref. 49).

Since Pb, Ti, Sr. Ba, Nb do not degrade the superconductor properties, some of the two phase systems such as 123:$BaTiO_3$, 123:$PbTiO_3$, 123:SBN were also studied. In all these cases the density of the ceramics improved (Fig. 5). Because the interfaces with dopants like Pb. Ti, Nb, etc. are still superconducting, it is quite possible that grain boundaries in such diphasic or composite systems are not unfavorable, instead those may be helping in improving the density of the 123 materials. The measurements on J_c and H_c are in progress and will be published in the forthcoming papers.

Although many impurities which we have used with the 123 compositions do not affect the superconducting T_c, some do affect the onset of the T_c. For example, in the case of K there was a continuous decrease in slope of ρ vs. T curves but the T_c of ~ 90K was not affected. Also the studies suggest that the magnitude of the resistance of the 123 compounds at room temperature, the ρ vs. T behavior and the resistance at the onset of the transition can be tailored simply by adjusting the impurities and their concentration in the 123 compositions. Such studies could also help in resolving some of the claims of higher temperature transitions present in 123 compounds which probably are not of the superconductor type but could be one of the insulator-metal or ferroic in nature.

In general, for the preparation of single crystal thin films of the 123 materials, the crystallographic lattice parameters of the substrate should be matched with the film and the (epitaxial) film should be deposited on a heated substrate at ~ 500°C. A post deposition annealing step at temperatures around 600-800°C is also required to produce a good superconducting film. Our studies provide the necessary information for the chemical compatibility of the 123 materials with the possible or at least the most important substrate materials.

The effects of the impurities on the 123 compounds suggest some interesting device design possibilities on the thin film superconductors. By the proper choice of the impurities, the localized passivation effects on the superconducting surfaces can be envisaged. Also, by using the laser beam heating, the pattern writing can be achieved, either as a result of diffusing the impurities in the 123 films or by selected area decomposition of the 123 compounds to modify the behavior of the 123 compounds to superconductor, semiconductor or insulator.

Effects of Oxygen Substitution by Fluorine and Sulphur

Substitution by higher atomic weight isotopes have not made any noticeable effects on the T_c of YBC compounds (ref. 50). In the recent reports of partial substitution of oxygen by fluorine, apparently it was claimed that the T_c of a YBC compound increased dramatically. The source of fluorine in this work was BaF_2 and the replacement amount was between one and two fluorine atoms in the formula $Y_1Ba_2Cu_3O_{7-x}F_{2x}$. Most of the ceramics studied were multiphasic and always had BaF_2 as one of the phases present in the powder. A T_c as high as 155K has been reported (ref. 51). In our studies we have used BaF_2/YF_3 or CuF_2 as the source of fluorine for the

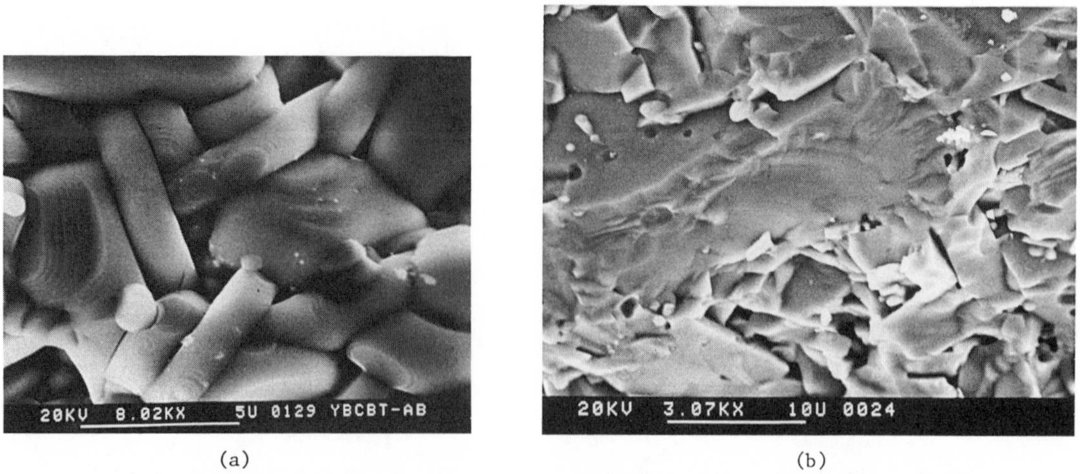

(a) (b)

Fig. 5. Microstructure of densified (>90%) ceramics - a) as sintered surface and b) fractured surface.

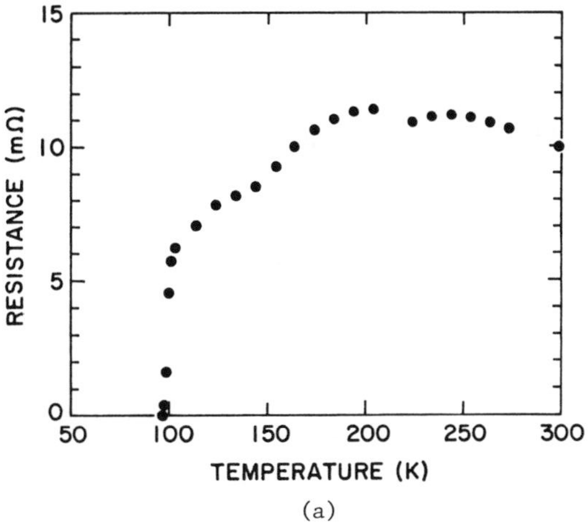

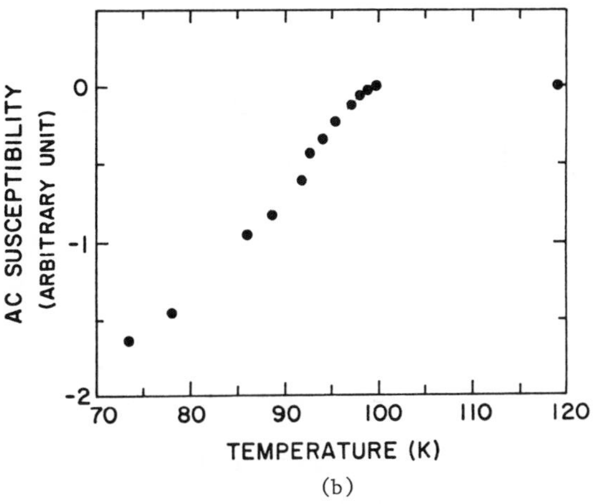

Fig. 6. a) Resistance vs. T and b) A.C. susceptibility vs. temperature measurements in $YBa_2Cu_3O_{7-x}F_{2x}$ prepared by using CuF_2.

synthesis of 123 compositions. The best results were obtained by using CuF_2 as the source for replacement of F for oxygen. Higher T_c's than 90K were obtained and in the best samples we could produce T_c up to 97-100K and with a sharp transition of ΔT_c›1K (Fig. 6). Reports on high T_c's ~155K could not be reproduced though some abnormalities and anomalous features in the ρ vs. T measurements in the vicinity of 155K were frequently observed in our results (as reported above). Recently the Philips group has reported a very sharp phase transition at ~240K in the oxides and fluorine-doped compounds and showed that by annealing the samples at ~240K the 90K superconducting T_c can be raised in the range of 150K (ref. 52). Our attempts to produce such increases in the T_c of oxide materials by cycling the samples between 90-240K are shown in Fig. 7. There is 5-6K increase in T_c in the case of YBC during the two cycles but further cycling did not change the T_c much further than 92K (Fig. 7). However, such samples invariably gave T_c lower than the standard 90K before the temperature cycling and also no indications of the upper T_c were observed in these samples, the two important criterias which were used for the selection of samples for such studies by the Philips research group. Therefore, the change in ΔT_c observed in our studies could be due to some factors which need more investigation.

In the case of low T_c superconductor oxide compositions, small concentrations of S at the O site improved the T_c by a few degrees. For the studies along similar lines, we prepared several compounds in the system $Y_1Ba_2Cu_3O_{5.5+\delta}S$ and measured their superconducting characteristics. In most cases the samples prepared were multiphasic and the properties were poor. The strong Meissner effect and the usual T_c of 90K were observed in the samples with compositions close to

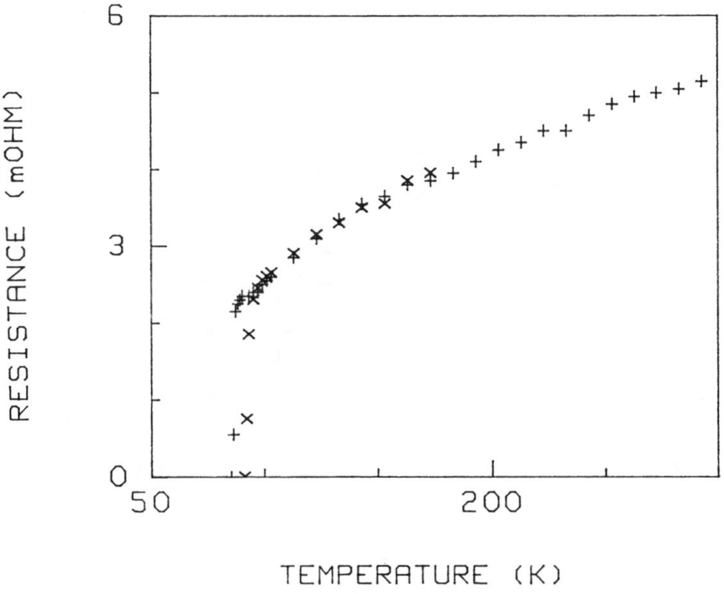

Fig. 7. Improvement in T_c of $YBa_2Cu_3O_7$ ceramic as a result of cycling to T = 240K.

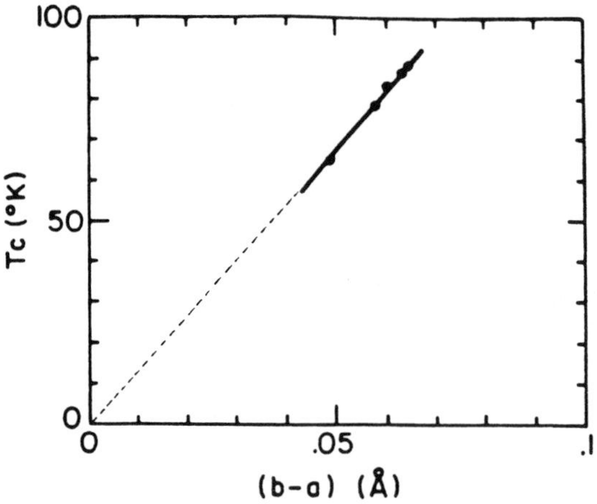

Fig. 8. Rhombicity (b-a) vs. T_c of $GdBa_2Cu_3O_7$ ceramic. (Setter, et al., ref. 53)

$Y_1Ba_2Cu_3O_6S$. The Meissner effect was stronger than that observed in the pure oxide 123 materials. Further studies on the effects of S in the oxide 123 compositions and on the Meissner effect in sulphides are in progress.

In the course of our studies on the addition of impurities in the 123 compositions, it is noticed that structurally pure orthorhombic phase showed the low room temperature resistivity, $T_c \geq 90K$ and strong Meissner effect. Recently, the studies of Setter, et al. in this laboratory (ref. 53), showed a strong correlation between the higher rhombicity and the high T_c in the $GdBa_2Cu_3O_{7-\partial}$ compositions. The various degrees of rhombicity in the samples were obtained by precise control of calcination and sintering conditions. Figure 8 shows that the T_c reduces to zero as the rhombicity

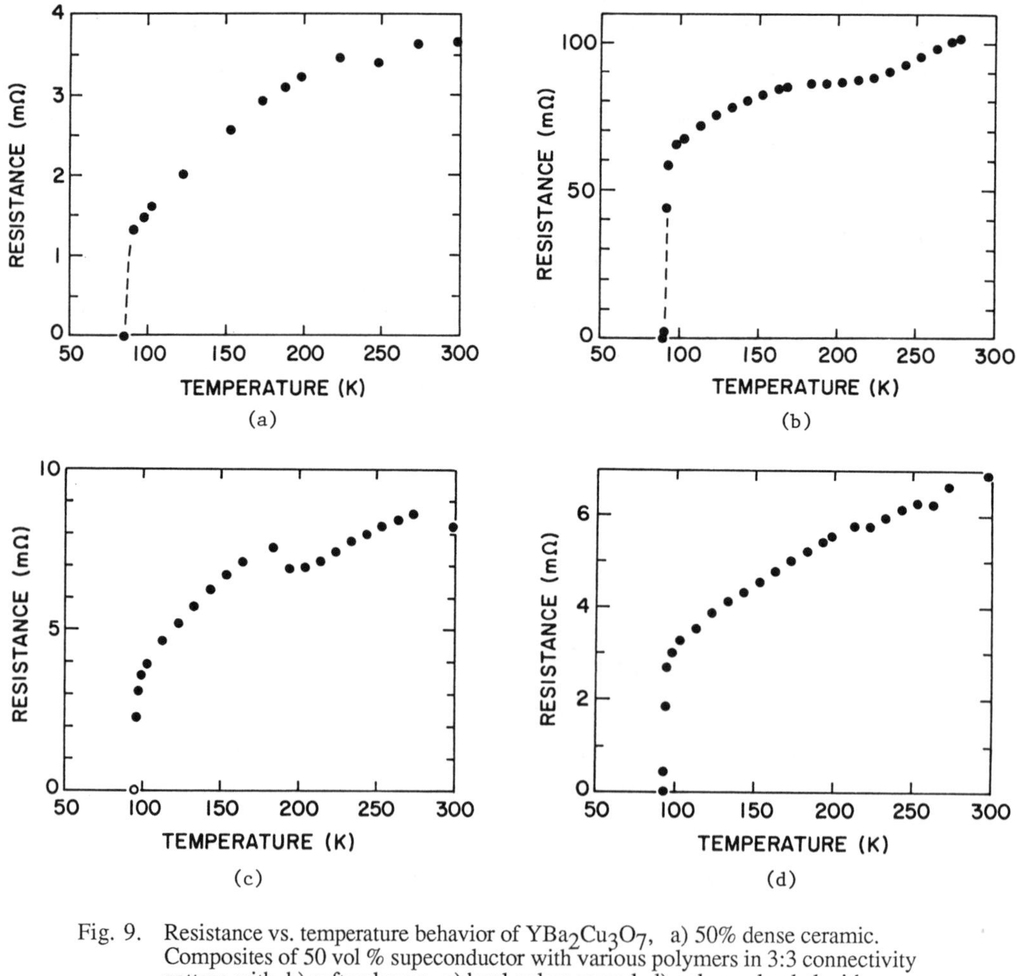

Fig. 9. Resistance vs. temperature behavior of $YBa_2Cu_3O_7$, a) 50% dense ceramic. Composites of 50 vol % supeconductor with various polymers in 3:3 connectivity pattern with b) soft polymer, c) hard polymer, and d) polymer loaded with Cu-particles.

disappears i.e. the material becomes tetragonal. One should be careful in interpreting such data as the samples with the low rhombicity always had a glassy phase which could be responsible for the poorer superconducting properties.

Similar correlations have been observed on $YBa_2Cu_3O_{7-\partial}$ by the Philips group.

COMPOSITES

There are many reasons for working with composites of high T_c-superconductor oxides. We follow here the principles and the terminology of composite design described in detail by Newnham (ref. 54):

i) It is relatively easier to prepare various disk and block-shaped samples of the high T_c-superconductors by the solid state reaction method. However, it will not be easy to prepare wires, spiral coils, or large area sheets because of the brittle nature of oxides,

ii) The high T_c superconductors are chemically sensitive to various solvents such as alcohol, alkalies, acids and especially to water and even humid ambients. In several cases the material loses its superconductivity but it is recoverable in some instances after certain annealing procedures. Thus, the superconducting bulk devices need special protection from the ambient,

iii) In general bulk high T_c superconductors are of much lower density compared to their theoretical density and hence are mechanically weaker.

To overcome these problems, the composite route is the simplest way for using these superconducting materials for larger scale applications. The multiphasic composites also provide remarkable coupled effects and in the case of high T_c-superconductors those help in controlling the sensitive chemical properties and improving the mechanical properties without the loss of the characteristics of the active superconducting phase. Figures 9(b) and (c) show some of the ρ vs. T results on the composites of 123 superconductors prepared with the soft and hard polymers in the 3:3 connectivity pattern. In some cases Cu-fine particles were loaded in the polymer (Fig. 9d). The composites showed very strong Meissner effects. Detailed results from these studies are described in forthcoming publications (ref. 55).

COMMENT ON THE HIGHER TEMPERATURES ›100K SUPERCONDUCTIVITY IN OXIDE SUPERCONDUCTORS

Though there are several reports on the superconductivity in oxide perovskite related compounds in the multiphase samples there are no proven reports satisfying the necessary requirements such as Meissner effect in these materials (ref. 56). Most of these reports prove to be anomalies within ρ vs. T data and the superconducting phase present is claimed to be ‹‹1% of the total material. Such anomalies have also been measured by us in several instances and are shown in Figures 10a-10c. Invariably these anomalies occurred in the case of the multiphase samples and such samples failed to show the Meissner effect within the accuracy of our measuring systems.

Figure 10a shows the ρ vs. T (μV reading for a constant current of +10mA through the sample) measurement on a sample prepared by the mixed oxide method. Such results could not be reproduced in repeated measurements on the same samples. Some of these artifacts could be resolved by taking the data point each time with positive and negative current passing through the sample. Even with such data collection procedure, the anomalies in the ρ vs. T measurements have been confirmed here as at many other laboratories at ~200K in several instances. These changes could be due to some subtle structural phase transition of magnetic, electric or semiconductor to metal nature and as a result changing the domain or electrical state of the sample.

Our measurement so far on the oxide superconductors though, failed to prove the superconductivity at higher temperatures than 100K. The BHK theory does see the possibilities of higher T_c, provided the superconducting T_c in these materials are associated with the ferroic twinning and such twinning states can be modified suitably. There are no published reports disfavoring such linkage of the high T_c with domain states in these oxide supeconductors. Therefore, more careful studies are needed to understand the nature of the twinning states in these new class of 'ferroics.'

Our present view on the anomalies of high T_c's are based on the BHK theory (refs. 37-39) which -- in grossly oversimplified terms -- correlates the enhancement of T_c to the fineness of the twinning in the microstructure. Hence we contend that in the anomalously high T_c samples, the other phases present mechanically constrain small volumes of the 1:2:3 phase to persist with a very, very fine twin structure and hence stay in a maximally enhanced T_c state. The "disappearance" of the effect is due to the relaxation of the artificial and serendipitous mechanical constraint.

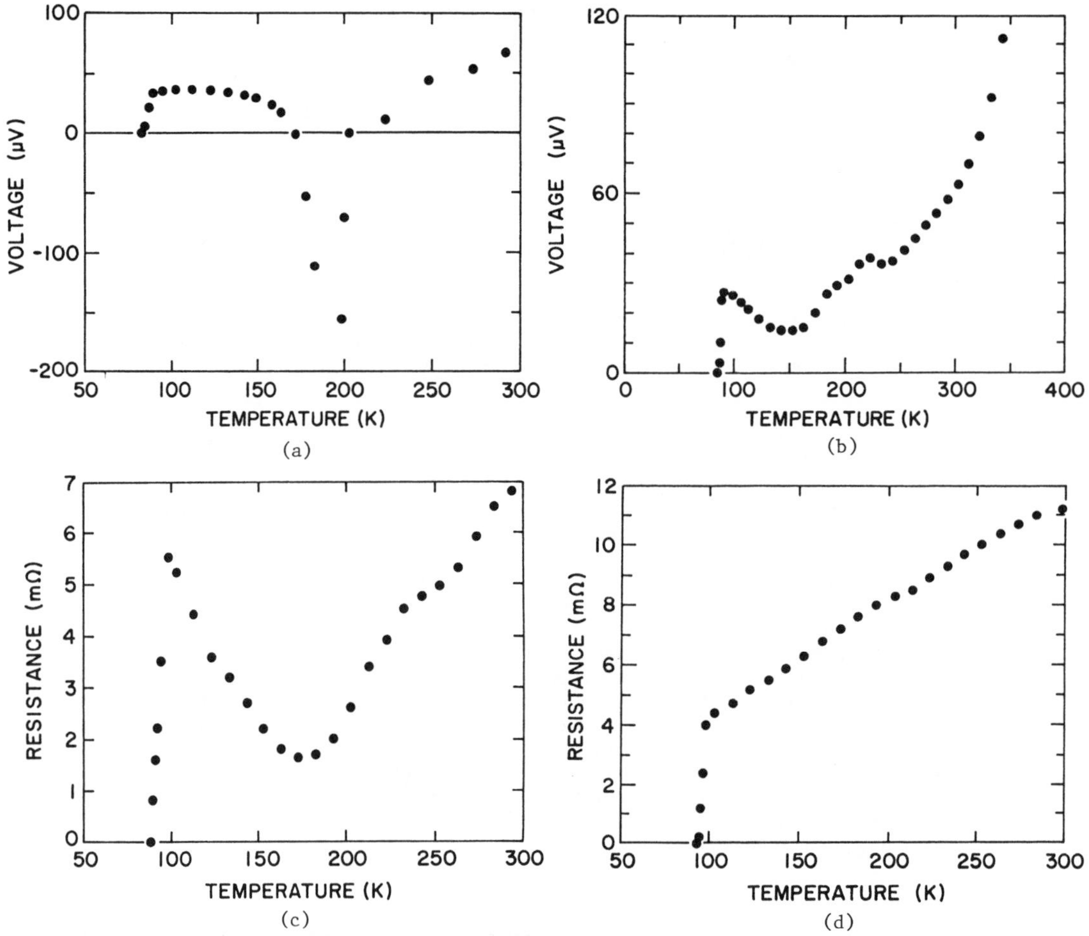

Fig. 10. Four terminal ρ vs. T behavior of a) mixed rare earth oxide superconductor, $(YYb)_1Ba_2Cu_3O_{7-\partial}$. Note that at about T = 205K and T = 175K voltage drops to zero and finally into the superconducting state at T - 90K. b) In some cases very sharp and reproducible drop in resistivity near room temperature has been observed. The sharp drop in resistance at 180K is introduced by the sample processing, c) sample prepared by excess oxygen pressure, and d) samples prepared under optimum conditions show no such strong abnormalities in ρ vs. T. measurements.

ACKNOWLEDGEMENTS

The authors are thankful to Profs. R.E Newnham, S.K. Kurtz and co-workers for the useful discussions. The work has been supported by the ONR-DARPA grant under the contract No. NOO14-86-K-0767.

REFERENCES

1. R.A. Hein, J.W. Gibson, R. Mazelsky, R.C. Miller and J.K. Hulni, Phys. Rev. Lett. 12, 320 (1964).
2. Ch.J. Raub, A.R. Sweedler, M.A. Jensen, S. Broadston and B.T. Matthias, Phys. Rev. Lett. 15 108 (1965).
3. J.P. Remeika, T.H. Geballe, B.T. Matthias, A.S. Cooper, G.W. Hull and E.M. Kelly, Phys. Lett. 24A, 565 (1867).
4. A.W. Sleight, T.A. Bither and P.E. Bierstedt, Sol. State Commun. 7, 299 (1969).
5. D.C. Johnston, H. Prakash, W.H. Zachariasen and R. Viswanathan, Mat. Res. Bull. 8, 777 (1973).
6. A.W. Sleight, J.L. Gillson and P.E. Bierstedt, Sol. State Commun. 17, 27 (1975).
7. L.R. Gilbert, R. Messier and R. Roy, Thin Solid Films 54, 129 (1978).
8. N.J. Doyle, J.K. Hulm, C.K. Jones, R.C. Miller and A. Taylor, Phys. Lett. 26A, 604 (1968).
9. H.E. Barz, A.S. Cooper, E. Corenzwit, M. Marezio, B.T. Matthias and P.H. Schmidt, Science 175, 884 (1972).
10. B.T. Matthias, M. Marezio, E. Corenzwit, A.S. Cooper and H.E. Barz, Science 175, 1465 (1972).

11. N.H. Van Maaren, G.M. Schaeffer and F.K. Lotgering, Phys. Letts 25A, 238 (1967). J.B. Goodenough, Mat. Res. Bull. 8, 423 (1973).
12. B.T. Matthias, Mat. Res. Bull 5, 665 (1974).
13. P. Ganguby and C.N. R. Rao, Mat. Res. Bull. 8, 408 (1973).
14. P. Ganguby and C.N.R. Rao, J. Sol. State Chem. 53, 193 (1984).
15. J.S. Shaplygin, B.G. Kakhan and V.B. Lazarov, Russ. J. Inorg. Chem. 24, 820 (1979).
16. C. Michel and B. Raveau, Revue de Chimie Minerals 21, 407 (1984).
17. C. Michel and L. Ev-Rakho and B. Raveau, Mat. Res. Bull. 20, 667 (1985).
18. J.G. Bednorz and K.A. Muller, Z. Phys. B64, 189 (1986).
19. H. Takagi, S. Uchida, K. Kitazawa and S. Tanaka, Jap. J. Appl. Phys. 26, L123 (1987).
20. J.G. Bednorz, M. Takashige and K.A. Muller, Europhys. Lett. 3, 379 (1987).
21. S. Uchida, H. Takagi, K. Kitazawa and S. Tanaka, Jap. J. Appl. Phys. 26, Ll (1987).
22. K. Kishio, K. Kitazawa, S. Kanbe, I. Yasuda, N. Sugii, H. Takagi, S. Uchida, K. Kucki and S. Tanaka, Chem. Lett. (Japan), 429 (1987).
23. R.J. Cava, R.B. Van Dover, B. Batlogg and E.A. Rietman, Phys. Rev. Lett. 58, 408 (1987).
24. M.K. Wu, J.R. Ashburn, C.J. Torng, P.H. Hos, R.L. Meng, L. Gao, Z.J. Huang, Y.Q. Wang and C.W. Chu, Phys. Rev. Lett. 58, 908 (1987).
25. Z. Shao, L. Chen, Q. Yang, Y. Huang, G. Chen, R. Tang, G. Liu, C. Cui, L. Chen, L. Wang, S. Gup, S. Li and J. Bi, Kexue Tong 32, 661 (1987).
26. S. Hikami, T. Hirai and S. Kogashima, Jap. J. Appl. Phys. 26, L314 (1987).
27. C.N.R. Rao, P. Ganguli, A.K. Raychaudhuri, K. Sreedhar and R.A. Mohan Ram, Nature 326, 856 (1987).
28. S.V. Bhat, P. Ganguly and C.N.R. Rao, Pramana 28, 425 (1987).
29. P. Ganguly, R.A. Mohan Ram, K. Sreedhar and C.N.R. Rao, Pramana J. Phys. 28, L321 (1987).
30. R.J. Cava, B. Batlogg, R.B. Van Dover, D.W. Murphy, S. Sunshine, T. Seigrist, J. P. Remeika, E.A. Rietman, S. Zahurak and G.P. Espinosa, Phys. Rev. Lett. 581676 (1987).
31. Y. de Page, W.R. McKinnon, J.M. Tarascon, L.H. Greene, G.W. Hull and D.M. Hwang, Phys. Rev. B35, 7245 (1987).
32. T. Siegrist, S. Sunshine, D.W. Murphy, R.J. Cava and S.M. Zahusak, Phys. Rev. B35, 7137 (1987).
33. Y. Le Page, W.R. McKinnon, J.M. Tarascon, L.H. Greene and G.W. Hull,Phys. Rev. B35, 7115 (1987).
34. J.M. Tarascon, L.H. Greene, W.R. McKinnon and G.W. Hull, Phys. Rev. B35, 7115 (1987).
35. S. Kogoshima, S. Hikami, Y. Nogami, T. Hirai and K. Kubo, Jap. J. Appl. Phys. 26, L318 (1987).
36. H. Takagi, S. Uchida, K. Kishio, K. Kitazawa, K. Fucki and S. Tanaka, Jap. J. Appl. Phys. 26, L320 (1987).
37. G.R. Barsch, B. Horovitz and J.A. Krumhansl, Phys. Rev. Lett. 59, 1251 (1987).
38. B. Horovitz, G.R. Barsch and J.A. Krumhansl, "Proceedings of the Adriatico Research Conference on High-T_c Superconductors," Trieste, June 1987.
40. A.S. Bhalla, R. Roy, et al. (to be published).
41. C.N.R. Rao (personal communication).
42. Ravindranathan, et al. (to be published).
43. R.R. Neurgaonkar, G. Shoop, J.R. Oliver, I. Santha, A.S. Bhalla and L.E. Cross, Mat. Res. Bull. (to be published, March, 1988).
44. D.E. Farrell, B.S. Chandrasekhar, M.R. De Guire, M.M. Fang, V.G. Kogan, J.R. Chen, D.K. Finnemore, Phys. Rev. B36, 4025 (1987).
45. A.S. Bhalla, N. Setter, R. Guo and S.K. Kurtz (to be published).
46. A.S. Bhalla, T.R. Shrout, A. Kumar and B. Jones (to be published).
47. J.P. Franck, J. Jung and M.A.K. Mohamed, Phys. Rev. B36, 2308 (1987).
48. R. Escudero, L.E. Rendon-Diazmiron, T. Akachi, J. Heiras, C. Vazquez, L. Banos, F. Estrada and G. Gonzalez, Jap. J. Appl. Phys. 26, L1020 (1987).
49. A.S. Bhalla, N. Setter and L.E. Cross (to be published).
50. L.C. Bourne, M.F. Crommie, A. Zettl, H. Loye, S.W. Keller, K.L. Leary, A. Stacy, K.J. Chang, M.L. Cohen and D. Morris, Phys. Rev. B52, 2337 (1987).
51. R.S. Ovshinsky, R.T. Young, D.D. Alfred, G. DeMaggio and G.A. Van der Leeden, Phys. Rev. Lett. 58, 2579 (1987).
52. R.N. Bhargava, S.P. Herko and W.N. Osborne, Phys. Rev. Lett. (Sept. 28, 2987).
53. N. Setter, C.A. Randall, W. Cao and A.S. Bhalla, Mat. Res. Bull. (to be published, Feb., 1987).
54. R.E. Newnham, D.P. Skinner and L.E. Cross, Mat. Res. Bull. 13, 525 (1978).
55. A.S. Bhalla, T.T. Srinivasan, R.E. Newnham and L.E. Cross (to be published).
56. J.C. Chen, L.E. Wenger, C.J. McEwan and E.M. Logothetis, Phys. Rev. Letts. 58, 1972 (1987).

Thermal decomposition studies of $YBa_2Cu_3O_{7-x}$ using beam modulation mass spectrometry

M. A. Frisch, F. Holtzberg and D. L. Kaiser

IBM T. J. Watson Research Center,
P. O. Box 218, Yorktown Heights, NY 10598-0218

Abstract Using Knudsen Effusion Mass Spectrometry we have studied the thermal decomposition kinetics of $YBa_2Cu_3O_{7-x}$. On heating to 1200 K, the material undergoes a loss of oxygen via a two step process. The first is a third order reaction with an activation energy of 188 kJ/mole over a range of oxygen concentration from 7.0 to 6.4. This reaction is inhibited by surface carbon dioxide which is removed by 640 K. The second reaction, covering the oxygen composition 6.4 to 6.0, has an activation energy of 350 kJ/mole. The sample decomposes to several phases at about 1200 K. Two different batches of material yielded identical kinetic parameters except for differences in the preexponential factor which we attribute to particle size variations.

INTRODUCTION

The discovery of high temperature superconducting copper oxides (ref. 1) has led to an overwhelming number of publications many of which have focussed on the higher transition temperature $YBa_2Cu_3O_{7-x}$ compound (ref. 2). It was demonstrated rather early that the superconducting transition temperature, the oxygen coordination and the structure of the compound depend critically on the amount of oxygen in the unit cell (ref. 3). A number of techniques have been used to monitor the loss of oxygen including thermogravimetric analysis (ref. 4), full structure neutron diffraction analysis (ref. 5), temperature dependent resistivity (ref. 6), x-ray structural analysis (ref. 7) and transmission electron microscopy (ref. 3). The results of these studies show that the $YBa_2Cu_3O_{7-x}$ compound is superconducting in the orthorhombic phase and becomes tetragonal and non-superconducting with loss of oxygen. In this paper we detail quantitatively, as a function of temperature, the desorption of oxygen and other vaporizable species including unintentional contaminants in the material.

The thermal stability of the high temperature superconductor, $YBa_2Cu_3O_{7-x}$, was studied by Knudsen Effusion Mass Spectrometry (KEMS). This technique is a powerful method for examining decomposition reactions because it is possible to analyze both equilibrium and activated reactions. The order and activation energy of kinetic processes and heats and entropies of equilibrium reactions can be derived directly from the mass spectroscopic data (ref. 8). The instrumentation of the high temperature mass spectrometer is described in detail elsewhere (ref. 9). Several key features are incorporated in this system which give us an exceptionally high sensitivity. These include: 1) a differentially pumped ultra high vacuum environment for both the mass spectrometer and furnace, 2) the use of digital modulation techniques and 3) a fully automated instrumentation controlled by an IBM Series/1 computer.

EXPERIMENTAL

The $YBa_2Cu_3O_{7-x}$ compound was made by mixing appropriate amounts of Y_2O_3, $BaCO_3$ and CuO. The mixture was pressed into pellets and heated to 950°C for 12 hours on a layer of powder of the same mixture, spread on a platinum sheet. After cooling, the pellets were ground, repressed and fired at 950°C for 16 hours and cooled at 50°C/hour to 400°C in flowing oxygen. X-ray powder diffraction measurements showed the samples to be orthorhombic, single phase with lattice constants $a=3.826\pm0.001$, $b=3.892\pm0.001$ and $c=11.665\pm0.003$ Å. Two different batches, designated as 202 and 203, were examined by KEMS and their similarities and differences are reported here.

The samples were heated in an all platinum Knudsen cell whose interior dimensions were 1.8 cm diameter and 3.0 cm high, giving a volume of 2.75 cc. The orifice, 1 mm diameter by 5 mm long, was on the cylindrical surface of the cell. The conductance of this aperture for O_2 at 700 K is 26 cc/sec (ref. 10), resulting in a time constant of about 0.1 second. The aperture can be accurately aligned with the entrance to the cross beam ionizer of the quadrupole mass spectrometer. For temperature measurement and control, a Pt vs 90Pt-10Rh thermocouple was welded to the surface of the cell near the sample position. The samples, 20 to 50 mg in size, were weighed to 10 μg accuracy into a platinum cup and reweighed after heating. Weight losses were correlated with the integrated ion signals from the mass spectrometer for pressure calibration.

The Knudsen cell was heated radiatively with a tungsten mesh element from room temperature up to some maximum temperature, using linear ramps of either 1 or 4 K/min. For exploratory studies the maximum temperature was 1600 K. For the majority of the studies, where less than one oxygen atom per formula unit was removed from the lattice, 900 K was the maximum. As an aid to identifying the principal reactions, the samples were also quenched from key temperatures and the product material examined by x-ray powder diffraction. The molecular beam from the Knudsen cell was modulated and the time dependent gas composition measured with an automated quadrupole mass spectrometer. The gases were ionized using 30 eV electrons, which minimized the contributions of fragment ions to the ion intensity for the simpler species in the spectrum.

RESULTS

The rate of oxygen loss, the major gas released in the decomposition of $YBa_2Cu_3O_{7-x}$, is shown in Fig. 1 for sample 202 heated to 1600 K at 4 K/min. The thermal desorption spectrum of oxygen is complex with several major peaks, indicating a series of discrete decomposition steps. The thermal spectrum shows two distinct regions. The first feature has a maximum rate at a temperature of about

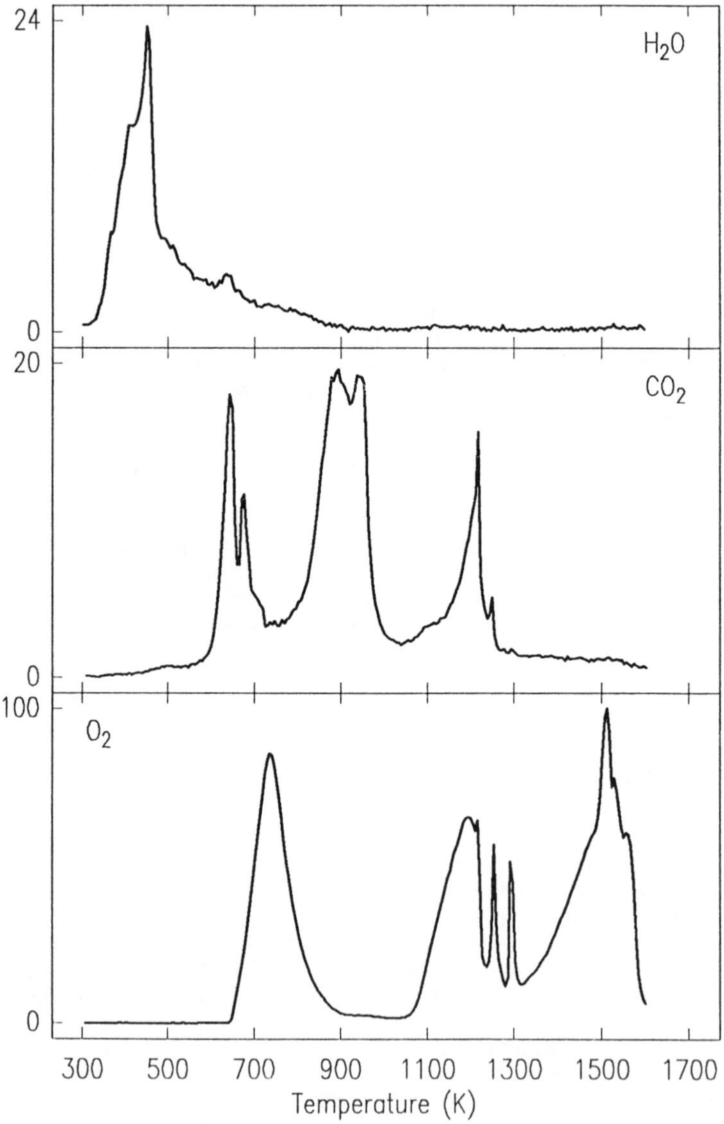

Fig. 1 Principal gases desorbing from $YBa_2Cu_3O_{7-x}$ (202) heated at 4 K/min.

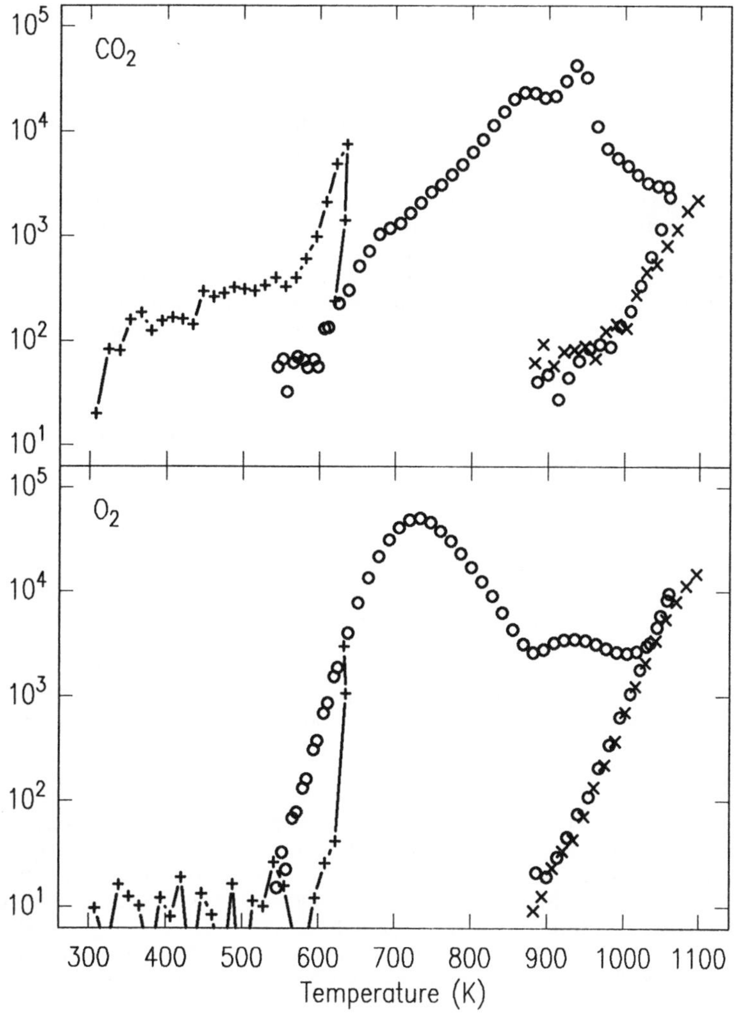

Fig. 2 Desorption of oxygen and carbon dioxide (arb. units) from $YBa_2Cu_3O_{7-x}$ (203), heated at 1 K/min. The +—+ curve is the first heating to 640 K.

740 K followed by a rather complex region which results in the complete decomposition of the compound into a number of phases. In this paper we focus primarily on the reactions involving the first well defined peak below 900 K.

In addition to the oxygen release, significant amounts of water and carbon dioxide were also observed but on the order of ten times less than that for oxygen. The desorption rates for these two gases are also plotted in Fig. 1 with their importance relative to oxygen shown on the vertical scale. The major water loss is in the early stages of heating and is essentially complete before the onset of the $YBa_2Cu_3O_{7-x}$ decomposition. However, the carbon dioxide desorption overlaps the oxygen signal. A detailed examination on the log scale of the decomposition rates reveal that the oxygen release is inhibited until 640 K. At this temperature the rate suddenly increases by about 2 orders of magnitude. This dramatic increase is coincident with the completion of a significant precursor in the CO_2 spectrum.

To examine the possible correlation between O_2 and CO_2, sample 203 was first heated to 640 K, then cooled to 540 K before the final heating to 1100 K. The desorption rates of oxygen and carbon dioxide for this heating schedule are shown on the log scale in Fig. 2. The first heating to 640 K is represented by dashed curves. The carbon dioxide desorption in this region appears to be a first order process, terminating about 640 K. The corresponding O_2 desorption is completely inhibited until 620 K. However, upon cooling to 540 K and reheating to 640 K, the O_2 desorption follows a well-defined rate curve whereas there is no measurable release of CO_2. This correlation in the two desorption spectra suggests that CO_2 molecules occupy surface sites which prevent the release of oxygen from the bulk. In subsequent experiments, we followed this heating schedule to insure that the correct reaction rates were being measured in this temperature regime.

The second CO_2 doublet peak (870 and 940 K) for sample 203 is about ten times larger than for sample 202 (see Fig. 1). This difference in carbon dioxide evolution could result from a small excess of barium carbonate in sample 203 or other subtle variations in processing conditions. However, it is unlikely that the excess CO_2 resulted from the decomposition of unreacted barium carbonate since supplemental measurements on pure barium carbonate showed that CO_2 is not released until well above 950 K. The double peak indicates more than one source for CO_2 which is not observed for pure barium carbonate. Additionally, the x-ray measurements do not show the presence of a minor phase to which we could attribute the large CO_2 loss. However, the sensitivity of the x-ray data does not preclude a small amount of a carbonate containing phase. This issue is being studied further both to identify the sources and control their incorporation in the ceramic material.

The oxygen curve is well behaved up to 900 K where an abrupt change in the rate is observed. This inflection point probably corresponds to an ordering of the remaining oxygen atoms in the basal plane. X-ray measurements on samples heated to 900 K show the tetragonal structure, with lattice parameters $a=3.860\pm0.001$ and $c=11.852\pm0.002$ Å. The small bump after this inflection point is most likely the release of the remaining oxygen freed up in the transformation. We are continuing studies to determine more precisely the orthorhombic to tetragonal transformation temperature.

The temperature was cycled up to 1060 K and down to 880 K to remove the oxygen from the loosely bound sites. Upon reheating to 1250 K, the O_2 reaction curve retraced exactly the rates measured during the cooling ramp. This duplication verifies that this reaction is a well-defined activated process. However, beginning at about 1200 K the sample undergoes a decomposition to several phases which were difficult to identify by x-ray diffraction. At this decomposition onset temperature, the presence of the copper and barium gas species is first detected in the mass spectrum. Yttrium vapor was not detected even up to 1600 K. The annealing of samples above 900°C under O_2 may not totally suppress the loss of copper and barium and could alter the stoichiometry of the cation species.

The data for oxygen, converted to pressure in atmospheres calculated from weight loss measurement (ref. 11), are plotted in Fig. 3 for sample 203 as a function of reciprocal temperature (1/K). The least squares fit to the linear portion of the two reaction peaks are also shown in this figure. In the fitted region, since the bulk oxygen composition remains essentially constant, the slope yields an activation energy independent of the reaction order. The first reaction, which peaks at 730 K for a heating rate of 1 K/min, has an activation energy of 188±3 kJ/mole. The second major oxygen peak begins about 900 K with an activation energy of 350±10 kJ/mole.

The mass spectrometric data provides a continuous monitoring of the oxygen loss from the Knudsen cell and consequently we are able to compute by integration the oxygen remaining in the sample. Thus, the first decomposition reaction proceeds with a loss of approximately 0.6 oxygen atom per formula unit, yielding a compound of the nominal composition $YBa_2Cu_3O_{6.4}$. The

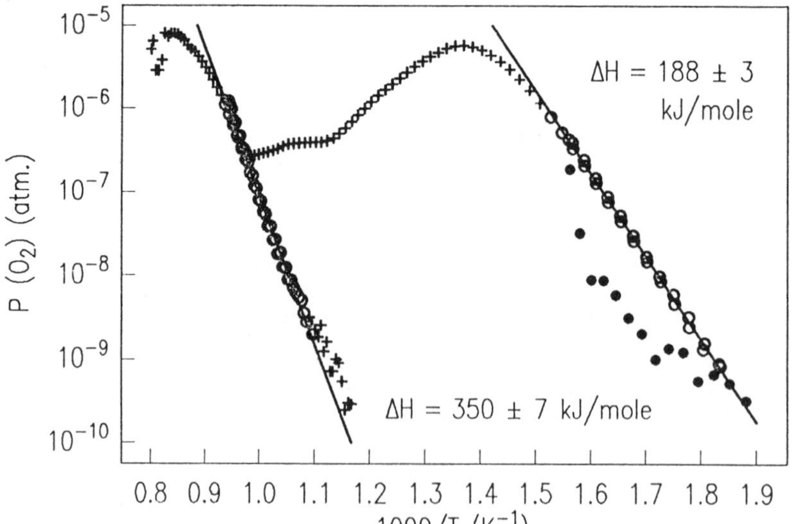

Fig. 3 Desorption of oxygen from $YBa_2Cu_3O_{7-x}$ (203), heated at 1 K/min.

stoichiometry of the residue in the cell up to the second peak (1200 K) was about $YBa_2Cu_3O_{6.0}$. These assignments are based on precise weight loss measurements and assuming a starting compound $YBa_2Cu_3O_{7.0}$. Substantial corrections to the total weight loss were required for the other major gases, (water, carbon monoxide and carbon dioxide) evolving from sample. There is also some uncertainty to the weight loss due to variable amount of gases desorbing from the cell walls which add to the mass spectrometric intensities but are not included in the weight loss. The total uncertainty in the composition is about 0.1 oxygen atom.

The first decomposition reaction was fitted to a kinetic model for first, second and third order processes (ref. 8), using a concentration independent preexponential rate constant of 1.7×10^{10} sec^{-1}. The theoretical curves overlaying the experimental data ($\log_{10} P(O_2)$ vs 1/T) are plotted in Fig. 4. We note that the third order fit is in remarkably good agreement with the data. In Fig. 5, the experimental data and the third order model are presented on a linear scale as a function of temperature. Even in this critical high pressure region the data and theory are essentially identical.

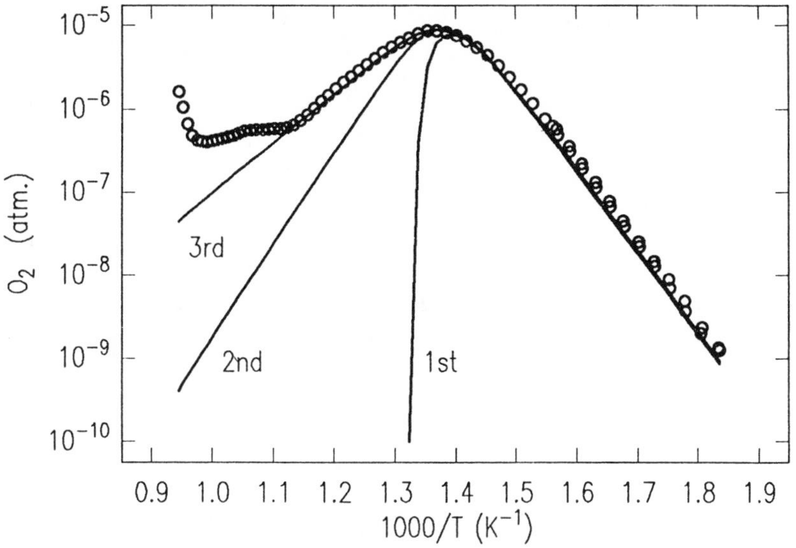

Fig. 4 Desorption of oxygen from $YBa_2Cu_3O_{7-x}$ (sample 203) heated at 1 K/min. Comparison of experimental data fitted with 1st, 2nd and 3rd order reactions using an activation energy of 188 kJ/mole.

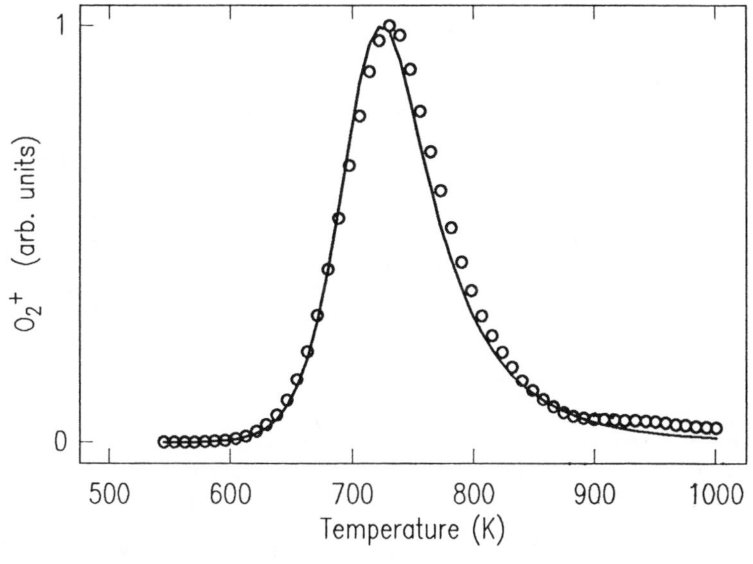

Fig. 5 Desorption of oxygen from $YBa_2Cu_3O_{7-x}$ (202), heated at 1 K/min. Data fitted with a 3rd order reaction using an activation energy of 188 kJ/mole.

One possible mechanism for a third order process is the formation of a peroxide species bonded to a surface oxygen atom followed by decomposition to oxygen gas.

The data for the second reaction also present a well-behaved activated process in the early stages, with an activation energy almost twice that of the first reaction. This suggests a dramatic stability increase in oxygen bonding in the tetragonal phase. However, we could not find any acceptable parameters to fit a kinetic model to describe this reaction. Beyond the linear region, the reaction rates fall more rapidly than those calculated from a kinetic model, indicating that diffusion may play a role in the final stages before decomposition at 1200 K.

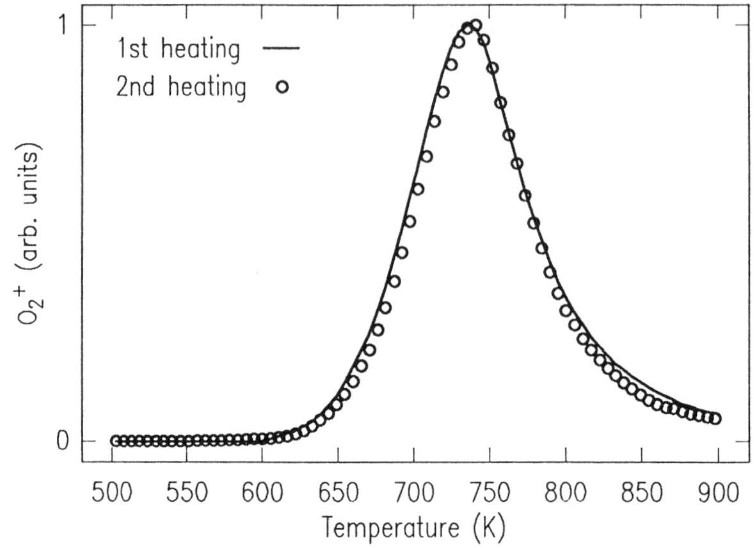

Fig. 6 Desorption of oxygen from $YBa_2Cu_3O_{7-x}$ (202). Sample heated to 900 K at 4 K/min followed by reoxygenation for 16 hours at 450°C and a second heating.

Sample 203 was also heated at 4 K/min and the temperature of the first peak shifted to 765 K with same activation energy obtained for 1 K/min. The third order kinetic model, using the same preexponential factor, fits the experimental data with very good precision as for the lower heating rate of 1 K/min. The fact that all three fitting parameters, (namely activation energy, reaction order and preexponential factor), are identical for the two heating rates shows that for this grain size ($<10\mu$) the rate limiting step is not diffusion controlled. Also, samples from this batch, stored under normal atmospheric conditions, were measured over a two month period and their thermal spectra were identical in all respects, indicating very good long term stability and exceptional homogeneity.

For duplicate measurements on sample 202, only a different preexponential rate constant ($5.0\text{x}10^{10}$ sec^{-1}) was required to fit the experimental data. The peak temperature was lower than that obtained for sample 203 with values of 705 and 737 K, for heating rates of 1 and 4 K/min, respectively. The observed differences in the absolute rate constant between the two samples probably results from a difference in particle size. This difference corresponds to a factor of three times smaller grain size for 202 which is equivalent to an increase in the surface to volume ratio by the same factor.

Another important consideration regarding the stability of oxygen in the lattice is the reversibility of the process. We examined this reaction by first heating sample 202 to 900 K at 4 K/min. About half the residue was then annealed at 450°C for 16 hours in flowing oxygen at ambient pressure. This reoxygenated sample was heated to 900 K again, in the exact manner as the first time. In Fig. 6, the oxygen desorption spectra for the two experiments are plotted as a function of temperature, normalized to the peak rate. The two curves overlay each other identically. The peak temperature of 737 K for the two runs is in agreement and is evidence that the annealing process does not alter the grain size of the sample. In effect, the ceramic material can probably be recylced several times without altering the basic crystal structure.

CONCLUSIONS

This study has provided considerable information pertaining to the thermal decomposition kinetics of $YBa_2Cu_3O_{7.0}$. On heating to 1200 K, the material undergoes a loss of oxygen via a two step process. The first is a third order reaction with an activation energy of 188 kJ/mole which encompasses the oxygen composition from 7.0 to 6.4. This reaction is inhibited by surface carbon dioxide which is removed by 640 K. The second reaction, covering the oxygen composition 6.4 to 6.0, has an activation energy of 350 kJ/mole. The sample decomposes to several phases at about 1200 K. Two different batches of material yielded identical kinetic parameters except for differences in the preexponential factor which we attribute to particle size variations.

ACKNOWLEDGEMENTS

The authors are indebted to George Pigey who meticulously performed the mass spectrometric experiments. The special platinum Knudsen cell was expertly constructed by Lou Posposil. The figures have been made with the graphic facility created by Richard F. Voss.

REFERENCES

1. J. G. Bednorz and K. A. Mueller, Z. Phys. **B64**,189(1986).
2. M. K. Wu, J. R. Asburn, C. J. Torng, P. H. Hor, R. L. Meng, L. Gao, Z. J. Huang, Y. Q. Wang and C. W. Chu, Phys. Rev. Lett. **58**, 908 (1987).
3. R. Beyers, G. Lin, E. M. Engler, R. J. Savoy, T. M. Shaw, T. R. Dinger and W. J. Gallagher, Appl. Phys. Lett. **50**,1918(1987).
4. R. J. Cava, B. Batlogg, R. B. van Dover, D. W. Murphy, S. Sunshine, T. Siegrist, J. P. Remeika, E. A. Rietman, S. Zahurak, and G. P. Espinosa, Phys. Rev. Lett. **58**,1676(1987).
5. M. Kikuchi, Y. Syono, A. Tokiwa, K. H. Oh-Ishi, H. Arai, K. Hiraga, N. Kobayashi, T. Sasaoka and Y. Muto, Jap. J. Appl. Phys. **26**,1066(1987).
6. J. D. Jorgensen, B. W. Veal, W. K. Kwok, G. W. Crabtree, A. Umezawa, L. J. Nowicki and A. P. Paulikas, (to be published).
7. P. P. Freitas and T. S. Plaskett, Phys. Rev. B **36**, 5723(1987).
8. M. A. Frisch, *Studies on Non-Equilibrium Reactions by Knudsen Effusion Mass Spectrometry*, ***Advances in Mass Spectrometry***, **8A**, 391, ed. by A. Quayle, (Heyden, London, 1980).
9. M. A. Frisch and W. Reuter, J. Vac. Sci. Technol. **16** 1020(1979).
10. S. Dushman, ***Scientific Foundations of Vacuum Technique***, John Wiley & Sons, NY (1949).
11. R. E. Honig, J. Chem. Phys. **22**,126(1954).

Kinetics of oxygen uptake in $YBa_2Cu_3O_x$

I. HALLER, M. W. SHAFER, R. FIGAT, and D. B. GOLAND

IBM Thomas J. Watson Research Center
P.O. Box 218, Yorktown Heights, NY 10598, U.S.A.

Abstract - The oxygen uptake rates of sintered $YBa_2Cu_3O_x$ were measured by isothermal thermogravimetric analysis. The uptake is an activated process with $E_a = 111 \pm 3$ kJ/mole. Diffusion control of the oxygen uptake rate is supported by identical E_a for uptake and loss, by the similar diffusion coefficients in the orthorhombic and tetragonal phases, and by a realistic jump frequency. Data at early stages of the reaction are inconsistent with diffusion, and surface oxidation reaction is likely rate determining at least in that regime. The equilibration of the superconducting transition characteristics is consistent with the oxygen uptake kinetics.

INTRODUCTION

The importance of oxygen in the superconductor $YBa_2Cu_3O_x$ is well known[1-4] and a rather adequate understanding of the relationship between the oxygen concentration, x, and many of the superconducting properties has been extablished. For example, for x values between 6.85 - 7.0 superconducting transition temperatures (T_c) 90 - 92K are normally seen. Decreasing x below about 6.8 results in broader transitions which decrease non-linearly[5-7]. The processing conditions have been empirically determined and the optimum temperatures and oxygen pressures to obtain the various x values are known. However, for technological applications, particularly for optimization of the processing times and temperatures, it is also important to know the rate at which oxygen is added or removed from these materials. In this paper we address this question by studying the reaction kinetics of oxygen with $YBa_2Cu_3O_x$. We show that oxygen uptake is an activated process and propose a mechanism in which oxygen taken up in a surface oxidation step is distributed into the bulk via a vacancy diffusion mechanism.

EXPERIMENTAL

The ceramic samples were made from a stoichiometric mixture of yttrium and copper oxides and barium carbonate by repeated grinding and calcining at 950 °C. Pellets (12.6 mm diameter) were cold pressed at 4200 kg/cm^2, then fired in air at 950 °C for 12 hours. Ramp rates were 5 °C/min heating and 1 °C/minute cooling. The pellets were sliced into approximately 0.7 x 3.5 x 11 mm bars with a diamond saw cooled with kerosene lubricant, which was subsequently removed by multiple rinses in n-hexane.

Batches of the $YBa_2Cu_3O_x$ slabs with lower oxygen content were prepared by annealing at 525 °C in a fused silica tube at reduced presure. The base pressure of the system was about 10^{-6} torr. In one batch the heating was continued, while continuously pumping, until the pressure above the samples fell to about 3×10^{-3} torr, which took about 5 hours. This procedure resulted in a weight loss of 1.80 percent and a tetragonal phase with a composition of $YBa_2Cu_3O_{6.25}$, (see Table I). In another batch of samples the annealing was carried out in flowing oxygen (p=7.5 torr, flow rate 1.6 sccm), which lead to an oxygen content slightly

TABLE I. Physical properties of the $YBa_2Cu_3O_x$ samples used.

x =	6.97	6.66	6.25
ρ_{298} (Ω-cm)	0.0017	0.005	5
$d\rho/dT$	metallic	metallic	semiconductor
T_c (°K)	91	61	none
XRD Structure	orthorhombic	orthorhombic	tetragonal
a (Å)	3.817	3.831	3.856
b (Å)	3.884	3.880	--
c (Å)	11.670	11.714	11.785

above that required to be on the orthorhombic side of the structural phase transition (x_0=6.66, Table I).

The density of the as prepared samples was found to be 4.20 and 4.40 (in 2 separate batches) implying about 33 percent pore volume. The metal ratio of the samples were confirmed by inductively coupled plasma atomic emission spectroscopy (ICP-AES) to be 1:2:3 with an accuracy of 5 percent[8]. The oxygen contents of the samples were determined thermogravimetrically by heating in flowing O_2 at 3 degree/min from room temperature to 950 °C. The oxygen content at the endpoint was taken as x=6.25 which value was established earlier by wet chemical analysis[9]. X-ray powder patterns have established that all samples (3 different oxygen concentrations) were single phase. The physical properties are summarized in Table I.

Oxygen uptake rates were measured by isothermal thermogravimetric analyses performed on a Perkin-Elmer Model 7 instrument. The temperature was stepped to the desired value at 200 °C/min, and we estimate that complete temperature equilibration of the sample took less than 3 minutes. The rates were low enough to ensure that no difficulties arose, except perhaps at the highest temperature used, in determining the effective onset time. All measurements were carried out at 1 atm pressure, generally in 100 % O_2 flowing at a rate of 80 sccm, but argon (<1 ppm O_2) and mixtures of the two gases were also used. The collected data were transferred to a mainframe computer for further analysis.

Sample resistances were measured in four-terminal measurements using a lock-in amplifier at 27.0 Hz with spring loaded contacts spaced at 1.8 mm. We define three quantities to characterize the superconducting transition: (1) T_{onset}, the intersection of a line extended from the linear R vs T region above the transition with a line through the steepest resitance drop, (2) T_{compl}, the temperature at which R becomes less than 1 % of that at T_{onset}, and (3) the slope of the line (in units of K^{-1}) of the line going through the 10 and 80 % resistance points of the transition.

RESULTS and DISCUSSION

Isothermal thermogravimetric measurements of the oxygen uptake by the vacuum preannealed (x_0=6.25) samples in pure O_2 were performed at four temperatures. A typical thermogravimetric trace, taken at 350 °C, and recomputed as mean oxygen content, x, versus time is shown as the top solid curve in Fig. 1. Also shown in Fig. 1, circles, are the numerically computed values of the uptake rate, dx/dt.

The oxygen uptake process in a porous solid is anticipated to consist of three consecutive steps, i.e. (1) gaseous oxygen transport through the pores, (2) oxidation reaction at the solid surface, and (3) solid state diffusion of oxygen into the bulk. Of these the latter two are expected to be activated processes, while the first one would be expected to show a much weaker temperature dependence.

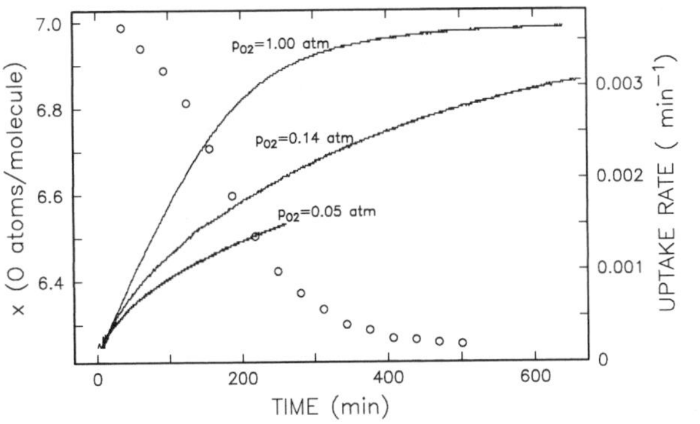

Fig. 1. Oxygen content of $YBa_2Cu_3O_x$ measured thermogravimetrically at 350 °C in pure oxygen (top trace) and in O_2 diluted with Ar (lower traces). Circles show derived uptake rates at 1 atm.

An Arrhenius plot of initial oxygen uptake rates at 1 atm O_2 pressure is shown as the circles in Fig. 2. It is immediately obvious that the rate-determining process is activated, hence gas diffusion is ruled out as the rate determining step.

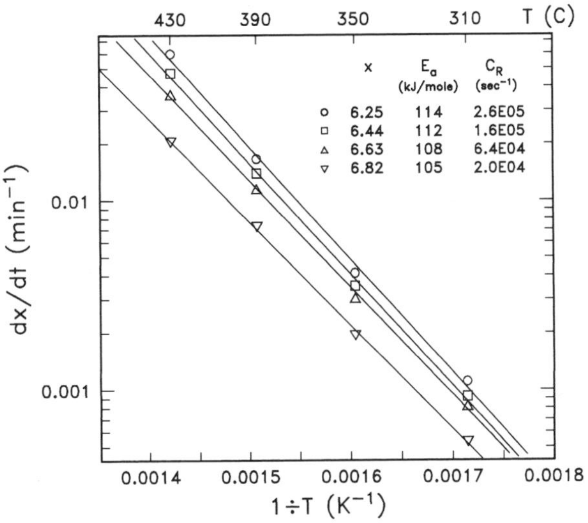

Fig. 2. Arrhenius plots of oxygen uptake rates, p_{O_2} = 1 atm, at various degrees of conversion. The fitting constants are tabulated in the inset.

Also shown in Fig. 2 are uptake rates observed at about 25, 50 and 75 percent completion of the reaction. The activation energies show a decreasing trend but no discontinuity as the conversion increases. Uptake rates were also measured at 350 °C with O_2 partial pressures of 0.14 and 0.05 atm (lower 2 solid lines of Fig. 1). A definitely sublinear pressure dependence (a factor of 2.5 decrease for a 20 fold decrease in p_{O_2}) was observed, but the available precision did not warrant further measurements to determine a more accurate functional form.

The oxidation step at the solid surface could include the following reactions where the * symbolizes empty adsorption sites, O_0 represents lattice oxygen and V_O oxygen vacancies.

$$O_2(gas) + * \rightleftarrows O_2(adsorbed) \qquad [Reaction\ 1]$$

$$O_2(adsorbed) + 2V_O \longrightarrow 2O_0 + * \qquad [Reaction\ 2]$$

$$O_2(adsorbed) \rightleftarrows 2O(adsorbed) \qquad [Reaction\ 3]$$

$$O(adsorbed) + V_O \longrightarrow O_0 + \tfrac{1}{2}* \qquad [Reaction\ 4]$$

Reaction 1 is written as equilibrium because the sublinear pressure dependence is incompatible with it being rate determining. If Reaction 2 is rate determining, the preexponential factor of the Arrhenius plots of Fig. 2 is given by $C_R = A_2[O_2(adsorbed)][V_O]^2$. As seen in the inset of Fig. 2, C_R indeed appears to vary quadratically with $\delta = 7.00 - x$. We caution, however, that the magnitude of the experimental error prevents us from ruling out mechanisms first order in V_O, such as the alternate path Reactions 3 and 4. Furthermore, as discussed below, diffusion of the lattice oxygen can be the rate determining step in other than the initial stages of the reaction.

If solid state (vacancy) diffusion is rate determining, the kinetics can be easily modeled[10] for particles of various shapes. For a semiinfinite solid, the solution of the diffusion equation with the boundary condition of constant surface concentration requires that the total amount of diffusing species taken up increases with the square root of time. For finite size solids the uptake of indiffusing species saturates as the bulk concentration approaches the surface concentration; the initial uptake, however, in all cases is again proportional to the 1/2 power of time. For example, for spheres of radius a the solution of the diffusion equation yields[11] Eqn. (1), where D is the diffusion coefficient and M_t/M_∞ the fractional mass uptake at time t.

$$M_t/M_\infty = 6\left(\frac{Dt}{a^2}\right)^{1/2}\left\{\pi^{-1/2} - 2\sum_{n=1}^{\infty} \text{ierfc}\, n\left(\frac{Dt}{a^2}\right)^{-1/2}\right\} - 3\frac{Dt}{a^2}. \qquad (1)$$

Since the errorfunction complement, ierfc, approaches zero rapidly as its argument increases above 1, for small $\frac{Dt}{a^2}$ Eqn. (1) reduces to

$$\frac{M_t}{M_\infty} = \frac{6}{\pi^{1/2}}\left(\frac{Dt}{a^2}\right)^{1/2}. \qquad (2)$$

Fig. 3a illustrates the expected fractional weight increase if the rate is diffusion controlled with constant surface concentration for various assumed geometries. The solid line is a plot of the solutions of the diffusion equation from Ref. 10 as it applies to the case of a set of monodisperse spheres of radius a. The dotted line is a recomputation representing a distribution of sizes of spheres, i.e. it assumes three subsets of equal numbers of spheres with radii $\frac{1}{2}a$, a,

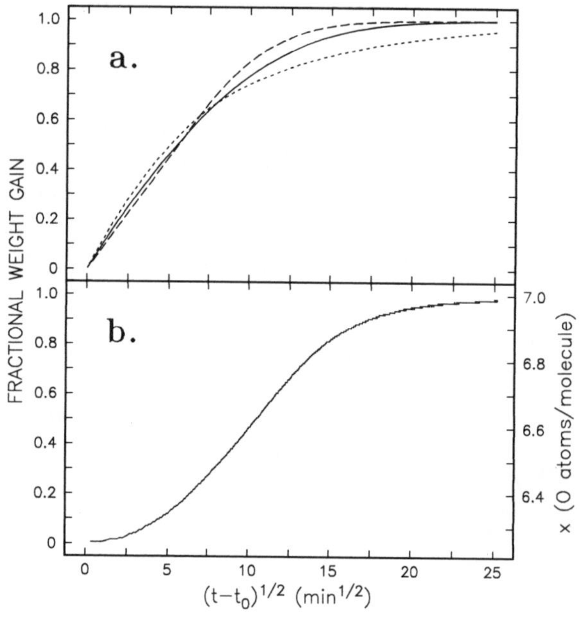

Fig. 3. a. Theoretical diffusion controlled gas uptake with constant surface concentration. Solid line: spheres $(D/a^2) = 1.71\times10^{-5}$, sec^{-1}, dotted line: equal number of spheres with (D/a^2) 0.43, 1.71, and 6.83 in units of 10^{-5} sec^{-1}, and dashed line: flat sheet, $(D/l^2) = 1.07\times10^{-4}$, sec^{-1}. b. Experimental curve, T=350 °C.

and 2*a* respectively. The dashed line represents the solution* of the diffusion equation for a plane sheet[10].

The experimental data of Fig. 1 are replotted in Fig. 3b as fractional weight uptake vs. the square root of time. At intermediate times this curve (as well as all other plots of this kind of data taken under different conditions) exhibits a linear section, from whose slope a diffusion coefficient can be extracted if one models the porous solid as a collection of spheres. But near the origin we observe a definite upward concave curvature which is inconsistent with theoretical diffusion controlled weight uptake curves (as long as constant surface concentration and constant diffusion coefficient is assumed) for any combination of shapes and size distributions. This deviation is typical for most of our thermogravimetric oxygen uptake measurements although not in all instances is the deviation as pronounced. In the lower temperature data, such as shown in Fig. 3b, the time required to reach the behavior consistent with the diffusion controlled process is much larger than what is needed for the sample to reach temperature equilibrium. Thus if we choose to interpret our data in terms of solid state diffusion, another process still has to be invoked for the observed induction period.

To test the possibility that a change in the diffusion coefficient at the tetragonal to orthorhombic phase transition might be responsible for the observed deviations, we repeated the measurements at the same four temperatures with $x_0=6.66$ starting compositions. These samples are orthorhombic at the outset and no structural phase transition takes place during the oxygen uptake process. We found the shapes of the M_t/M_∞ vs. $t^{1/2}$ curves to be essentially the same as in the lower x_0 samples. Assuming that the model of a set of spheres adequately represents diffusion into the porous solid, the diffusion coefficients are, according to Eqn. (2), proportional to the squares of the slopes of the linear sections of the M_t/M_∞ vs. $t^{1/2}$ curves. The circles in Fig. 4 are an Arrhenius plot for the squares of the slopes for the $x_0=6.25$ sample.

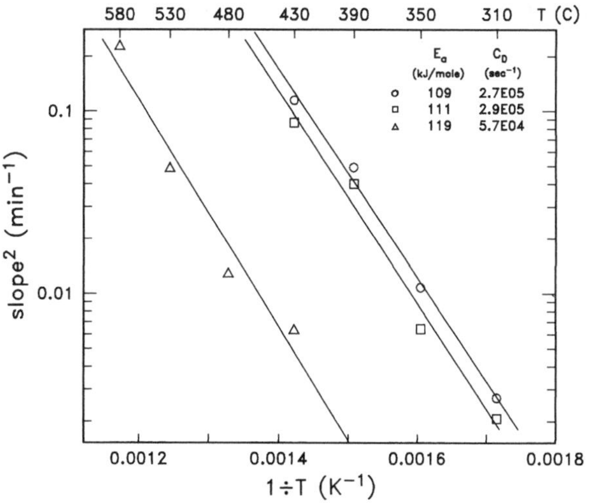

Fig. 4. Temperature dependence of the diffusion product (square of slope of Fig. 3b type curves) for oxygen uptake by $x_0=6.25$ and $x_0=6.66$ starting materials (circles and squares, resp.), and for oxygen loss (triangles).

We note that the activation energies are essentially the same as E_a obtained (in Fig. 2) from the uptake rates. The rectangles in Fig. 4 refer to the $x_0=6.66$ starting material and imply a diffusion coefficient only 15 % lower than in the $x_0=6.25$ samples. It is, of course, impossible to determine whether this minor difference is due to a change in the activation energy or in the preexponential factor, but in either case it is compatible with a vacancy diffusion model with either a minor change in the barrier or in the number of available jump sites at the structural phase transition. We conclude that the deviations of the experimental data from the straightforward diffusion model do not have their origin in the phase transition, and, if indeed diffusion

*The solution given by Ref. 10, p. 48, is an infinite series of errorfunction complements, somewhat similar to Eqn. (1), which for short times reduces to $\frac{M_t}{M_\infty} = \frac{6}{\pi^{1/2}}\left(\frac{Dt}{l^2}\right)^{-1/2}$, l being the half thickness of the sheet.

is the rate determining step, the diffusion coefficients of oxygen in the tetragonal and orthorhombic phases of $YBa_2Cu_3O_x$ are identical.

Thermogravimetric determination of the oxygen uptake rate is a relatively precise measurement because the driving force for the reaction, i.e. the difference between the thermodynamic equilibrium oxygen content and that of the starting material, $x_{eq}(T) - x_0$, can easily be made large and at 1 atm O_2 pressure is only slightly temperature dependent. For oxygen loss rate measurements, however, by isothermal thermogravimetry at low O_2 partial pressures, this is not the case. The driving force for the oxygen loss reaction is very sensitive to the experimental conditions for two reasons. Firstly, we find, in agreement with Ref. 12, that at low p_{O_2} the equilibrium oxygen content in the temperature range suitable for rate measurements, is a strong function of temperature. Secondly, the partial pressure of O_2 at the solid-gas interface inside the porous solid is much more poorly defined in flowing argon carrier gas than in pure O_2. Thus oxygen loss rate measurements by isothermal thermogravimetry have inherently larger experimental errors than uptake rate measurements. We have nevertheless measured oxygen loss rates from x_0=6.97 samples and have included these data as triangles in Fig. 4. If solid state diffusion is the rate limiting step both for oxygen loss and uptake, the activation energies of the coefficients of diffusion in both direction must be the same. The experimental data are not inconsistent with this requirement.

The activation energies obtained, E_D = 111 kJ/mole, is in the right range for vacancy diffusion of oxygen in other oxides[13-16]. It follows from Eqn. (2) that the preexponential constant of the diffusion coefficient can obtained from the preexponential coefficients, C_D, of the plots in Fig. 4 as $D_0 = (\pi a^2/36)C_D$. From the dimensions of grain sizes observed under the optical microscope, which frequently exceed 10 μm, we estimate the equivalent radius for diffusion a = 3 μm. The observed value of $C_D = 2.85 \times 10^5$ sec^{-1} leads then to the order-of-magnitude estimate of $D_0 = 2.2 \times 10^{-3}$ cm^2 sec^{-1}. The well-known[17] transition state theory of vacancy diffusion expresses this constant as $D_0 = x_{vac}\alpha a_0^2\chi\nu \exp(\Delta S^\ddagger/R)$. In $YBa_2Cu_3O_x$ the oxygen vacancy mole fraction, $x_{vac,}$ is determined by the oxygen content and is of the order of 0.5. Assuming the jump distance a_0 = 2 Å, with $\alpha = \chi = 1$ and $\Delta S^\ddagger = 0$, from the above D_0 we arrive to $\nu = 1.1 \times 10^{13}$ sec^{-1}. Considering the crudeness of the approximations, this is certainly the right order of magnitude for the jump frequency and supports the notion that diffusion is rate controlling.

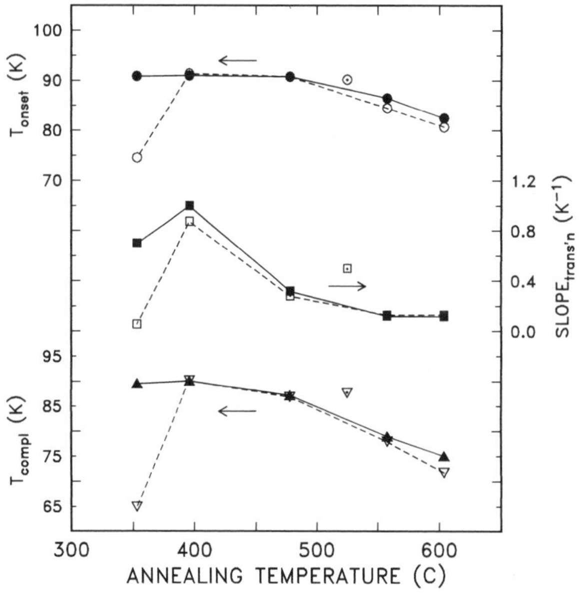

Fig. 5. Comparison of the characteristics of the superconducting transition as a function of annealing temperature (in 0.1 atm O_2, 7 hours) for x_0=6.97 starting material (full symbols) and vacuum annealed x_0=6.25 samples (open symbols). Half filled symbols show unannealed x_0=6.97 sample.

To ascertain that the oxygen uptake kinetics also control the equilibration of the superconducting transition characteristics, we carried out resistance vs temperature (<150 °K) measurements on samples isothermally annealed in oxygen under various conditions. At each annealing condition two samples were used simultaneously: one as grown (x_0=6.97) and another vacuum preannealed (x_0=6.25). As an example, Fig. 5 shows the onset and completion temperatures as well as the sharpness (slope) of the superconducting transition for samples annealed in 0.1 atm O_2 for 7 hours at various temperatures. There is no difference in any of the characteristics of samples annealed at or over 400 °C between starting compositions x_0=6.25 (open symbols) and x_0=6.97 (full symbols). But at 350 °C 7 hours is inadequate for complete equilibration. We note that the best characteristics are achieved in annealing at the lowest temperature that is adequate for equilibration in the alloted time; the decrease at higher than required temperatures is due to the lower equilibrium oxygen content. Entirely analogous results were obtained by carrying out the annealing at 1 atm oxygen pressure except the maxima in the characteristics are not as sharp.

SUMMARY

The oxygen uptake in porous ceramic $YBa_2Cu_3O_x$ is an activated process with E_a = 111±3 kJ/mole. Lattice oxygen diffusion by a vacancy mechanism is likely to control the O_2 uptake rate. This is supported by the observation of (a) the same E_D for uptake and loss, (b) essentially the same diffusion coefficient for tetragonal and orthorhombic starting materials, and (c) a realistic jump frequency. Data points at the early stages of the O_2 uptake are inconsistent with diffusion with constant surface concentration. A surface oxidation reaction is probably rate determining in that regime and may, indeed, control the rate until a high degree of completion is reached. The determination of where the cross-over is, must await the availability $YBa_2Cu_3O_x$ in well controlled particle sizes. The equilibration of the superconducting transition characteristics is consistent with the oxygen uptake kinetics.

ACKNOWLEDGEMENTS

We thank M. M. Plechaty for metal analyses and W. J. Gallagher for making available to us his apparatus for measurements of the superconducting transition characteristics.

REFERENCES

1. M.A. Beno, L. Soderholm, D.W. Capone, D.G. Hinks, J.D. Jorgenson, I.K. Schuller, C.V. Segre, K. Zhang and J.D. Grace, Appl. Phys. Lett. **51**, 57 (1987).
2. J.J. Capponi, C. Chaillout, A.W. Hewat, P. Lejay, M. Marezio, N. Nguyen, B. Raveau, J.J. Souberroux, J.L. Tholence, and R. Tournier, Euro. Phys. Lett. **3**, 1301 (1987).
3. J.E. Greedan, A.H. O'Reilly and C.V. Stager, Phys. Rev. B **35**, 8770 (1987).
4. D.W. Murphy, S.A. Sunshine, P.K. Gallagher, H.M. O'Bryan, R.J. Cava, B. Batlogg, R.B. van Dover, L.F. Schneemeyer, and S.M. Zahurak in Chemistry of High-Temperature Superconductors, ACS Symposium Series 351, edited by D.L. Nelson, M.S. Wittingham and T.F. George, p. 181, Am. Chem. Soc., Washington (1987).
5. B. Batlogg, Procedings of Int. Workshop on Novel Mechanisms of Superconductivity, Berkeley, CA June 22-26, (1987).
6. D.C. Johnson, A.J. Jacobson, J.M. Newsom, J.T. Lewandowski, D.P. Goshorn, D. Xie, and W.B. Yelon in Chemistry of High-Temperature Superconductors, ACS Symposium Series 351, edited by D.L. Nelson, M.S. Wittingham and T.F. George, p. 136 Am. Chem. Soc., Washington (1987).
7. P. Monod, M. Ribault, F. D'Yvoire, J. Jegoudez, G. Collin, and A. Revcolevschi, J. Physique **48**, 1369 (1987).

8. M.M. Plechaty, unpublished data.
9. M.W. Shafer, T. Penney, and B.L. Olson, Phys. Rev. B **36**, 4047 (1987).
10. J. Crank, The Mathematics of Diffusion, 2nd ed. Clarendon Press, Oxford, (1980).
11. Reference 10, pp. 90-91.
12. P.K. Gallagher, Advanced Ceram. Mater. **2**, 632 (1987).
13. R.L. Sproull, R.S. Bever, and G. Libowitz, Phys. Rev. **92**, 77 (1953).
14. M.F. Berard, C.D. Wirkus, and D.R. Wilder, J. Am. Ceram. Soc. **51**, 643 (1967).
15. I. Matsuura and M. Kirishiki, Chem. Letters **1986**, 1141.
16. G.W. Crabtree, J.W. Downey, B.K. Flandermeyer, J.D. Jorgensen, T.E. Klippert, D.S. Kupperman, W.K. Kwok, D.J. Lam, A.W. Mitchell, A.G. McKale, M.V. Nevitt, L.J. Nowicki, A.P. Paulikas, R.B. Poeppel, S.J. Rothman, J.L. Routbort, J.P. Singh, C.H. Sowers, A. Umezawa, B.W. Veal, and J.E. Baker, Advanced Ceram. Mater. **2**, 444 (1987).
17. See for example S. Mrowec, Defects and Diffusion in Solids p. 180, Elsevier, New York, (1980).

The role of oxygen in $YBa_2Cu_3O_{7-\delta}$

John B. Goodenough and A. Manthiram

Center for Materials Science & Engineering
The University of Texas at Austin, Austin, Texas 78712-1084 USA

Abstract - Five observations concerning the role of oxygen in the high-T_c superconductors $YBa_2Cu_3O_{7-\delta}$ are emphasized: (1) An internal electric field parallel to the c-axis constrains the redox reactions associated with the intercalation/disintercalation of oxygen to the $[CuO_{3-\delta}]^{3-}$ layers located between Ba^{2+}-ion layers. (2) Intercalation of O^{2-} ions into the tetragonal $YBa_2Cu_3O_6$ phase results in an ordering onto one of the basal-plane axes of the Cu(1) layer, the b-axis of the orthorhombic phase, before any copper is oxidized beyond the Cu^{2+} state. (3) Ordering of oxygen vacancies within or between b-axis chains gives rise to discrete phases predicted to occur at $O_{6.67}$, $O_{6.50}$, and $O_{6.33}$. (4) An equilibrium oxidation state for the intercalation layer depends not only on the oxygen partial pressure and temperature, but also on the cations of the structure. (5) At higher oxidation states of the intercalation layer, any oxygen atom on the a-axis interacts with a near-neighbor oxygen on the b-axis to trap out holes in a peroxide ion $(O_2)^{2-}$. The significance of these roles for the optimization and processing of the superconductor material is introduced.

INTRODUCTION

The system $YBa_2Cu_3O_{7-\delta}$, $0.04 \leq \delta \leq 1$, has the tetragonal structure of Fig. 1(a) in the limit O_6 (i.e. $\delta = 1$) (ref. 1); it has the orthorhombic structure of Fig. 1(b) in the ideal limit O_7 (i.e. $\delta = 0$) (ref. 2). By varying the temperature in the interval 350 - 950 °C and the partial pressure of oxygen over the range $0 \leq p_{O_2} \leq 1$ atm, it is possible to adjust the equilibrium oxygen concentration over the entire range $O_{6.96}$ to O_6 ($0.04 \leq \delta \leq 1$) (ref. 3). The mobile oxygen is rapidly inserted into/extracted from the Cu(1) planes at temperatures above 350 °C; at room temperature the oxygen is essentially immobile, so non-equilibrium oxygen concentrations can be quenched in. The full range of oxygen compositions $0.04 \leq \delta \leq 1$ has been obtained at room temperature; the structure was found (ref. 3) to change smoothly with oxygen concentration through an orthorhombic-tetragonal transition at $O_{6.27}$, see Fig. 2. At lower oxygen concentrations (O_6 to $O_{6.27}$), the inserted oxygen atoms randomly occupy the Cu(1)-bridging positions on the two equivalent axes of the tetragonal structure; at higher oxygen concentrations ($O_{6.27}$ to $O_{6.96}$) the inserted oxygen are ordered onto one of these axes, the b-axis of the orthorhombic structure.

On raising the temperature in air or 1 atm O_2, the $O_{6.96}$ composition loses oxygen, but the O_6 composition initially picks up oxygen. Fig. 3 illustrates the weight change due to oxygen insertion on heating an $O_{6.18}$ sample in O_2 at 1 °C/min; the equilibrium oxygen concentration is approached below 400 °C, and a continued increase in temperature results in a loss of oxygen as the equilibrium oxygen concentration decreases. On cooling from 920 °C at 1 °C/min, the sample regains oxygen, but the time at temperature is never sufficient for the full equilibrium concentration to be reached (ref. 3). The high-temperature x-ray data of Fig. 4 shows heating a fully oxidized sample $O_{6.96}$ in air; the orthorhombic-tetragonal transition occurs smoothly, but rather abruptly, near 600 °C (ref. 4a). In 1 atm O_2, it occurs near 680 °C (ref. 4b). The c-axis oxygens--O(1) of Fig. 1--neighboring the Cu(1) atoms are not disordered by either temperature or changing oxygen concentration (ref. 5).

On the other hand, the superconducting transition temperature T_c does not vary smoothly with oxygen concentration. Cava et al. (ref. 6) have shown that T_c varies stepwise from near 90 K at $O_{6.96}$ to about 60 K in the vicinity of $O_{6.67}$ and to a lower temperature ($0 < T_c < 20$ K) below $O_{6.5}$, see Fig. 5.

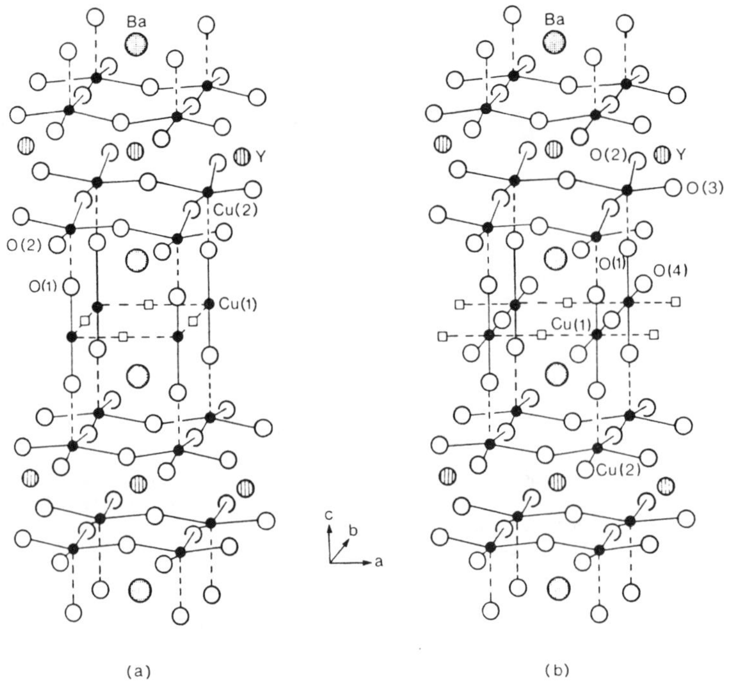

Fig. 1. Structure of (a) tetragonal $YBa_2Cu_3O_6$ and (b) orthorhombic, ideal $YBa_2Cu_3O_7$.

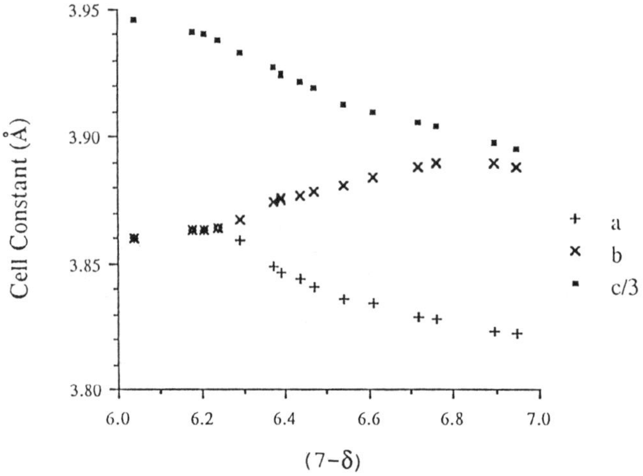

Fig. 2. Variation of room-temperature lattice parameters with oxygen concentration in the system $YBa_2Cu_3O_{7-\delta}$, after Swinnea (ref. 3).

Fig. 3. TGA Curve for a metastable, room-temperature $YBa_2Cu_3O_{6.18}$ sample heated in O_2 at 1 °C/min. Numbers refer to oxygen content.

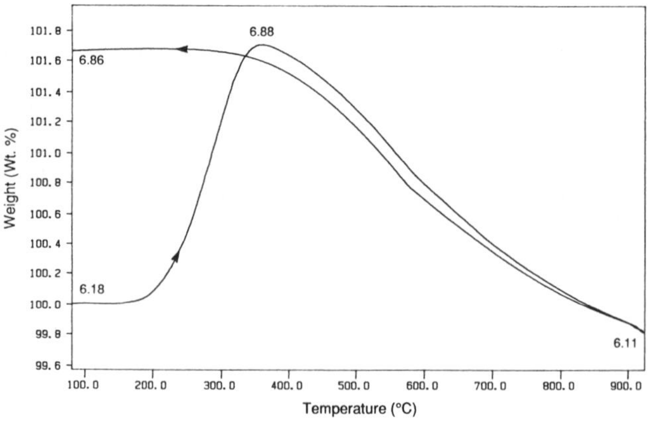

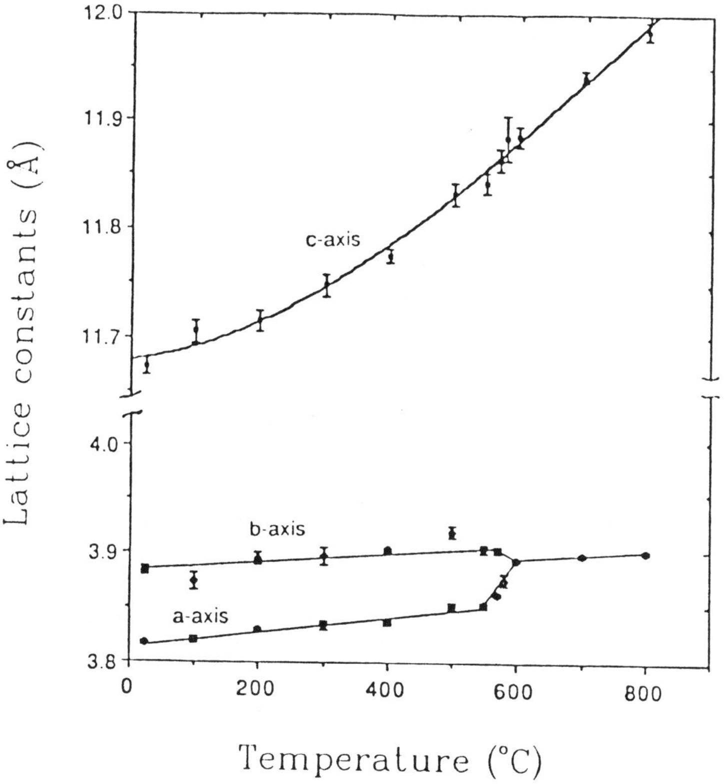

Fig. 4. Temperature variation of the equilibrium lattice parameters of $YBa_2Cu_3O_{7-\delta}$ heated in ambient atm (after ref. 4).

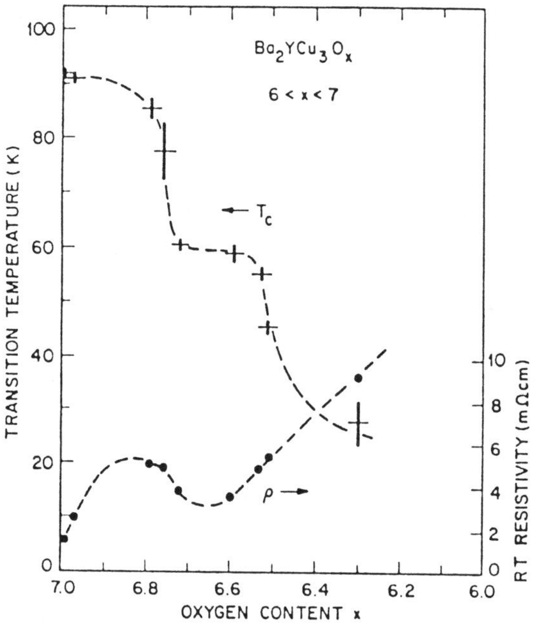

Fig. 5. Variation of T_c and room-temperature resistivity ρ with oxygen concentration (after ref. 6).

Moreover, the transitions are sharp and the room-temperature resistivity is lower in the center of a plateau region, the transitions are broad at intermediate compositions.

These data clearly indicate the presence of some additional ordering not readily detectable with diffraction techniques.

This chemical flexibility raises several important questions that can be addressed experimentally:

(1) Where in the structure does the redox reaction occur on insertion/extraction of oxygen?

(2) To what extent is electronic charge transferred from the host matrix to the inserted oxygen, and how meaningful is the assignment of formal valence states?

(3) In addition to ordering of the inserted oxygen on the orthorhombic b-axis, what other ordering of these oxygen can be expected?

(4) Ordering of the mobile oxygen onto the orthorhombic b-axis prevents direct contact between these oxygen atoms even at the higher oxygen concentrations. Does oxygen clustering (as in peroxide-ion $(O_2)^{2-}$ formation), occur if the intercalated oxygen are placed in direct contact under conditions of higher total oxidation?

(5) To what extent can the equilibrium oxygen concentration for a fixed p_{O_2} and temperature be varied by cation substitution?

(6) Can a fluctuating valence on the Cu(1) and bridging oxygen near neighbors be responsible for the enhanced superconducting transition temperature T_c in these materials?

THE REDOX REACTION

The structure of Fig. 1 represents an intergrowth of layers of different charge. As a result, an internal electric field is established parallel to the c-axis that determines not only the location of the oxygen vacancies in the structure, but also the region of the crystal where redox reactions occur on insertion/extraction of oxygen.

The crystallographic feature that builds in an internal electric field is the stabilization of eightfold coordination at the Y^{3+} ion instead of the twelvefold coordination found in the perovskite structure. The absence of O^{2-} ions in the Y^{3+}-ion plane forces the adjacent Cu(2) planes to contain the full complement of two O^{2-} ions per Cu(2) atom so as to approach local charge neutrality. In fact, the remaining oxygen vacancies are constrained to the Cu(1) planes, which are as far removed as possible from the Y^{3+}-ion planes.

In the tetragonal $YBa_2Cu_3O_6$ structure of Fig. 1(a), the Cu(1) atoms have only two near-neighbor oxygens in linear O-Cu-O units paralled to the c-axis. This coordination is characteristic of Cu(I) species; the local electrostatic forces destabilizing the Cu:3d manifold are sufficiently reduced relative to those at the fivefold-coordinated Cu(2) atoms that we may assign to the Cu(2) atoms a formal valence close to Cu(II) despite the c-axis axial field, which would stabilize a mixed Cu(I), Cu(II) valence at the Cu(2) sites in order to achieve local charge neutrality. It therefore follows that, omitting the Ba^{2+} ions, the Cu(1) atoms and their near-neighbor oxygen atoms form a layer having a formal charge approaching $[Cu(1)O_2]^{3-}$.

Introduction of oxygen into the Cu(1) plane must oxidize the Cu(1) atoms, not the Cu(2), if the internal electric field is to be minimized. Therefore insertion/extraction of oxygen into/from the Cu(1) plane introduces a redox reaction between Cu(1) and inserted oxygen that is essentially constrained to the Cu(1) plane, which means that the Cu(1) layer retains its formal charge $[Cu(1)O_{3-\delta}]^{3-}$ for all values of δ. As a corollary, the Cu(2) atoms must retain a formal valence Cu(II), which is consistent with the small change in basal-plane area on oxygen insertion (Fig. 2).

Identification of the location of the redox reaction in the structure does not specify the extent of charge transfer from the Cu(1) atoms to the inserted oxygen, the second of our original questions. Before addressing this question, it is convenient to consider first the simpler problem of oxygen ordering.

OXYGEN ORDERING

A remarkable feature of Fig. 2 is the appearance of ordering onto the b-axis of the inserted oxygen with less than 15 percent occupancy of the Cu(1)-bridging oxygen sites. Two factors contribute to the ordering, the electrostatic repulsion between oxide ions and the preference of formal-valence Cu(II) and Cu(III) for coplanar anion coordination.

As indicated in Fig. 6(a), introduction of an oxygen atom into the Cu(1) plane of the host O_6 structure would induce an electron transfer from the two neighboring Cu(1) atoms to create an inserted oxide ion O^{2-} bridging two neighboring Cu(II) atoms in the Cu(1) plane. The electrostatic interactions between such inserted oxide-ion clusters are strong and long-range, which is why b-axis ordering occurs at $O_{6.27}$. It follows that these same forces must induce either interchain or intrachain ordering.

At the composition $O_{6.5}$, the intrachain ordering of Fig. 6(b) would minimize the electrostatic forces between inserted oxygen and at the same time be compatible with a coplanar anion configuration at Cu(1) atoms uniformly oxidized to the formal valence state Cu(II). This order would be stabilized further by Cu(1)-Cu(1) bonding across the oxygen vacancy; trapping of two electrons per oxide-ion vacancy is commonly found in oxides (ref. 7). Alternatively, completely oxidized chains may alternate with unoxidized chains along the a-axis. In either case, weak coupling between (001) layers separated by more than 11.5 Å may make difficult detection of the order by x-ray or neutron diffraction.

Other intrachain orderings can also be envisaged. The next most stable would occur at $O_{6.67}$ and $O_{6.33}$. Fig. 6(c) illustrates the ordered arrangement to be anticipated for $O_{6.67}$; however, assignment of formal valences to the atoms has been withheld for the reasons discussed in the next section. Fig. 5 presents what evidence there is presently that such an intrachain ordering exists. The ordering at $O_{6.33}$ would be analogous, but with bridging oxygen and vacancies interchanged. In this case formal valences would be meaningful;

-Cu(II)-O-Cu(II)-□-Cu(I)-□-Cu(II)-O-Cu(II)

The fact that b-axis ordering extends to $O_{6.27}$, Fig. 2, indicates that intrachain ordering at $O_{6.33}$ probably exists also. However, interchain ordering of oxidized and non-oxidized chains alternating 2:1 or 1:2 along the a-axis are also possible.

The above considerations lead to the schematic phase diagram of Fig. 7. The solid curve is an experimental thermogravimetric oxygen analysis (TGA) like that of Fig. 3. The dashed orthorhombic-tetragonal transition temperature T_c is estimated from the room-temperature transition at $O_{6.27}$ and the equilibrium transition in 1 atm air near 650 °C. The shaded areas represent schematically, immiscibility domes between the ordered phases at $O_{6.33}$, $O_{6.5}$, $O_{6.67}$, and $O_{7-\delta'}$.

In the last of these phases, the δ' oxygen vacancies per formula unit ($\delta' \lesssim 0.15$) may be randomly distributed on the b-axis chain segments between twin planes; however, they would be attracted to twin-plane boundaries, as is illustrated in Fig. 8, so as to inhibit formation of close oxide-ion pairs across such a boundary. The electrostatic repulsive force between such a pair would increase the surface energy of a twin-plane boundary were such pairs to exist. The orthorhombic phase contains a high density of twin planes of the type illustrated, but the density is not great enough to attract all the oxygen vacancies at the equilibrium concentration $O_{6.96}$. Therefore ordering into b-axis chains prevents near-neighbor contact between inserted oxygen over the entire compositional range $O_{6.96}$ to O_6.

The fact that the phase $O_{6.67}$ is superconducting with a $T_c \approx 60$ K demonstrates that either tunneling between Cu(1) atoms is occurring across the oxygen vacancies as well as through the bridging oxygen atoms or that coupling occurs via Cu(2) planes.

FORMAL VALENCES

The formal valences Y^{3+} and Ba^{2+} are not at issue. However, the assignment of formal valences to the other atomic species does require definition.

The covalent hybridization between Cu-3d and O-2p orbitals is strong. In the oxygen coordinations occurring in the $YBa_2Cu_3O_{7-\delta}$ system, the copper d_{z^2} and $d_{x^2-y^2}$ orbitals (x,y,z axes are the crystallographic axes) σ-bond with the nearest-neighbor oxygen atoms; they are orthogonal to the O-p_π orbitals (we neglect the displacement of the Cu(2) atoms from the oxygen plane). Conversely, the Cu-3d orbitals d_{xy}, d_{yz}, d_{zx} π-bond with the nearest-neighbor oxygen atoms; they are orthogonal to the O-p_σ and O-s orbitals. Consequently the crystal-field orbitals at the copper atoms become

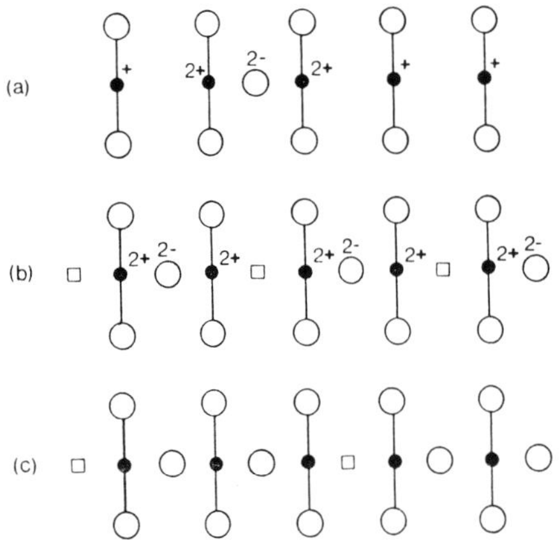

Fig. 6. Inserted bridging oxygen in Cu(1) plane: (a) isolated species, (b) b-axis ordering at $O_{6.5}$, (c) b-axis ordering at $O_{6.67}$.

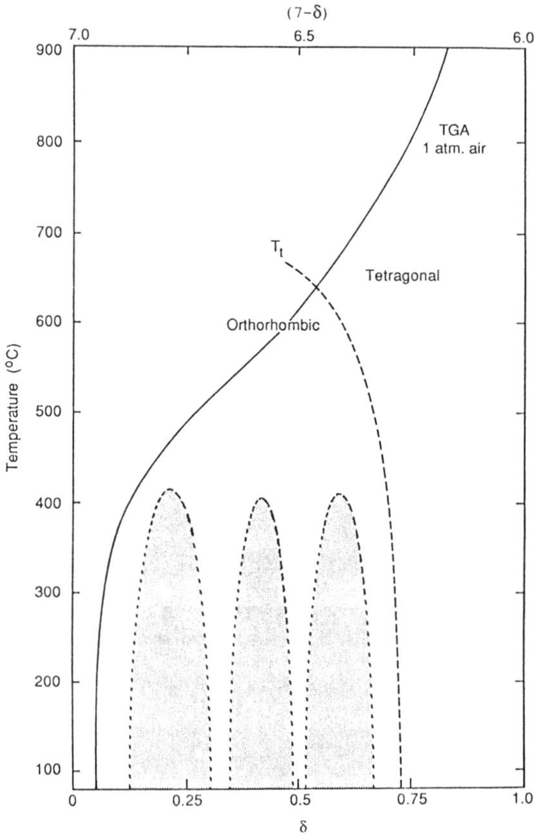

Fig. 7. Schematic phase diagram for the system $YBa_2Cu_3O_{7-\delta}$ and the TGA curve, solid line, for 1 atm air at 1 °C/min.

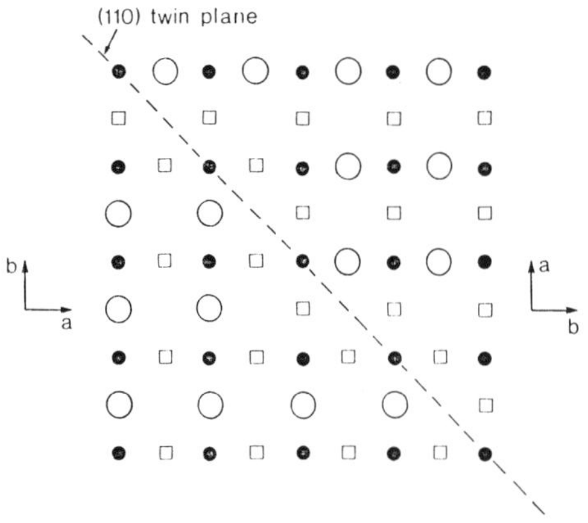

Fig. 8. Orthorhombic $YBa_2Cu_3O_{7-\delta}$ fully oxidized except at twin planes.

$$\Phi_{z^2} = N_\sigma(f_{z^2} - \lambda_s\Phi_s - \lambda_\sigma\Phi_\sigma)$$

$$\Phi_{x^2-y^2} = N_{\sigma'}(f_{x^2-y^2} - \lambda_{s'}\Phi_{s'} - \lambda_{\sigma'}\Phi_{\sigma'})$$

$$\Phi_{\mu\nu} = N_\pi(f_{\mu\nu} + \lambda_\pi\Phi_\pi),\ \mu\nu = xy,\ yz,\ \text{or}\ zx$$

where the N_i are normalization constants, the f_i are the atomic Cu-3d orbitals, the Φ_i are appropriately symmetrized O-2s, O-2p_σ, or O-2p_π orbitals, and the $\lambda_i \equiv |b_{ca}|_i/\Delta\varepsilon_i$ are the covalent-mixing parameters associated with a particular Cu-O transfer-energy matrix element (resonance integral) b_{ca} for a virtual electron transfer from an occupied to an empty state separated by an energy $\Delta\varepsilon$. The crystal-field orbitals at an oxygen atom have a corresponding admixture of Cu-3d character.

Interactions between crystal-field orbitals give rise to itinerant-electron bands. The dominant in-plane interactions between Cu(2) atoms are 180 ° Cu-O-Cu interactions having tight-binding bandwidths

$$w_{\sigma'} \sim \varepsilon_{\sigma'}\lambda_{\sigma'}^2, \quad \text{and} \quad w_\pi \sim \varepsilon_\pi\lambda_\pi^2$$

where $\varepsilon_{\sigma'}$ and ε_π are one-electron energies. A $\lambda_{\sigma'} > \lambda_\pi$ makes $w_{\sigma'} > w_\pi$.

What makes the copper oxides a special case is the positioning of the relative energies of the Cu-3d and O-2p_σ orbitals. The $Cu^{2+/+}$ redox couple lies above the O-2p energy, so covalent mixing raises the energy of the σ-bonding, copper crystal-field orbitals and lowers the energy of the corresponding O-2p_σ orbitals. The antibonding σ^* band is therefore composed of orbitals of primarily Cu-3d parentage; the top of the band derives primarily from the $\Phi_{x^2-y^2}$ orbitals where $\lambda_{\sigma'} >> \lambda_\sigma$ holds. On the other hand, the O-2p_π energies appear to lie above the π-bonding Cu-3d energies; in this event, covalent mixing lowers the energy of the π-bonding copper crystal-field orbitals and raises the energy of the corresponding O-2p_π orbitals. The antibonding π^* band would thus be composed of orbitals of primarily O-2p_π parentage. Nevertheless, this situation need not create a potential ambiguity in the definition of formal valence so long as a $w_{\sigma'} > w_\pi$ places all the holes associated with a formal $Cu^{2+/+}$ redox state in the antibonding σ^* orbitals of primarily Cu-3d parentage. This situation appears to hold so long as the oxidation involves only a $Cu^{2+/+}$ redox couple and not the $Cu^{3+/2+}$ redox couple. Therefore we refer to the Cu(I) and Cu(II) valence states in association with oxide ions O^{2-} wherever the operative redox couple is $Cu^{2+/+}$. Jahn-Teller distortions, magnetic moments, and spectroscopic data for copper oxides all attest to a meaningful Cu^{2+} formal valence, so we retain the formal valence Cu(II) for the Cu(2) atoms over the entire compositional range $O_{6.96}$ to O_6.

On the other hand, the apparent correlation splitting between the $Cu^{2+/+}$ and $Cu^{3+/2+}$ redox couples, as determined by photoelectron spectroscopy (ref. 8), is large. This observation suggests that the $Cu^{3+/2+}$ redox energy may lie below the O-2p_σ level, which would change the character of the antibonding σ^* orbitals from primarily Cu-3d parentage to primarily O-2p_σ parentage as the oxidation of a $[CuO_{3-\delta}]^{3-}$ layer increases in the range $O_{6.96}$ to $O_{6.5}$. In addition, the Fermi energy may begin to enter the band of π^* orbitals (ref. 9); the holes in the π^* orbitals are associated with orbitals of primarily O-2p parentage. In this situation, assignment of formal valences to the Cu(1) atoms and their near-neighbor oxygen atoms becomes totally ambiguous; it is more meaningful simply to identify the distribution of the holes between the two partially filled, overlapping π^* and σ^* bonds of a $(CuO_{3-\delta})^{3-}$ layer.

For $YBa_2Cu_3O_{7-\delta}$, the total hole concentration performula unit in the σ^* and π^* bands of the $(CuO_{3-\delta})^{3-}$ layers is $2(1-\delta)$ since each inserted oxygen atom has two holes in its 2p shell. An ambiguous distribution of the holes between the two bands forces introduction of the parameter $p(\delta)$ to describe the mole fraction of holes in the π^* bands; the hole distribution is then given by $\pi^{*p}\sigma^{*2(1-\delta)-p}$ per Cu(1) atom. For $\delta < 0.5$ a $p = 0$ and a $Cu^{2+/+}$ redox energy above the O-2p_σ level makes meaningful a conventional average formal valence state per Cu(1) atom of $[1 + 2(1-\delta)]+$. However, for $\delta < 0.5$ the definition of formal valence becomes ambiguous.

An experimental test of the formal-valence ambiguity would be the trapping out of holes at oxygen-atom clusters. A peroxide ion $(O_2)^{2-}$, for example, traps two holes in its antibonding orbitals. Trapping out of holes on oxygen-atom clusters would, in effect, increase p and cause transfer of electrons to the σ^* bands. With sufficient electron transfer, a formal valence on the copper may be meaningfully restored.

The apparent absence of peroxide-ion formation in $YBa_2Cu_3O_{6.96}$--as normally prepared from tetragonal $YBa_2Cu_3O_{6+\varepsilon}$, $\varepsilon < 0.25$, by oxygen insertion--demonstrates that there is no peroxide formation so long as the copper retain a coplanar coordination. We may conclude from this observation that the terminal oxygen atoms of a b-axis chain, the O(1) of Fig. 1, do not participate in O-O bonding; presumably they retain a formal valence O^{2-}. On the other hand, the bridging oxygen are prevented from making direct contact with one another by the ordering that occurs. The inserted, bridging oxygen are disordered only at low oxidation levels, $(7-\delta) < 6.5$, where formal valence states are meaningful.

This line of argument leads to the following testable prediction:

> Where the oxygen concentration in a $YBa_2Cu_3O_{7-\delta}$ phase--or in a cation-substituted isostructural phase--is high enough to create π^*-band holes ($p > 0$), there these holes will be trapped out as peroxide ions $(O_2)^{2-}$ wherever two Cu(1)-bridging oxygen atoms make direct contact with one another.

Two types of experiments have been performed that provide a confirmation of this prediction: low-temperature preparations of tetragonal phases containing higher oxygen concentrations, $(7-\delta) > 6.5$, and cation substitutions that induce oxygen concentrations in excess of O_7.

LOW-TEMPERATURE PREPARATIONS

In order to prepare $YBa_2Cu_3O_{7-\delta}$ phases in particle sizes < 0.5 μm, it is necessary to decompose precursors mixed in stoichiometric proportions at temperatures too low for sintering. This procedure has led to tetragonal phases having an oxygen content $(7-\delta) > 6.5$ (refs. 10 & 11). In view of the rapid diffusion of oxide ions O^{2-} inserted in $YBa_2Cu_3O_{7-\delta}$, the synthesis of a tetragonal phase having $(7-\delta) = 6.7$ at 780 °C (ref. 11) clearly indicates that the bridging oxygen may have a quite different character if present in high concentration and also disordered on the a-axis and b-axis bridging sites.

The system $YBa_2Cu_3O_{7-\delta}$ is commonly prepared from a mixture of $BaCO_3$, Y_2O_3, and CuO; this mixture must be fired above 900 °C to remove the CO_2 from $BaCO_3$. At this temperature a $(7-\delta) < 6.25$ represents an oxidation state where the formal valence O^{2-} is meaningful, and no oxygen clustering occurs. On lowering the temperature in air or oxygen, b-axis ordering of the bridging oxygen below an orthorhombic-tetragonal transition occurs at a $(7-\delta) \approx 6.5$, an oxidation state where formal valence states are still meaningful. At higher oxygen concentrations, ordering of the bridging oxygen prevents their contact even at twin planes as discussed above.

In a low-temperature preparation, on the other hand, the bridging oxygen are initially present in a disordered manner, which allows direct contact between them. If the initial concentration is high, $(7-\delta) > 6.5$, then O-2p holes may be trapped at oxygen clusters such as a peroxide $(O_2)^{2-}$ ion. A molecular species would be much less mobile than a monomeric O^{2-} ion, so the oxygen is retained to higher temperatures. Transfer of electrons back to the σ^* bands reduces the copper to a meaningful Cu(II) valence state, and the system is semiconducting rather than superconducting.

Transformation of a tetragonal, semiconducting phase prepared at temperature $T \leq 780$ °C to the orthorhombic superconducting phase can be achieved by first removing the "paired" oxygen at temperatures $T > 810$ °C in air or, to prevent sintering, at 750 °C in N_2--and then, by reannealing in O_2 at 400 °C, reinserting monomeric oxygen atoms. The monomeric oxygen order--as in a conventional preparation--before their concentration is high enough for O-2p holes to be introduced (ref. 11).

CATION SUBSTITUTIONS

Three types of cation substitutions have been investigated: (1) isovalent substitutions for Y or Ba, (2) aliovalent substitutions for Y or Ba, and (3) substitutions for copper.

Isovalent substitutions for Y or Ba.

Immediately after identification of the $YBa_2Cu_3O_{6-\delta}$ phase, substitution of the Ln^{3+} lanthanides for Y^{3+} was investigated (refs. 12 & 13). Even the magnetic lanthanides were found to have a negligible effect on the superconducting transition temperature $T_c \approx 90$ K. In view of the sensitivity of T_c to the oxygen concentration, this observation leads to two conclusions: (1) the equilibrium oxygen concentration is not appreciably changed by these substitutions and (2) the rare-earth ions are isolated from the Cooper pairs responsible for superconductivity. This latter conclusion is consistent with confinement of both the redox reactions and the metallic conduction to the Cu(1) planes.

On the other hand, substitution of the non-magnetic Sr^{2+} ion for Ba^{2+} causes T_c to decrease with increasing Sr^{2+}-ion concentration (ref. 14). An analysis of the oxygen concentration in the system $YBa_{2-x}Sr_xCu_3O_{7-\delta}$ showed that δ is independent of x for a given thermal treatment (ref. 15). This observation shows that the counter cations adjacent to the $[Cu(1)O_{3-\delta}]^{3-}$ layer have an influence on the band structure where Cooper pairs are formed and that a higher T_c is associated with the more basic (ionic) counter cation.

Aliovalent substitutions for Y or Ba.

Investigations of M^{4+}-ion substitutions for Y^{3+} need to be repeated and refined.

An investigation (ref. 16) of the system $Y_{1-x}Ca_xBa_2Cu_3O_{7-\delta}$ $(0 \leq x < 0.3)$ showed that, after annealing in oxygen in the range $350 < T < 450\ ^\circ C$, the oxygen concentration of the Cu(1) layer varies as $(CuO_{3-\delta_0-0.5x})^{(3-x)-}$ for all x, where δ_0 is the equilibrium value for x = 0 at a particular temperature and partial pressure of oxygen. This observation shows that the "formal oxidation state" of the Cu(1) atoms--if all oxygen are taken to be O^{2-} ions and the Cu(2) remain "Cu^{2+}"--is independent of x; only the oxygen-vacancy concentration $[V_O]$ within the $(CuO_{3-\delta})^{3-}$ chains is changing. Measurement of T_c gave a decrease with x that varied as

$$(\partial T_c/\partial(0.5x))_{[h]=1.9} = (\partial T_c/\partial[V_O])_{[h]=1.9} \approx -180\ K$$

for a fixed hole concentration $[h] \approx 1.9$, i.e. $\delta_0 \approx 0.05$, in the conduction bands of the Cu(1) plane.

An investigation (ref. 17) of La substitution for Ba in the system $YBa_{2-x}La_xCu_3O_{7\pm\delta}$ $(0 \leq x < 0.5)$ showed a similar tendency to retain compositional domains over which the equilibrium "formal oxidation state" of the Cu(1) atoms remains constant. In this system, the oxygen concentration of the Cu(1) layers varies as $(CuO_{3-\delta_0+0.5x})^{(3+x)-}$ for a given temperature and partial pressure of oxygen. Three domains could be distinguished: Domain I $(0 \leq x < 0.05)$ was narrow; both T_c and the orthorhombic b/a ratio increased with increasing x, and a $\delta_0 \approx 0.05$ is similar to that found in the $Y_{1-x}Ca_xBa_2Cu_3O_{7-\delta}$ system. In this domain the oxygen atoms remain monomeric. Domain III $(0.10 \leq x < 0.5)$ was extensive; both T_c and the orthorhombic b/a ratio decreased with increasing x, see Fig. 9, and a $\delta_0 \approx 0.15$ indicated that the equilibrium oxygen concentration had been changed. Domain III can therefore represent a second crystallographic phase with Domain II $(0.05 \leq x < 0.1)$ corresponding to a two-phase region separating the phases in Domain I and Domain III. A straightforward interpretation of the properties of the phase of Domain III could be obtained by introducing the possibility of peroxide-ion formation between a-axis and b-axis oxygen. With this assumption, the Cu(1) layer is represented formally as

$$[Cu^{n+}O^{2-}_{3\pm\delta-y}(O_2)^{2-}_{0.5y}]^{(3+x)-}$$

$$n = 3-y-x\pm2\delta$$

where a $|\delta_0| = 0.15$ makes $n = 2.7-y$. If x_o is the critical concentration for the generation of peroxide ions, then $y = (x-x_o)$ and

$$n = (2.7+x_o) - x$$

In the high-T_c superconducting copper oxides, the appearance of superconductivity correlates well with a formal oxidation state on the copper that is greater than Cu^{2+}. Extrapolation of T_c vs x, Fig. 10, gives $T_c = 0$ at $x_c = 0.77$. Substitution of $x_c = 0.77$ for x at n = 2 in the above expression gives $x_o = 0.07$, which is in the middle of the two-phase region, Domain II. Were the samples prefectly homogeneous, the upper bound of the y = 0 phase would, according to the calculation, be $O_{6.98}$, which is consistent with what is permissible according to the sparse data for the twin-plane density.

The significance of this analysis is that it offers clear evidence for a T_c that varies linearly with the "formal oxidation state" on the Cu(1) atoms; it is independent of an orthorhombic-tetragonal transition at x = 0.3, Fig. 9. Moreover, the data provide striking evidence for the formation of peroxide ions where a-axis and b-axis oxygen make contact.

Substitutions for copper

Numerous investigations of cation substitutions for copper have been reported. All such substitutions decrease T_c whether or not the substituent carries an atomic magnetic moment.

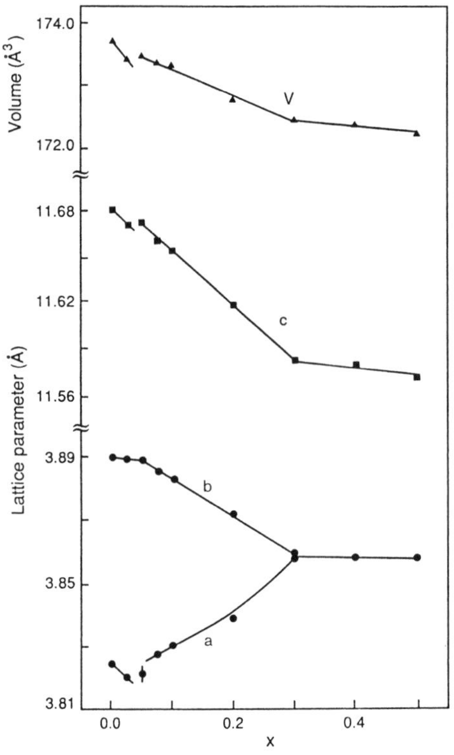

Fig. 9. Variation of lattice parameters and unit-cell volume with composition x of oxygen-annealed $YBa_{2-x}La_xCu_3O_{7\pm\delta}$.

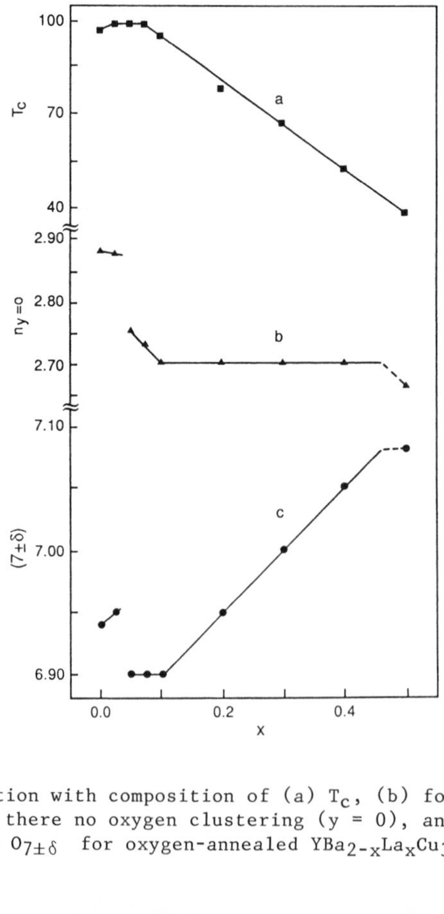

Fig. 10. Variation with composition of (a) T_c, (b) formal oxidation state of Cu(1) were there no oxygen clustering (y = 0), and (c) the total oxygen content $O_{7\pm\delta}$ for oxygen-annealed $YBa_{2-x}La_xCu_3O_{7\pm\delta}$.

An investigation (ref. 18) of the compounds representing the limiting substitution of Co or Fe for Cu in $YBa_2Cu_3O_{7-\delta}$ gives additional evidence of peroxide (or oxygen clustering) formation where a-axis and b-axis oxygen make contact.

One cobalt per formula unit can be substituted for copper. After annealing at 400 °C and slow cooling in O_2, the composition is $YBa_2Cu_2CoO_{7.25}$; it is tetragonal and semiconducting. Single-crystal x-ray structural analysis located the Co at the Cu(1) positions of the structure. Comparison of the Cu(2)-O(2) bond lengths in the Cu(2) plane with those observed for $YBa_2Cu_3O_6$ indicates that the Cu(2) atoms remain Cu^{2+}, which is consistent with the semiconducting character of the compound. The tetragonal structure shows that the 1.25 inserted oxygen per formula unit randomly occupy a-axis and b-axis sites, which places them in direct contact with one another. Large thermal parameters associated with the cobalt atoms and their nearest-neighbor oxygen atoms were consistent with random adjustments of the atomic positions about cobalt in fourfold or fivefold oxygen coordination as well as oxygen-atom clustering. If all the inserted oxygen form peroxide ion pairs $(O_2)^{2-}$, then the chemical formula may be written as $Ba_2Cu_2CoO_6(O_2)_{0.625}$, which would allow all the octahedral-site cobalt to have the formal valence state Co^{3+}, all the tetrahedral-site cobalt to be Co^{2+}; these coordinations are satisfactory from the point of view of site-stabilization energies. Although such a cobalt layer has mixed valence, the Co^{3+} and Co^{2+} ions occupy inequivalent lattice sites, which inhibits charge transfer.

A similar analysis for $YBa_2Cu_{2.5}Fe_{0.5}O_{7.19}$ leads to the formulation $YBa_2Cu_{2.5}Fe_{0.5}O_6(O_2)_{0.6}$ with Fe^{3+} ions in tetrahedral sites in the Cu(1) plane.

COOPER PAIRING

The identification of oxygen-atom clustering, or peroxide-ion formation, between inserted oxygen atoms at higher oxidation states signals the presence of holes in orbitals of primarily O-2p character. As pointed out above, such a situation should be anticipated first for the π-bands. An important correlation splitting between the $Cu^{2+/+}$ and $Cu^{3+/2+}$ levels is implied by the semiconducting, antiferromagnetic character of the $YBa_2Cu_3O_6$ composition (ref. 18); it has Cu(2) atoms with a Cu^{2+} valence state, and increasing the "formal oxidation state" of the Cu(1) atoms beyond Cu^{2+}, i.e. $(7-\delta) > 6.5$ in $YBa_2Cu_3O_{7-\delta}$, would cause the Fermi energy to move into the π^* band of the $[Cu(1)O_{3-\delta}]^{3-}$ layer (ref. 9). As argued above, this situation leads to the hole configuration

$$\pi^{*p}\sigma^{*2(1-\delta)-p}$$

for the Cu(1)-layer π^* and σ^* bands, where here p is the number holes per formula unit in the π^* bands. Note that an increase in the hole concentration p in the π^* bands of primarily O-2p parentage results in a corresponding increase in the electron concentration in the σ^* bands of primarily Cu(1)-3d parentage. This reciprocal relationship is capable of introducing polarization fluctuations that enhance the Bardeen-Cooper-Schrieffer (BCS) pairing potential

$$V_{BCS} = V_C - U$$

where V_C is the Cooper-pair attractive potential and U is the electrostatic repulsive energy between paired electrons. An increase in V_{BCS} increases the BCS

$$T_c \approx 1.14\, \Theta_D \exp(-1/V_{BCS}N(0))$$

where Θ_D is the Debye temperature and N(0) is the density of one-electron states at the Fermi energy at T = 0 K.

The enhancement arises from the fact that the relative strengths of the π and σ components of a Cu-O bond vary with the bond distance R, which makes p = p(R). Therefore, coupling of a pair of σ^* electrons to a phonon may be enhanced by an associated increase in p, thus increasing V_C. Moreover, the interrelationship between σ^* electrons and π^* holes would serve to reduce the electrostatic repulsion between the paired electrons, thus decreasing U.

The band calculation of de Groot et al. (ref. 19) indicates the presence of "valence" fluctuations that can enhance T_c, and their view of the O-2p character of the states at the top of the π^* band appears to be similar to the one presented here. Yu et al. (ref. 9) have also emphasized the role of charge fluctuations between σ^* and π^* bands, but without introducing an O-2p parentage for the states at the top of the π^* band.

The authors gratefully acknowledge the R.A. Welch Foundation, Houston, Texas, for partial support of this research.

REFERENCES

1. J.S. Swinnea and H. Steinfink, J. Mat. Res. 2, 424 (1987).
2. W.I.F. David, W.T.A. Harrison, J.M.F. Gunn, O. Moze, A.K. Soper, P. Day, J.D. Jorgensen, D.G. Hinks, M.A. Beno, L. Soderholm, D.W. Capone II, I.K. Schuller, C.U. Segre, K. Zhang, J.D. Grace, Nature 327, 310 (1987).
3. A. Manthiram, J.S. Swinnea, Z.T. Sui, H. Steinfink, and J.B. Goodenough, J. Am. Chem. Soc. 109, 6667 (1987).
4. (a) M.O. Eatough, D.S. Grimley, B. Morosin, E.L. Venturimi, Appl. Phys. Lett. 51, 367 (1987).
(b) P.P. Fereitas and T.S. Plaskett, Phys. Rev. B36, 5723 (1987).
5. J.D. Jorgensen, M.A. Beno, D.G. Hink, L. Solderholm, K.J. Volin, R.L. Hitterman, J.D. Grace, I.K. Schuller, C.U. Segre, K. Zhang, M.S. Kleefisch, Phys. Rev. B36, 3608 (1987).
6. R.J. Cava, B. Batlogg, C.H. Chen, E.A. Rietman, S.M. Zahurak, D. Werder, Nature 329, 423 (1987); Phys. Rev. B36, 5719 (1987).
7. J.B. Goodenough, Progress in Solid State Chemistry 5, 145 (1972).
8. J.C. Fuggle, P.J. W. Weijs, R. Schoorl, G.A. Sawatzky, J. Fink, N. Nücker, P.J. Durham, W.M. Temmerman, Phys. Rev. B. (in press).
9. J. Yu, S. Massidda, and A.J. Freeman, Phys. Lett. A 122, 203 (1987).
10. C.C. Torardi, E.M. McCarron, M.A. Subramanian, M.S. Horowitz, J.B. Michel, A.W. Sleight, and D.E. Cox, Chemistry of High-Temperature Superconductors, ACS Symposium Series 351, D.L. Nelson, M.S. Whittingham, and T.F. George, eds., p. 152, Am. Chem. Soc., Washington, D.C., (1987).
11. A. Manthiram and J.B. Goodenough, Nature 329, 701 (1987).
12. P.H. Hor, R.L. Meng, Y.Q. Wang, L. Gao, Z.J. Huang, J. Bechtold, K. Forster, and C.W. Chu, Phys. Rev. Lett. 58, 1891 (1987).
13. J.M. Tarascon, L.H. Greene, B.G. Bagley, W.R. McKinnon, P. Barboux, and G.W. Hull, in Novel Superconductivity, S.A. Wolf and V. Kreshnin, eds., p. 705, Plenum Press, N.Y., London, (1987).
14. B.W. Veal, W.K. Kwok, A. Umezawa, G.W. Crabtree, J.D. Jorgensen, J.W. Downey, L.J. Nowicki, A.W. Mitchell, A.P. Panlikas, C.H. Showers, Appl. Phys. Lett. 51, 279 (1987).
15. A. Manthiram, unpublished.
16. A. Manthiram, S.J. Lee, and J.B. Goodenough, J. Solid State Chem. (in press).
17. A. Manthiram, X.X. Tang, and J.B. Goodenough, submitted.
18. Y.K. Tao, J.S. Swinnea, A. Manthiram, J.S. Kim, J.B. Goodenough and H. Steinfink, J. Mat. Res. (in press).
19. R.A. de Groot, H. Gutfreund, and M. Weger, Solid State Commun. 63, 451 (1987).

Are one-dimensional structural features essential for superconductivity at 90 K?

A.Simon, H.Borrmann, Hj.Mattausch, W.Bauhofer and R.Kremer

Max-Planck-Institut fuer Festkoerperforschung,
D-7000 Stuttgart 80, FRG

Abstract - We propose a model for the electron pair formation in high-T_c superconducting oxocuprates which is based on Cu^+/Cu^{3+} valence fluctuations made possible by the configurational instability of O^{2-}. The mechanism seems optimized for the special arrangement of Cu and O in the CuO_{3-x} ribbons occurring in $YBa_2Cu_3O_{7-x}$ and related "123" phases, as they contain (chemically) highly oxidized Cu in a (structurally) preformed coordination of the reduced state. $La_{3-x}Ba_{3+x}Cu_6O_{14+\delta}$ ("336") also contains such ribbons. $La_{2-x}Sr_{1+x}Cu_2O_6$ which only contains twin layers (as in "123" phases) but no CuO_3 ribbons does not become superconducting.

The crystal structures of the novel high-T_c oxocuprates belong to different types which have certain features in common, but also exhibit essential differences. $La_{1.8}M_{0.2}CuO_4$ (M=Ba,Sr; T_c 30 to 40 K) (ref.1,2) contains copper in an average oxidation state +2.2. The coordination octahedron around Cu is tetragonally elongated (d(Cu-O)=190 pm (4x) and 240 pm (2x)). Taking the drastic differentiation of the Cu-O distances into account, the structure contains CuO_2 layers formed from corner-sharing CuO_4 squares according to $^2_\infty[CuO_{4/2}]$ which are sandwiched between rigid slabs of composition $^2_\infty[(La,M)_2O_2]$. Obviously, the layers are only weakly coupled to the surrounding slabs.

In the structure of $YBa_2Cu_3O_{7-x}$ ($T_c \approx 90$ K for $x \to 0$) (ref.3,4) the same kind of CuO_2 layers pairwise surround the Y atoms. The CuO_4 squares are orthorhombically distorted and the layers are buckled. These structural details might indicate that the CuO_2 layers in $YBa_2Cu_3O_{7-x}$ are not responsible for the 90 K superconductivity, in particular, as the replacement of yttrium near the layers by magnetic rare earth ions does not suppress superconductivity. Between the slabs of composition $^2_\infty[YCu_2O_4]$ structural units Ba_2CuO_{3-x} are inserted which contain parallel ribbons $^1_\infty[CuO_2O_{2((1-x)/2)}]$. We associate the origin of the high-T_c of $YBa_2Cu_3O_{7-x}$ with these ribbons. They consist of linear CuO_2 units (d(Cu-O) = 185 pm) which are bridged by oxygen atoms (O1) resulting in square planar Cu coordination with a rhomboidal distortion of the squares (d(Cu-O1) = 194 pm). It seems important that the ribbons exhibit crystallographic m m m symmetry and are again weakly coupled to the slabs via O-Cu contacts near 230 pm. The oxygen deficiency in the structure is restricted to the O1 atoms and determines the oxidation state of Cu in the ribbons. For x=0 these Cu atoms are essentially trivalent. Removing the loosely bound O1 atoms leads to a lowering of T_c. For x=0.5 corresponding to Cu^{2+} the phase becomes semiconducting. Finally, for x=1.0 an array of isolated linear CuO_2 units with Cu^+ is left. This oxygen removal can be performed in a reversible topochemical reaction (ref.5).

Our hypothesis for the spin-pairing mechanism (ref.6) in superconducting $YBa_2Cu_3O_{7-x}$ is based on a dynamical simulation of the topochemical redox process in terms of a phonon mediated valence fluctuation of Cu between the oxidation states +1 and +3. For that purpose a lattice vibration is needed which changes the environment of the Cu atoms between the characteristic coordinations for Cu^+ and Cu^{3+}.

All established structures of oxocuprates contain Cu^{3+} (isoelectronic with Ni^{2+}!) in a square (or rectangular) planar coordination (ref.7,8,9) and Cu^+ in a linear coordination (ref.10,11). It is important that the Cu-O distances in both coordinations are essentially the same (d(Cu-O) $\approx$ 184 pm) and slightly shorter than in oxocuprates(II) (d(Cu-O) $\geq$ 189 pm). Simple valence bond theory takes account of the bonding in terms of a dsp^2 hybrid for the (low-spin) d^8 system of Cu^{3+} and in terms of a sp hybrid for the d^{10} system of Cu^+.

The fully occupied CuO_3 ribbon in $YBa_2Cu_3O_7$ chemically contains Cu in an oxidation state near +3. Yet, the CuO_2 dumbbell characteristic for Cu^+ is preformed structurally (d(Cu-O) = 185 pm (2x) and 194 pm (2x)). The pairwise increase of d(Cu-O1) around one Cu site enhances the tendency towards a d^{10} configuration simultaneously favoring a d^8 configuration for the adjacent Cu atom in the ribbon (Fig.1). The indicated zone edge mode should result in a pairwise attraction of electrons via strong electron-phonon coupling. The valence fluctuation of copper is assisted by the configurational instability of the oxygen dianion.

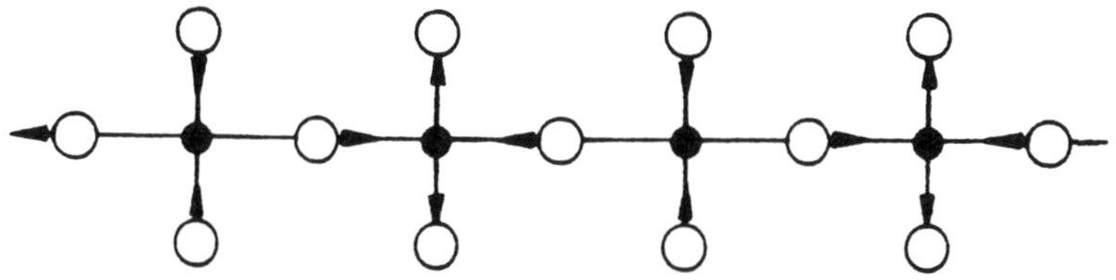

Fig. 1. CuO_3 ribbon in $YBa_2Cu_3O_7$ and indicated lattice vibration which leads to a tendency for d^{10} and d^8 configurations of adjacent Cu atoms.

It has been shown that the polarizability of the O^{2-} ion in oxides changes dramatically with the anion-cation distance ($\Delta\alpha \sim \Delta d^{12}$ in case of pronounced anisotropy) (ref.12) which in case of the CuO_3 ribbon leads to a dynamic delocalization of an oxygen 2p-electron to the more distant Cu atom. Simultaneously this Cu atom has the tendency to localize two electrons in the d^{10} configuration. In the case of the CuO_2 layer in $La_{1.8}M_{0.2}CuO_4$ the resulting electronic resonant state can be associated with a mode that leads to a quadrupolar distortion of the Cu environment, changing c.n.=4 to 2+2. The CuO_3 ribbon in $YBa_2Cu_3O_7$ meets the structural requirements for the discussed valence fluctuation of Cu more efficiently. In fact, recent ^{63}Cu nuclear spin-lattice relaxation measurements on $YBa_2Cu_3O_{7-x}$ (ref.13) reveal substantially different pairing energies for electrons in the layers ($2\Delta=2.4\ kT_c$) and ribbons ($8.3\ kT_c$).

From the above the onset of superconductivity around 90 K seems intimately related to the presence of the one-dimensional CuO_{3-x} ribbons in $YBa_2Cu_3O_{7-x}$ and related "123" systems, where yttrium is replaced by other rare earth metals (ref.14).

It is claimed (ref.15) that a new class of high-T_c superconductors, $La_{3-x}Ba_{3+x}Cu_6O_{14+\delta}$ (abbreviated "336") shows an onset of superconductivity near 90 K, yet does not contain the characteristic ribbons of the "123" phases. In the following we present arguments against the structural conclusions drawn by Mitzi et al. (ref.15).

Furthermore, we report on investigations of the system $La_{2-x}Sr_{1+x}Cu_2O_{6-x/2}$ (ref.16) whose structure is characterized by the presence of CuO_2 twin layers as in "123" compounds but without the ribbons being present.

<u>Comments on the structure of "336" superconductors</u>

Mitzi et al. (ref.15) show that samples of $La_{3-x}Ba_{3+x}Cu_6O_{14+\delta}$ exhibit an onset of superconductivity at approx. 45 K (x=0), 55 K (x=0.25), 60 K (x=0.5), 85 K (x=0.75) and 90 K (x=1.0) (Note a). X-ray powder investigations reveal the samples with x=0.25 and 0.75 to be single phase, whereas the sample with x=1.0 contains $BaCuO_2$. The compositions of the single-phased samples

(a) $La_{1.38}Ba_{1.63}Cu_3O_{7+\delta/2}$ and

(b) $La_{1.13}Ba_{1.88}Cu_3O_{7+\delta/2}$

are rather close to the composition of the corresponding "123" phase, in particular for sample (b). From the X-ray powder diagrams of (a) and (b) it is concluded that essential differences exist for the structure types of "336" and "123". Whereas "123" contains CuO_2 twin layers and (O-deficient)

Note a: Others (ref.17) report T_c's around 40 K. Our sample $La_3Ba_3Cu_6O_{14+\delta}$ prepared by heating La_2O_3, $BaCO_3$ and CuO in a stoichiometric ratio at 1000°C (48 h, flowing O_2) and slowly cooling to 500°C under O_2 also shows a resistivity anomaly but no superconductivity at 40 K. According to Guinier diagrams the sample is single phase.

CuO_3 ribbons, "336" is said to contain only layers linked via additional oxygen atoms following the earlier suggestion of Er-Rakho et al. (ref.18).

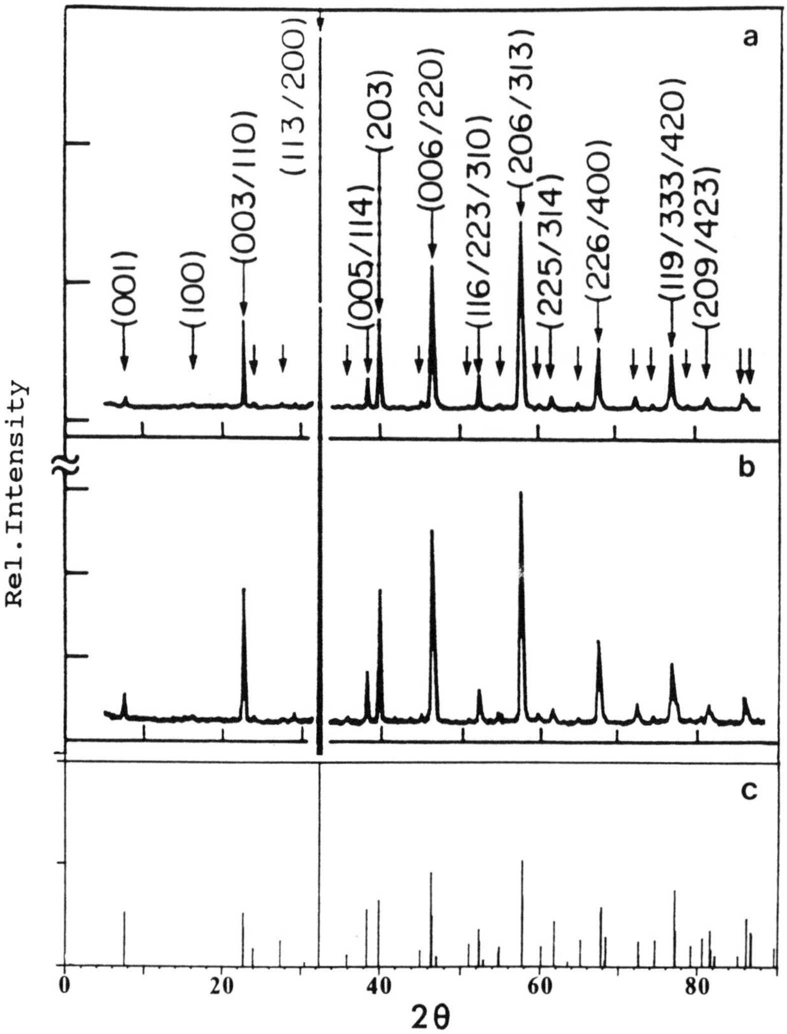

Fig.2 Published X-ray powder diagrams for $La_{3-x}Ba_{3+x}Cu_6O_{14+\delta}$ a) x = 0.25, b) x = 0.75. The arrows indicate the calculated positions of the peaks for the "336" structure (ref.15). c) Calculated powder diagram (ref.19) for $La(La,Ba)_2Cu_3O_7$. To enhance the intensity of the weak reflexions $(I(hkl))^{1/2}$ is plotted.

The oxygen content of the samples has not been analysed. The proposed model holds for δ=0 and the assumptions about the arrangement of oxygen atoms is entirely based on the X-ray powder diffraction results. One should remark that the contribution of the one oxygen atom in question to the total scattering intensity of the samples is less than 1%!

The X-ray diagrams of samples (a) and (b) are shown in Fig.2a,b. Very weak reflections, especially the reflection near $2\Theta \approx 16^{o}$ have been interpreted in terms of a superstructure calling for a unit cell twice as large as that of the "123" structure. The doubling is achieved by a special rearrangement of oxygen atoms around the Cu atoms near barium in the "123" structure. The relations between the "123" and the "336" structures are shown in Fig.3a,b. In "123" all Cu atoms in the z=0 plane are equivalent in which case the small unit cell with lattice vectors $\vec{a}$ and $\vec{b}$ is sufficient to describe the structure (Fig.3a). In the "336" structure the same Cu atoms are made non-equivalent by placing four and six oxygen atoms (square planar and octahedral coordination, respectively) around adjacent Cu atoms. Thus, the doubled unit cell with vectors $\vec{a}'$and $\vec{b}'$ is created (Fig.3b).

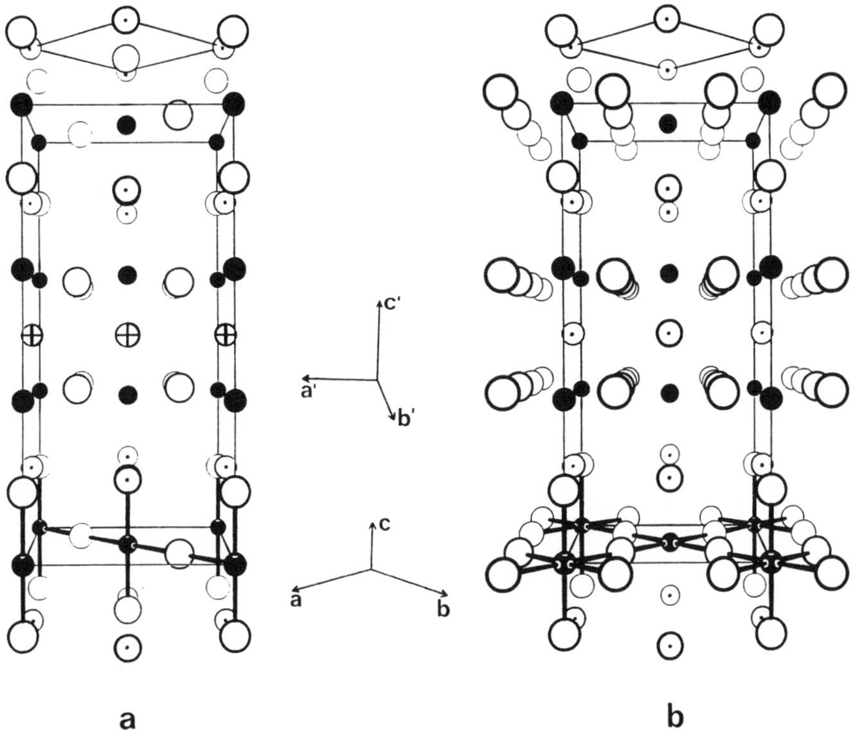

Fig.3(a) Doubled unit cell of $YBa_2Cu_3O_7$ with lattice vectors $\vec{a}'$, $\vec{b}'$, $\vec{c}'$.The lattice vectors of the conventional cell ($\vec{a}$, $\vec{b}$, $\vec{c}$; $\vec{c}=\vec{c}'$) are indicated. (b) Proposed crystal structure for $La_{3-x}Ba_{3+x}Cu_6O_{14}$ (ref.15). Cu: black, O: open circles, La,Ba: dotted circles and Y marked by a cross.

The calculation of scattering intensities using the "123" structure with Y replaced by La (orthorhombic symmetry, P m m m, and atomic coordinates taken from (ref.4); $a = b = a' \cdot 2^{1/2} = 391$ pm, $c = 1172$ pm taken from (ref.15)) leads to the result plotted in Fig.2c. Ba^{2+} and La^{3+} are indistinguishable to X-rays and therefore the deviations of the Ba:La ratio from two in samples (a) and (b) do not lead to an ambiguity for the calculated intensities.

It is evident that all weak reflections assigned to the "336" structure model by Mitzi et al. (ref.15) actually belong to the diagram of the "123"structure with just one exception. The small peak near $2\Theta \approx 16^o$ indexed as 100 should not occur with the "123" diagram. Unfortunately, the experimental conditions are not stated in (ref.15). Suppose the diagrams were taken with crystal-monochromatized radiation, then the $\lambda/_2$-reflections of the strongest 113/200 lines occur in exactly the observed position of 100.

In conclusion, from the experiments of (ref. 15) there is no experimental evidence for a new "336" structure type. In fact, recent neutron diffraction data of $La_3Ba_3Cu_6O_{14.39}$ confirm the isomorphism with the tetragonal variant of $YBa_2Cu_3O_{7-x}$ (ref.20).

$La_{2-x}Sr_{1+x}Cu_2O_6$, a phase with only CuO_2 twin layers.
H.Nguyen et al. describe a series of compounds with the general composition $La_{2-x}Sr_{1+x}Cu_2O_{6-x/2}$, for $0 \le x \le 0.14$ (ref.16).

The structure (Fig.4) results from an "intergrowth" of slabs of composition $^2_\infty[(La,Sr)_2O_2]$, as they occur in $La_{1.8}M_{0.2}CuO_4$, and twin layers $^2_\infty[(La,Sr)Cu_2O_4]$ which are one structural feature of $YBa_2Cu_3O_7$. The geometrical details of the twin layers, i.e. slight displacement of the Cu atoms out of the oxygen planes and the distances d(Cu-O) = 193 pm, are similar to those in $YBa_2Cu_3O_7$.

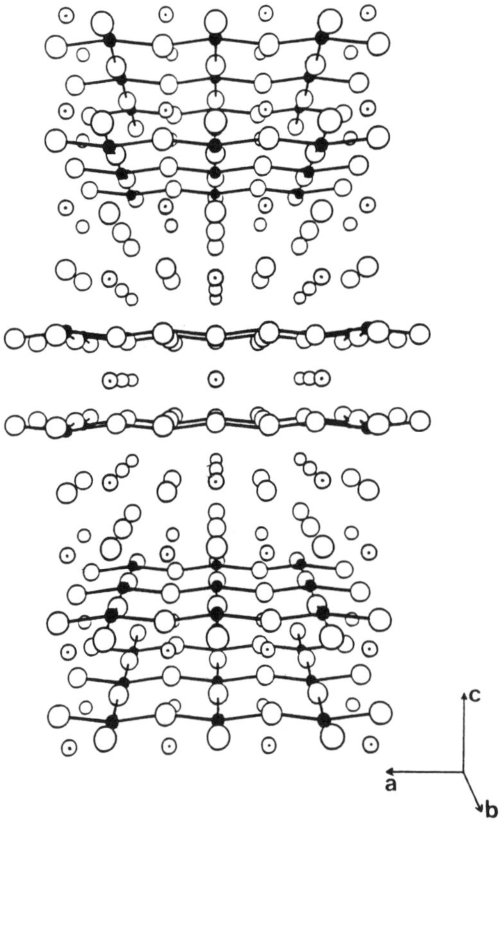

Fig. 4. Crystal structure of $La_{2-x}Sr_{1+x}Cu_2O_6$ (ref.16). The CuO_2 twin layers around La(Sr) are outlined.

In fact, the tetragonal and near 90 K superconducting $YBa_2Cu_3O_{7-x}$ prepared by heating single crystals of $YBa_2Cu_3O_6$ at 450°C (ref.21) exhibits identical Cu-O distances within the Cu-O twin layers. As in $La_{2-x}Sr_{1+x}Cu_2O_{6-x/2}$ no CuO_3 ribbons but only twin layers exist, the compounds can be taken to test whether these twin layers are responsible for 90 K superconductivity.

We have prepared $La_{2-x}Sr_{1+x}Cu_2O_{6-x/2}$ phases within the range $0.2 \leq x \leq 0.5$, see Table 1 (note b). All products are black and highly conducting which indicates the presence of Cu^{3+} in agreement with the results of iodometric titration for such samples (ref.16). For a sample with x=0.1 a content of 3% Cu^{3+} is found which nearly corresponds to the composition $La_{1.9}Sr_{1.1}Cu_2O_6$. Strangely, the formula $La_{2-x}Sr_{1+x}Cu_2O_{6-x/2}$ does not take account of any Cu^{3+}, as the replacement of La^{3+} by Sr^{2+} is compensated by a corresponding number of oxygen vacancies to keep all copper in the oxidation state +2. Our analyses of samples with x=0.2 and 0.3 (by heating in H_2 at 950°C and measurements of both weight loss and amount of formed H_2O) show that the general formula $La_{2-x}Sr_{1+x}Cu_2O_6$ holds. Annealing the samples at 350°C in 30 bar O_2 does not change their oxygen content.

TABLE 1 Characterization of samples aimed at a general composition $La_{2-x}Sr_{1+x}Cu_2O_6$ (see Note b)

x	0.2	0.2	0.3	0.5	1.0
Guinier diagram	single phase: a= 385.85 (2)pm c= 2000.2 (2)pm	single phase: a= 385.92 (2)pm c= 1997.4 (1)pm (prepared under 30 bar O_2)	single phase: a= 385.05 (4)pm c= 2010.2 (5)pm	contamination by $SrCuO_2$ and other unidentified phases a= 385.58(2)pm c= 2001.4(4)pm	mainly $SrCuO_2$ and other unidentified phases; traces of $La_{2-x}Sr_{1+x}Cu_2O_6$
specific resistivity $\rho_{300}[\Omega cm^{-1}]$	~ 10^{-2}	~ 10^{2-}	~ 10^{-2}	~ 10^{-2}	10^2

Note b: Stoichiometric amounts of La_2O_3 (4N, preheated at 600°C, 10^{-5} Torr), $SrCO_3$ (p.a. quality) and CuO (p.a.) were reacted according to (ref.16). Finally the samples were kept at 500°C in 1 bar O_2 for 3h and a different one at 350°C in 30 bar O_2 for 18h.

The temperature dependencies of the electrical resistivity and magnetic susceptibility of $La_{1.8}Sr_{1.2}Cu_2O_6$ are shown in Fig.5. With decreasing temperature the resistance increases monotonically. Only at approx. 30 and 20 K small discontinuities are apparent. Due to the magnetization only the 30 K effect could be attributed to a superconductivity minority phase e.g. $La_{2-x}Sr_xCuO_4$ at an impurity level of approximately 100 ppm. The measurements on $La_{1.8}Sr_{1.2}Cu_2O_6$ as well as the other samples of Table 1 reveal no transition to bulk superconductivity down to 5 K.

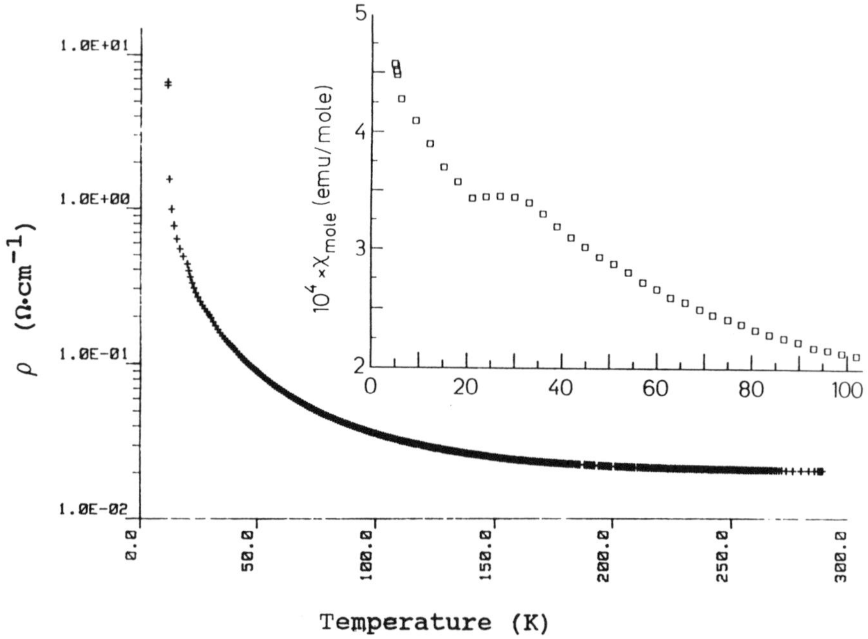

Fig.5. Electrical resistivity $\rho[\Omega{\cdot}cm^{-1}]$ and magnetic susceptibility χ_{mol} $(emu{\cdot}mol^{-1})$ of $La_{1.8}Sr_{1.2}Cu_2O_6$.

Conclusion

The oxidation state of Cu in the twin layers of (metallic) $La_{2-x}Sr_{1+x}Cu_2O_6$ can be widely varied and therefore adjusted to the value for Cu in the twin layers of $YBa_2Cu_3O_{7-x}$. Yet, the phases $La_{2-x}Sr_{1+x}Cu_2O_6$ do not become superconducting. Obviously, additional Cu and O atoms between the twin layers in "123" phases are essential for superconductivity which becomes optimized when these atoms order in as complete as possible CuO_{3-x} ribbons. These ribbons are also present in "336" phases.

REFERENCES

1. J.G.Bednorz and K.A.Müller, Z.Phys. B64, 189 (1986)
2. R.J.Cava, R.B.v.Dover, B.Batlog, and E.A.Rietmann, Phys.Rev.Lett. 58, 408 (1978)
3. M.K.Wu, J.R.Ashburn, C.J.Torug, P.H.Hor, R.L.Meng, L.Gao, Z.J.Huang, Y.Q.Wang, and C.W.Chu, Phys.Rev.Lett. 58, 908 (1987)
4. M.A.Beno, L.Soderholm, D.W.Capone II, D.G.Hinks, J.D.Jorgenson, I.K.Schuller, C.U.Segre, K.Zhang, J.D.Grace, Appl.Phys.Lett. 51, 57 (1987)
5. H.Eickenbusch, W.Paulus, E.Gocke, J.F.March, H.Koch and R.Schöllhorn, Angew.Chem.Int.Ed.Engl.26, 1182 (1987)
6. A.Simon, Angew.Chem.Int.Ed.Engl.26, 579 (1987)
7. Hk.Müller-Buschbaum, Angew.Chem.Int.Ed.Engl. 16, 674 (1977)
8. K.Hestermann and R.Hoppe, Z.anorg.allg.Chem. 367, 249 (1969)
9. K.Wahl and W.Klemm, Z.anorg.allg.Chem. 270, 69 (1952)
10. C.L.Teske and Hk.Müller-Buschbaum, Z.anorg.allg.Chem. 379, 113 (1970)
11. R.W.G.Wykoff, Crystal Structures, 2nd Ed., Vol.1, p.331, Interscience publ., J.Wiley, New York (1963)
12. A.Bussmann, H.Bilz, R.Roenspiess, and K.Schwarz, Ferroelectrics 25, 343 (1980)
13. W.W.Warren, R.E.Walstedt, G.F.Brennert, G.P.Espinosa, and J.P.Remeika, Phys.Rev.Lett. 59, 1860 (1987)
14. H.P.Hor, R.L.Meng, Y.Q.Wang, L.Gao, Z.J.Huang, J.Bechtold, K.Forster, and C.W.Chu, Phys.Rev.Lett. 58, 1891 (1987)
15. D.B.Mitzi, J.Z.Sun, D.J.Webb, M.R.Beasley, T.H.Geballe, A.Kapitulnik, Adriatico Research Conference on High Temperature Superconductors, Trieste 1987
16. N.Nguyen, L.Er-Rakho, C.Michel, J.Choisnet, and B.Raveau, Mat.Res.Bull. 15, 891 (1980)
17. D.M.de Leeuw, C.A.H.A. Mutsaers, C.Langereis, J.W.C.de Vries, P.C.Zalm, and P.F.Bongers, European Materials Research Meeting, Strasbourg, June 3, 1987
18. L.Er-Rakho, C.Michel, J.Provost, and B.Raveau, J.Solid State Chem. 37, 151 (1981)
19. K.Yvon, W.Jeitschko, and E.Parté, J.Appl.Crystallogr. 10, 73 (1977)
20. W.I.David, W.T.A.Harrison, R.M.Ibberson, M.T.Weller, J.R.Grasmeder, and P.Lanchester, Nature, 328, 328 (1987)
21. S.Takekawa and N.Iyi, Jap.J.Appl.Phys. 26, 851 (1987)

Isoelectronic analogs of the high temperature oxide superconductor

J. Thiel*, S. N. Song+, J. B. Ketterson+, and K. R. Poeppelmeier*

*Department of Chemistry, +Department of Physics and Astronomy,
and Materials Research Center, Northwestern University
Evanston, Illinois 60208

Abstract - High temperature superconductivity in $YBa_2Cu_3O_{7-\delta}$ is dependent on the mixed oxidation states of copper (Cu^{3+}/Cu^{2+}). Replacement of the d^8 Cu^{3+}-ion with the isoelectronic d^8 Ni^{2+}-ion and Ba^{2+} with La^{3+} for charge compensation has been studied. In the composition series $YBa_{2-x}La_xCu_{3-x}Ni_xO_{7+x/2-\delta}$; x = 0.0, 0.1, 0.2, 0.3, 0.4, 0.5, 0.7, and 1.0 a systematic change from orthorhombic to tetragonal symmetry was observed. There is a corresponding decrease in the resistive transition temperature from 92K → 72K → 29K and a decrease in the gram-susceptibility. The series $YBa_{2-x}La_xCu_{3-x}Ni_xO_{7+x/2-\delta}$; $0 \leq x \leq 1$ is related in structure to $YBa_2Cu_3O_{7-\delta}$, but contains Ni^{3+} and more than seven oxygen atoms per unit cell.

INTRODUCTION

The concept that a significant fraction of oxygen atoms may be removed by reduction from metallic oxide lattices with little or no structural change was first studied (ref. 1,2) and debated (ref. 3) over twenty-five years ago. In particular the perovskite lattice has been shown to persist over the limits ABO_{3-x}; $0 \geq x \geq 0.5$ for many transition and some main group metal cations. The first compounds with the composition $ABO_{2.5}$ and ordered oxide ion vacancies that were recognized (ref. 4,5) to be related to perovskite were $Ca_2Fe_2O_5$ (ref. 6) and the mineral brownmillerite, Ca_2AlFeO_5 (ref. 7). In general the superstructure formed depends upon the electronic configuration (d^n) and preferred coordination of the smaller B-site cation and the ionic radius of the larger, electropositive A-cation (ref. 8,9). Complex pseudocubic and oxygen-deficient structures based on perovskite are known for manganese (ref. 10), iron (ref. 11), cobalt (ref. 12), nickel (ref. 13) and copper (ref. 14). Oxygen deficient compounds based on the related K_2NiF_4 structure, e.g. $Ca_2MnO_{3.5}$, are also known (ref. 15) although they have not been as thoroughly investigated.

In addition to pseudocubic structures, ABO_3 compounds form an extraordinary number of structures based on mixed cubic (c) and hexagonal (h) close-packed AO_3 layers (ref. 16). Some of this polytypism has been shown to be associated with the oxygen composition (ref. 17,18). The combination of mixed sequences (c,h), cation composition, and oxygen stoichiometry can give rise to a large number of complex structures and a range in properties (ref. 19,20).

Following the recent report of Bednorz and Müller (ref. 21) on possible high T_c (>30K) superconductivity in the Ba-La-Cu-O mixed phase system, the compound $La_{2-x}M_xCuO_{4-x/2+\delta}$ (M = Ba^{2+}, Sr^{2+}, or Ca^{2+}) with the tetragonal K_2NiF_4 structure was identified (ref. 22,23) to be the superconducting phase. Soon thereafter Chu et al. (ref. 24,25) reported superconductivity above 90K in the Ba-Y-Cu-O system. The compound has been identified by numerous groups (ref. 26–29) and has the composition $YBa_2Cu_3O_{7-\delta}$ and a structure (ref. 30, Fig. 1) to the tetragonal structure reported for $La_3Ba_3Cu_6O_{14.1}$ (ref. 31).

Substitution of 10 mole% nickel for copper has been shown to dramatically reduce the superconducting transition temperature (ref. 32). In this paper we report on our efforts to replace the d^8 Cu^{3+}-ion in $YBa_2Cu_3O_7$ with the isoelectronic d^8 Ni^{2+}-ion. Lanthanum was substituted in equimolar amounts with nickel to provide the necessary charge compensation. However, the stability of Ni^{3+} and the ability of the structure adopted by $YBa_2Cu_3O_7$ (ref. 30,31) to contain excess oxygen become important factors in the solid-state chemistry of nickel.

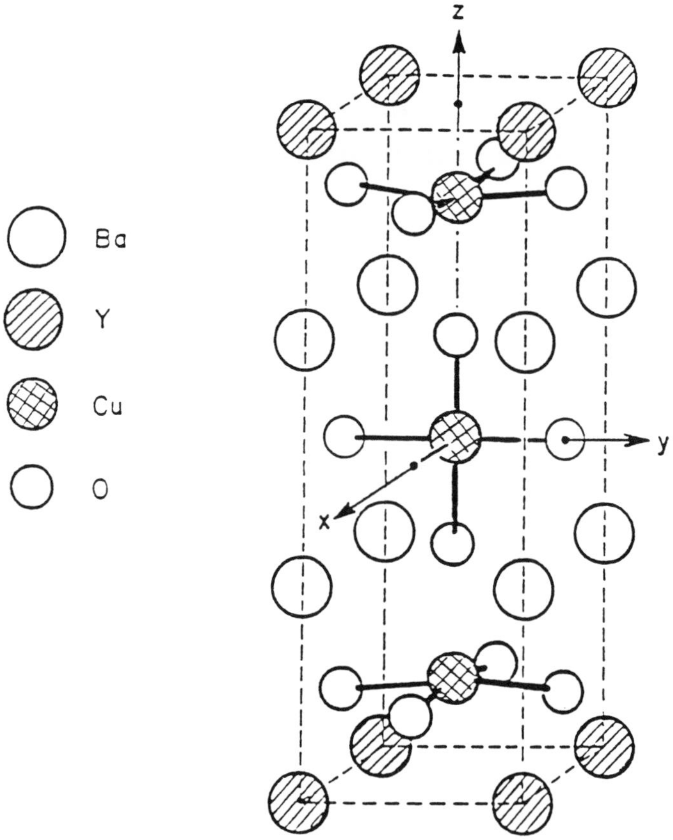

Fig. 1. Structure of $YBa_2Cu_3O_7$ with square planar Cu^{+3} ion centered at 1/2, 1/2, 1/2.

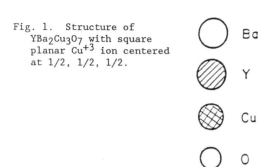

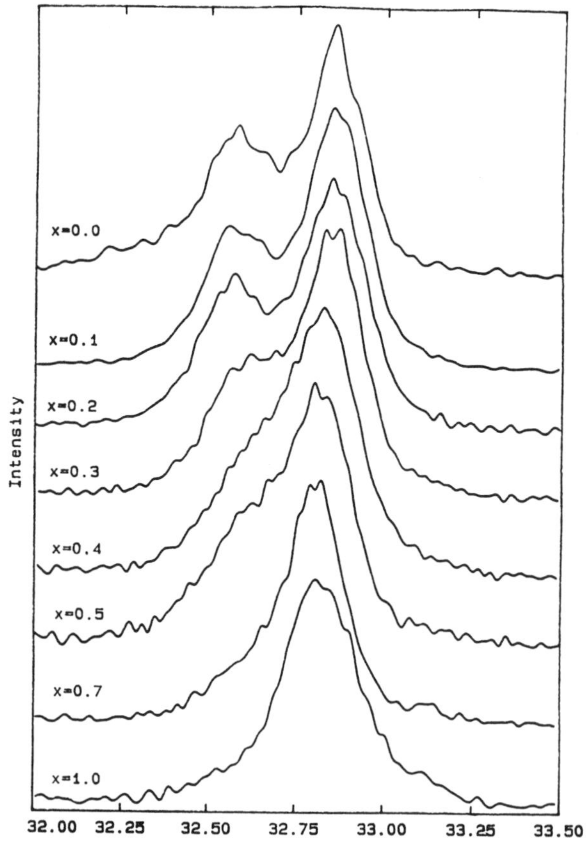

Fig. 2. The X-ray powder diffraction pattern for the region $32^\circ \leq 2\Theta \leq 33.5^\circ$ using Cu Kα radiation.

EXPERIMENTAL

Samples of composition $YBa_{2-x}La_xCu_{3-x}Ni_xO_{7+x/2-\delta}$ with x = 0.0, 0.1, 0.2, 0.3, 0.4, 0.5, 0.7 and 1.0 were prepared by solid state reaction of Aldrich yttrium oxide (99.999%), cupric oxide (99.999%), barium carbonate (99.999%) lanthanum oxide (99.999%) and nickel oxide (99.999%). Powders were ground with a mortar and calcined in air at 900°C for 8 hours. The samples were reground daily and calcined for 16 hours at 950°C in air for 5 days. The compounds were ground and heated in an oxygen atmosphere first for 6 hours at 900°C and then for 12 hours at 700°C. Thermogravimetric studies with a Du Pont Thermal Analysis System by reduction in hydrogen were used to determine the oxygen composition. For resistivity measurements disc-shaped specimens 1.25 cm in diameter and 1.5 mm thick were isostatically pressed at 7.2 kbar at room temperature. The oxygen treatment at 900°/700°C was repeated to anneal the pellet. After cooling, the discs were cut into rectangular specimens with cross section of 1.5 x 3 mm^2 and four leads were attached with silver paint for 4-point resistivity measurements.

X-ray diffraction (XRD) and magnetic susceptibility measurements were carried out on polycrystalline materials that had been calcined in oxygen. XRD powder patterns were recorded with Cu Kα radiation using a Ni filter on a Rigaku diffractometer. For reference all patterns were recorded with an internal NBS Si standard. After correcting the peak position based on the observed silicon line positions, the lattice constants were then refined by a least squares method weighted proportional to the square root of the height and inversely with the square of the width of the peak.

A VTS-50 susceptometer was used for the magnetization measurements. When equipped with a specially designed transport probe it could also be used for resistivity measurements. Temperature was controlled in the range 1.7-400K and magnetic fields in the range 0-5T. The susceptometer was calibrated with NBS aluminum and platinum standards. For the transport measurements temperatures could be read with either the thermometer in the VTS or with a carbon glass thermometer mounted on the probe; both were calibrated to an accuracy of 0.010K. Field-cooling (H = 100 Oe) was used to trace the temperature dependence of the susceptibility. The samples were cooled slowly, passing through the transition point to lower temperature at a fixed field. The contribution from the quartz basket was carefully subtracted. The DC resistivity was measured with care to insure thermal equilibrium. Currents of 1 mA were employed and voltages were determined with a Keithley 181 nanovoltmeter.

RESULTS

Figure 2 shows the X-ray powder diffraction of the compositions x = 0.0, 0.1, 0.2, 0.3, 0.4, 0.5, 0.7 and 1.0 from 32.00 ≤ 2Θ (deg) ≤ 33.50. In this region the 013 (d calcd./(Å) 2.750) and 103/110 (d calcd./(Å) 2.727, 2.726) diffraction peaks occur for the orthorhombic phase $YBa_2Cu_3O_{7-\delta}$. The equimolar substitution of lanthanum/nickel for Ba^{2+}/Cu^{3+} results in a systematic change from orthorhombic to tetragonal symmetry. Small amounts of a phase, green in color, with a diffraction pattern similar to Y_2BaCuO_5 were observed for x > 0.5. The data are summarized in Table 1. Oxygen analyses are summarized in Table 2.

TABLE 1. Lattice constants as a function of lanthanum/nickel substitution.

	a/Å	b/Å	c/Å
x=0.0	3.823(1)	3.888(1)	11.670(2)
0.1	3.821(1)	3.886(1)	11.666(2)
0.2	3.819(1)	3.891(1)	11.678(2)
0.3	3.824(1)	3.885(1)	11.653(2)
0.4	3.840(2)	2.879(1)	11.638(3)
0.5	3.837(2)	3.879(2)	11.642(4)
0.7	3.863(1)	-	11.592(4)
1.0	3.858(1)	-	11.574(2)

TABLE 2. Thermogravimetric analyses: $YBa_{2-x}La_xCu_{2-x}Ni_xO_{7+x/2-\delta}$

	$x/2_{-\delta}$ Observed	$x/2_{-\delta}$ for Ni^{3+}
x = 0.2	0.11	0.10
x = 0.5	0.22-0.27	0.25
x = 0.7	0.33	0.35
x = 1.0	0.47-0.54	0.50

Susceptibility results are summarized in Fig. 3 where normalized signal intensity at 5.2K and H = 100 Oe are plotted for the compositions x = 0.1, 0.2, 0.3, 0.4, 0.5, 0.7 and 1.0. As we have reported earlier (ref. 33,34), the magnetic susceptibility is particularly sensitive to the presence or absence of a superconducting phase. The intensity of the diamagnetic signal should scale with the volume fraction and therefore could be used to locate phase boundaries. Substitution of lanthanum/nickel greatly reduces the diamagnetic signal, and for x = 0.7, 0.8 and 1.0 only a small remnant diamagnetic signal was detected.

In Fig. 4 R(T)/R(300K) plots show two distinct regions. For x = 0.0, 0.1, 0.2, and 0.3 the midpoint of the resistive transition decreases from 92K to 72K in a nearly linearly fashion. For x = 0.4 and 0.5 a drop in resistance is detected between 70-75K and a resistive transition occurs near 28-29K. In Fig. 5 the much larger resistance observed for the x = 0.7 composition and semiconducting behavior is plotted. Similar behavior was observed for x = 0.8 and x = 1.0.

DISCUSSION

In orthorhombic $YBa_2Cu_3O_{7-\delta}$ when $\delta = 0$ ideally all oxygen sites along the b-axis are fully occupied. When $0.0 < \delta \leq 1.0$ there are oxygen vacancies, but these are thought to be located on oxygen sites along the b-axis and/or disordered on both a and b-axes as a function of the oxygen composition. Additional oxygen could potentially be incorporated although probably not in $YBa_2Cu_3O_7$ because this would require further oxidation of Cu^{3+} to Cu^{4+}. We have attempted to form the isoelectronic compound $YBaLaNiCu_2O_7$ and solid solutions with $YBa_2Cu_3O_7$. However we have prepared all our compositions in oxygen and additional oxygen in the structure up to $YBaLaNiCu_2O_8$ is entirely possible. Given that $LaNiO_3$ (Ni^{3+}) and $BaNiO_3$ (Ni^{4+}) are known compounds, and are more stable than $LaCuO_3$, this possibility seems all the more likely. In our thinking we assume that the planes that contain yttrium will not allow additional oxygen into the lattice, and therefore nine oxygen atoms per unit cell are not possible and the limiting oxygen stoichiometry is eight. We also assume that any additional oxygen incorporated into the lattice along the a-axis will result in the oxidation of Ni^{2+} (substituting for Cu^{3+}).

The change from orthorhombic to tetragonal symmetry, shown in Fig. 2, is unusual and informative. Orthorhombic $YBa_2Cu_3O_7$ has a c/a ratio greater than three (c/a = 3.06) but has a c/b ratio equal to three (pseudocubic) which explains the accidental overlap of the 103 and 110 reflections in the X-ray powder diffraction pattern. However tetragonal $Y(BaLa)(Cu_2Ni)O_{7+x/2-\delta}$ has a c/a ratio of three and the 013, 103 and 110 reflections appear as a single diffraction peak. This is a contrast to the oxygen-deficient, tetragonal compound $YBa_2Cu_3O_{7-\delta}$ ($\delta \approx 1$) where c/a is greater than three and two peaks, the 110 and the 013 overlapping the 103, with a relative intensity of 1:2 are observed. These data are consistent with the thermogravimetric data which demonstrations that the substitution of lanthanum/nickel results in La^{3+}/Ni^{3+} and a corresponding oxygen excess around the square-planar site. The oxygen-excess, tetragonal phase $YBaLaCu_2NiO_{7.5}$ probably consists of square-pyramidal Cu(II) sites as in the nickel-free compound, but in place of square-planar Cu(I) sites (Cu^{3+}) there are square-pyramidal Ni^{3+} ions due to the 0.5 excess oxygen atoms per unit cell. This excess oxygen is probably sited in one-half the sites along the a-axis, or disordered between the a-and b-axis, resulting in either case in square-pyramidal nickel sites (Ni^{3+}, d^7) rotated 90° from the two square-pyramidal copper sites (Cu^{2+}, d^9).

From the X-ray diffraction data, resistivity measurements, and susceptibility measurements we observe three regions of distinct behavior:

Region I: x = 0.0 to x = 0.3
Region II: x = 0.3 to x = 0.7
Region III: x = 0.7 to x = 1.0

In region I (x = 0.0 to x = 0.3) the midpoint of the resistive transition changes from greater than 90K to approximately 72K. The X-ray diffraction patterns of these compositions show an orthorhombic structure from which we infer that the oxygen occupancy along the a- and b-axes are still very unequal in at least one of the phases present. At most five phases can coexist at equilibrium in the BaO-Y_2O_3-La_2O_3-CuO-NiO system when the temperature

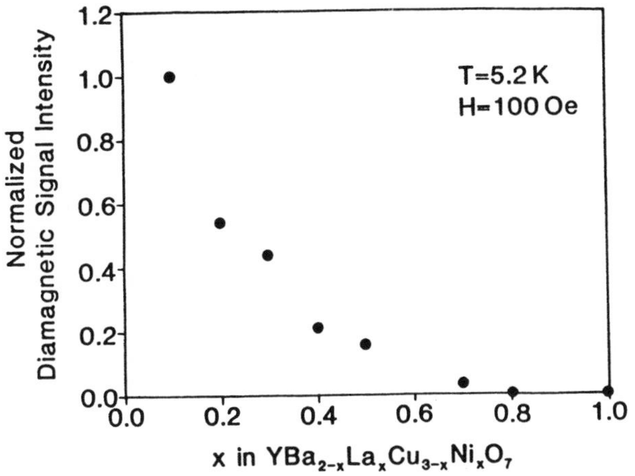

Fig. 3. The normalized shielding signal intensity vs. composition x in $YBa_{2-x}(LaNi)_xCu_{3-x}O_{7+x/2-\delta}$ The signal intensity was normalized with respect to that from $YBa_{1.9}(LaNi)_{0.1}Cu_{2.9}O_{7.05}$

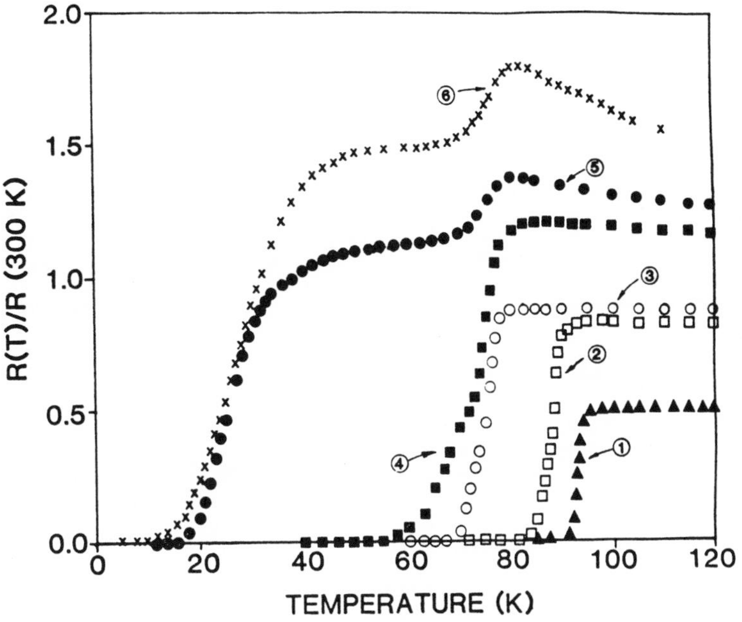

Fig. 4. The normalized resistance vs. temperature for $YBa_{2-x}(LaNi)_xCu_{3-x}O_{7+x/2-\delta}$

(1) x = 0.0; (2) x = 0.1;
(3) x = 0.2; (4) x = 0.3;
(5) x = 0.4; (6) x = 0.5.

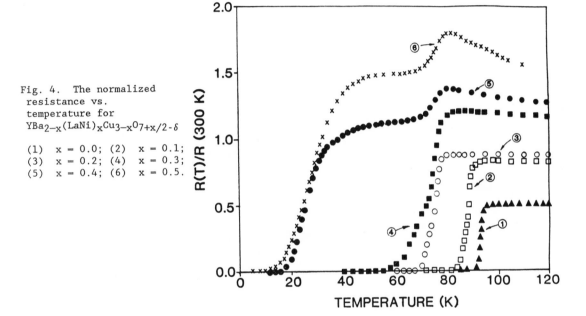

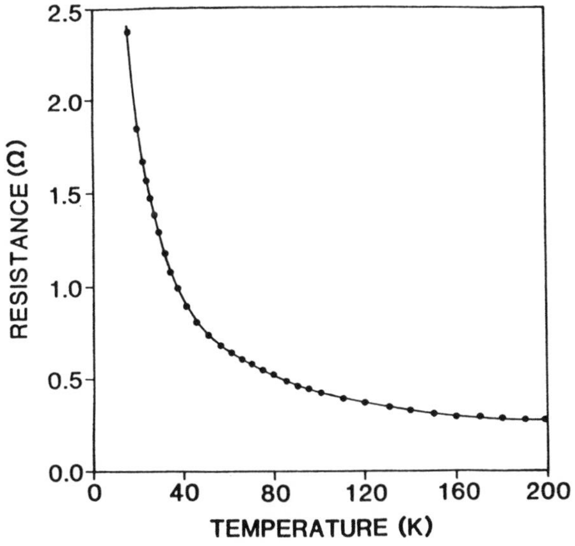

Fig. 5. The temperature dependence of the resistance of $YBa_{1.3}(LaNi)_{0.7}Cu_{2.3}O_{7.35}$. The solid curve is a guide to the eyes.

and oxygen partial pressure (1 atm) are fixed, although fewer are likely due to the chemical similarity of the components. The fraction of oxide superconductor drops significantly with substitution of La^{3+}/Ni^{3+} for Ba^{2+}/Cu^{3+} and indicates that at least one additional non-superconducting component is present, however, structural and compositional variances are subtle.

In region II the resistive behavior of x = 0.4 and 0.5 show a small fraction of the 72K material and a new transition near 29K. Similarly the X-ray diffraction patterns show the major phases present have a much smaller orthorhombic lattice distortion. Small amounts of $La_{2-x}Ba_xCuO_{4-x/2+\delta}$ could be responsible for the transition near 30K, but this phase was not observed in the diffraction patterns. The susceptibility data indicate that these samples may be largely bi- or triphasic depending on the amount of non-superconducting oxide present.

Region III (x = 0.7, 0.8 and 1.0) does not show any evidence for the presence of a superconducting phase based on resistance measurements. A weak diamagnetic signal is found for all three samples up to approximately 60K which indicates that a small amount of the oxide superconductor from Region I still remains. This implies that the Region II superconductor has reacted which is consistent with the fairly sharp X-ray diffraction patterns observed of a tetragonal phase. In this case the tetragonal phase is probably due to an excess of oxygen and Ni^{3+} e.g. $YBaLaNi^{3+}(Cu^{2+})_2O_{7.5}$. The intermediate region II material may contain small clusters (dimers) of d^7-Ni^{3+} alternating with d^8-Cu^{3+} ions, e.g. $YBa_{1.5}La_{0.5}Ni_{0.5}Cu_{2.5}O_{7.25}$.

CONCLUSION

Our attempts to prepare $YBaLaCu_2NiO_7$, the isoelectronic analog of $YBa_2Cu_3O_7$, and to replace the d^8 Cu^{3+}-ion with the d^8 Ni^{2+}-ion in the series $YBa_{2-x}La_xCu_{3-x}Ni_xO_{7+x/2-\delta}$; $0 \leq x \leq 1$ resulted in stable Ni^{3+} compounds due to the ability of the oxygen-deficient structure of $YBa_2Cu_3O_7$ to incorporate additional oxygen. The excess oxygen in the lattice effectively oxidizes the filled $(d_{z^2})^2$ states and creates bridges between the one-dimensional chains in the structure (see Fig. 1). We also find that reduction of these compounds when $0.0 \leq x \leq 0.5$ results in semiconducting behavior (versus metallic), which is most likely due to reduction of the copper chain sites (Cu(I)).

ACKNOWLEDGEMENTS

We acknowledge support for this research from NSF (DMR-8610659) and from the Materials Research Center of Northwestern University, NSF (MRL-8520280). We are grateful to S. Massidda, J. J. Yu and A. J. Freeman for helpful discussions.

LITERATURE CITED

1. G.H. Jonker, Physica. 20, 1118 (1954).
2. M. Kestigian, J.G. Dickenson, R. Ward, J. Am. Chem. Soc. 79, 5598 (1957).
3. S. Anderson, A.D. Wadsley, Nature 187, 499 (1960).
4. H. Watanabe, M. Sugimoto, M. Fukase, T. Hirone, J. Appl. Phys. 36, 988 (1965).
5. S. Geller, R.W. Grant, U. Gonser, H. Wiedersich, G.P. Espinosa, Phys. Lett. 20, 115 (1966).
6. E.F. Bertaut, P. Blum, A. Sagnieres, Acta Crystallogr. 12, 149 (1959).
7. W.C. Hansen, L.T. Brownmiller, Amer. J. Sci. 15, 224 (1928) .
8. J.C. Grenier, J. Parriet, M. Pouchard, Mat. Res. Bull. 11, 1219 (1976).
9. K.R. Poeppelmeier, M.E. Leonowicz, J.M. Longo, J. Solid State Chem. 44, 89 (1982).
10. K.R. Poeppelmeier, M.E. Leonowicz, J.C. Scanlon, J.M. Longo, W.B. Yelon, J. Solid State Chem. 45, 71 (1982).
11. J.C. Grenier, M. Pouchard, P. Hagenmuller, Structure and Bonding 47, 1 (1981).
12. K. Vidyasagar, J. Gopalakrishnan, C.N.R. Rao, Inorg. Chem. 23, 1206 (1984).
13. K. Vidyasagar, A. Reller, J. Gopalaknishnan, C.N.R. Rao, J. Chem. Soc., Chem. Commun. 7 (1985).
14. C. Michel, B. Raveau, Revue de Chimie Minerale 21, 407 (1984).
15. M.E. Leonowicz, K.R. Poeppelmeier, J.M. Longo, J. Solid State Chem. 59, 71 (1985).
16. L. Katz, R. Ward, Inorg. Chem. 3, 205 (1964).
17. T. Negas, R.S. Roth, J. Solid State Chem. 1, 409 (1970).
18. T. Negas, R.S. Roth, J. Solid State Chem. 3, 323 (1971).
19. J.B. Goodenough, J.M. Longo, Landolt-Bornstein, Group III/Vol.4a, Springer-Verlag, New York/Berlin (1970).
20. A.J. Jacobson, A.J.W. Horrox, Acta Crystallogr. Sect. B 32, 1003 (1976).
21. J.G. Bednorz, K.A. Müller, Z. Phys. B64, 189 (1986).
22. H. Takagi, S. Uchida, K. Kitazawa, S. Tanaka, Jpn. J. Appl. Phys. Lett. 26, L123 (1987).

23. R.J. Cava, R.B. van Dover, B. Batlogg, E.A. Rietman, Phys. Rev. Lett. 58, 408 (1987).
24. M.K. Wu, J.R. Ashburn, C.J. Torng, P.H. Hur, R.L. Meng, L. Gao, Z.J. Huang, Y.Q. Wang, C.W. Chu, Phys. Rev. Lett. 58, 908 (1987).
25. P.H. Hur, L. Gao, R.L. Meng, Z.J. Huang, K. Forster, J. Vassilious, C.W. Chu, C. W. Phys. Rev. Lett. 58, 911 (1987).
26. S.-J. Hwu, S.N. Song, J.P. Thiel, K.R. Poeppelmeier, J.B. Ketterson, A.J. Freeman, Phys. Rev. B. 35, 7119 (1987).
27. R.J. Cava, B. Batlogg, R.B. van Dover, D.W. Murphy, S. Sunshine, T. Siegrist, J.P. Remcika, E.A. Reitman, S. Zahurak, G.P. Espinosa, Phys. Rev. Lett. 58, 1676 (1987).
28. A.M. Stacy, J.V. Badding, M.J. Geselbracht, W.K. Ham, G.F. Holland, R.L. Hoskins, S.W. Keller, C.F. Millikan, H.-C. zur Loye, J. Am. Chem. Soc. 109, 2528 (1987).
29. H. Steinfink, J.S. Swinnea, Z.T. Sui, H.M. Hsu, J.B. Goodenough, J. Am. Chem. Soc. 109, 3348 (1987).
30. M.A. Beno, L. Soderholm, D.W. Capone, J.D. Jorgensen, I.K. Schuller, C.U. Segre, K. Zhang, J.D. Grace, Appl. Phys. Lett. 51, 57 (1987).
31. L. Er-Rakho, C. Michel, J. Provost, B. Raveau, J. Solid State Chem. 37, 151 (1981).
32. G. Xiao, F.H. Streitz, A. Gavrin, Y.W. Du, C.L. Chien, Phys. Rev. B, 35, 8782 (1987).
33. S.-J. Hwu, S.N. Song, J.B. Ketterson, T.O. Mason, K.R. Poeppelmeier, Commun. Am. Ceram. Soc. 70, C-165 (1987).
34. G. Wang, S.-J. Hwu, S.N. Song, J.B. Ketterson, L.D. Marks, K.R. Poeppelmeier, T.O. Mason, Adv. Ceram. Mater. 2, 313 (1987).

Substitutional chemistry of the 1:2:3 phase and studies of the Y_2O_3–SrO–CuO phase diagram

M. R. Harrison,* I. C. S. T. Hegedus, W. G. Freeman
GEC Research Limited, Hirst Research Centre, East Lane, Wembley, Middlesex, HA9 7PP, Great Britain.

R. Jones, P. P. Edwards*
National Centre for Superconductivity Research, University West Cambridge Site, Madingley Road, Cambridge, CB3 0HE, Great Britain and University Chemical Laboratory, Lensfield Road, Cambridge, CB2 1EW, Great Britain.

W. I. F. David, C. C. Wilson
Neutron Division, Rutherford Appleton Laboratory, Chilton, Didcot, OX11 0QX, Great Britain.

Abstract - Recent measurements have suggested that the superconducting mechanism in $YBa_2Cu_3O_{7-y}$ involves a coupling between the planes of Cu(2) atoms and chains of Cu(1) atoms, with the superconducting transition temperature varying with the chain length. Three possible methods of altering the chain length by changing the chemical composition of the 1:2:3 phase are discussed and our experimental results in this area are reviewed. Other phases with similar structural features are of interest in the search for improved superconducting materials. In this regard, phase relationships in the Y_2O_3-SrO-CuO system at 900-1000°C are reported and compared with those in the corresponding Y_2O_3-BaO-CuO system. The stable phases and tie-lines have been determined using a combination of X-ray powder diffraction and electron spin resonance spectroscopy. The latter is shown to be a powerful technique for detecting small amounts of copper-containing phases.

INTRODUCTION

In the short period of time since the discovery of superconductivity at 90K in $YBa_2Cu_3O_{7-y}$, an immense amount of work has been directed towards understanding the properties of this fascinating "1:2:3" phase. There is still no concensus on the mechanism of superconductivity in this material and it is vital that this mechanism is determined if new materials with improved superconducting properties are to be discovered. This paper is broadly divided into two sections. In the first we concentrate upon the chemistry of the 1:2:3 phase and describe the effects on the superconducting properties when some of the host ions are replaced by dopant ions. This allows us to identify the important structural and electronic features of this material. In the second section we describe our efforts to synthesise new superconducting phases with similar structural features but containing strontium instead of barium.

The crystal structure (ref. 1) of the 1:2:3 phase is shown in Fig. 1. The complexity of the properties is due in part to the presence of two chemically dissimilar copper sites: the square planar Cu(1) chains, containing predominantly Cu^{3+} ions, and the square pyramidal Cu(2) planes containing predominantly Cu^{2+} ions. It has been clearly demonstrated that the 1:2:3 phase exhibits superconductivity only if the oxygen ions in the basal plane are restricted to the sites along the <u>b</u> axis. This ordering results in the most interesting structural feature of the 1:2:3 phase; the parallel chains of copper and oxygen atoms running along the <u>b</u> axis.

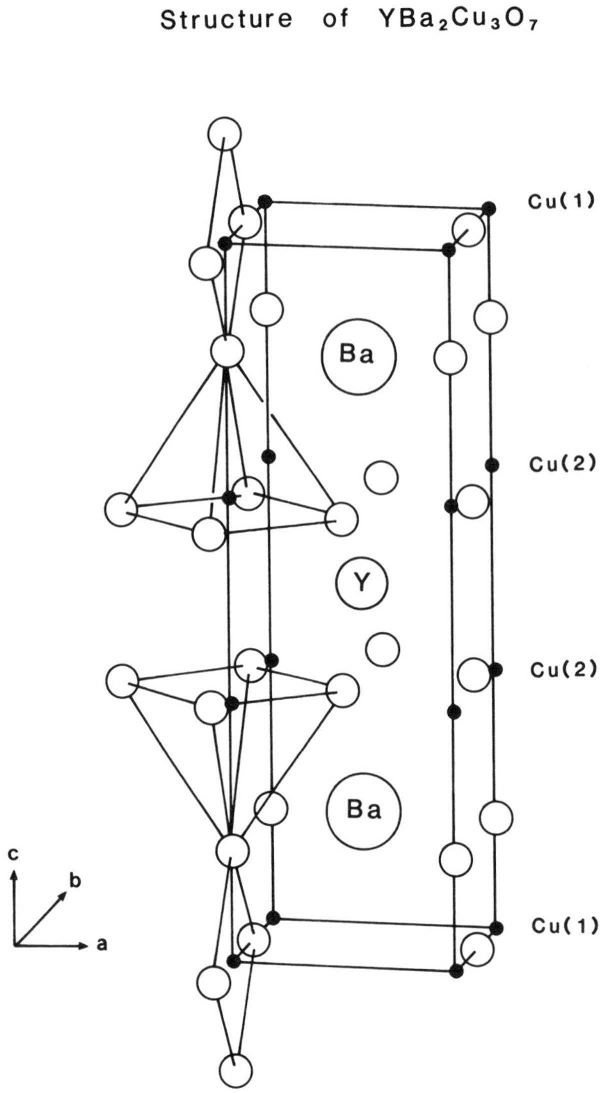

Fig. 1
The crystal structure of $YBa_2Cu_3O_{7-y}$, showing the Cu(1) and Cu(2)

Early substitution experiments concentrated upon replacing the Y^{3+} ions by isovalent members of the rare earth family and it was soon discovered that the majority of the rare earths could be successfully substituted with little effect either on the transition temperature or on the crystal structure. Isovalent substitution of the Ba^{2+} ions has proved more difficult, and only partial replacement by Sr^{2+} ions has been clearly demonstrated to date (refs. 2,3), although complete replacement has been claimed (refs. 4,5). The transition temperature appears to decrease with increasing strontium content, although this could also be caused by a simultaneous change in the oxygen content of the 1:2:3 phase (ref. 6).

Aliovalent substitution of the Y^{3+} ions by Pr^{4+} ions was found to rapidly reduce the transition temperature (ref. 7), and aliovalent substitution of the Ba^{2+} ions by La^{3+} ions was found to have a similar effect (ref. 8). These results could be explained in two ways. First, the higher charge of the dopant ions could cause the valence state of the copper ions to decrease. Second, the oxygen content of the phase could increase to compensate for the higher charge. Because of the different environments of the Y^{3+} and Ba^{2+} ions, it is plausible that Pr^{4+} substitution decreases the copper average charge while La^{3+} substitution increases the oxygen content.

Support for this hypothesis is provided by the structural studies of the compound $La_3Ba_3Cu_6O_{14+2y}$. Neutron diffraction studies showed that this compound was actually the end member of the solid solution $La(Ba_{1-x}La_x)_2Cu_3O_{7+y}$ with x = 0.25 (ref. 9). This system is therefore analogous to the $Y(Ba_{1-x}La_x)_2Cu_3O_{7+y}$ compositions with x > 0.25 studied in this paper. It was shown that the La^{3+} substitution indeed increased the oxygen content while the Cu^{3+}/Cu^{2+} ratio did not alter to a significant extent (refs. 9,10).

Nuclear spin-lattice relaxation measurements (refs. 11,12) have shown that the nuclear relaxation rate of both the Cu(1) and Cu(2) atoms changes at the transition temperature, and this suggests that the superconductivity mechanism involves a coupling between the planes and chains. Other recent measurements have hinted that the transition temperature varies with the chain length, and three possible methods of altering the chain length by changing the chemical composition of the 1:2:3 phase are discussed in this paper. These methods are shown schematically in Fig. 2 which shows only those ions in the vicinity of the basal plane.

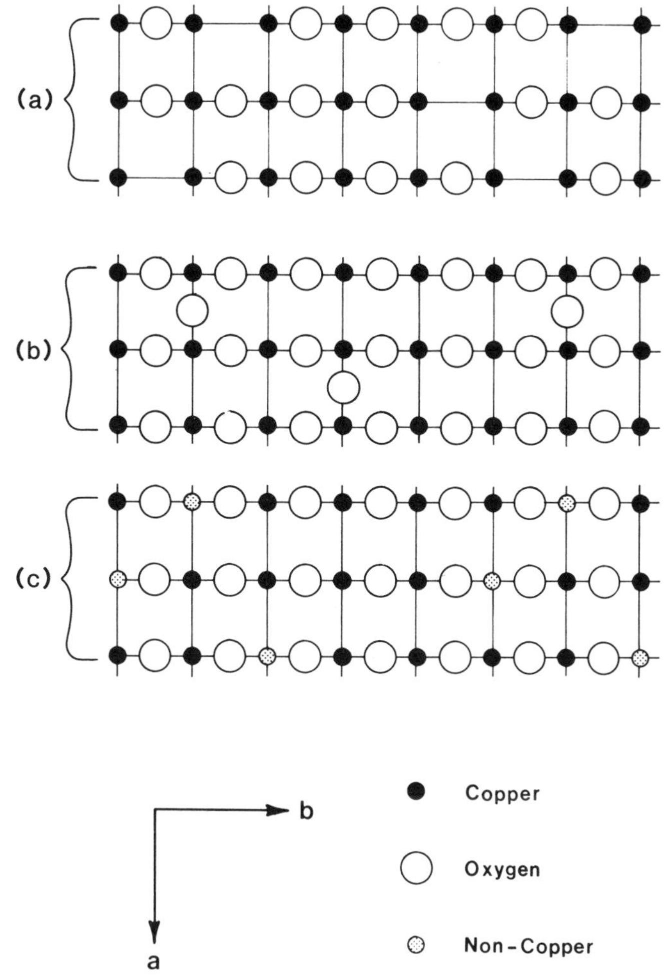

Fig. 2
The chains of copper and oxygen atoms in the basal plane can be interrupted by: (a) oxygen vacancies on the sites along the b axis; oxygen ions on the sites along the a axis; and (c) substitutional ions on the Cu(1) sites. The chain length is chosen to be four in each case.

The first method is to decrease the oxygen content of the 1:2:3 phase ($y > 0$) by producing oxygen vacancies along the b axis. This is readily achieved by annealing the material in a low pressure of oxygen above 400°C, and it has been clearly demonstrated that the transition temperature drops rapidly as the oxygen content decreases. The problem here is that the valence state of the copper ions also decreases as the oxygen content decreases, and the reduction in the transition temperature could arise from either or both of these mechanisms.

The second method is to increase the oxygen content of the phase ($y > 0$) by placing extra oxygen atoms in the sites along the a axis. The coordination of the copper atoms next to these oxygen atoms is then increased from four-fold to five-fold and this appears to interrupt the one-dimensional Cu-O chains. In priciple, this could be achieved by annealing the 1:2:3 phase in a high pressure of oxygen. In practice, chemical doping of the 1:2:3 phase is required to achieve a significant increase in the oxygen content. This can be done by replacing the Ba^{2+} ions by La^{3+} ions as discussed above (ref. 8). The virtue of this method is that the valence state of the copper ions apparently remains almost independent of the oxygen content, allowing the effect of truncating the chain length to be studied in isolation. In this paper, we describe our X-ray diffraction measurements, electrical resistance measurements and magnetic inductance measurements on samples with different lanthanum contents.

The third method is to replace some of the Cu(1) ions by dopant ions. Substitution of the copper ions in the 1:2:3 phase can indeed have a dramatic effect on the transition temperature (refs. 13,14). Interpretation of these doping experiments is difficult, however, since the dopant ions could substitute on the Cu(1) site, the Cu(2) site or both sites depending upon the radius, charge and electronic configuration of the ions. It is also possible that none of the dopant ions are incorporated into the 1:2:3 phase, instead forming impurity phases at the grain boundaries and surface.

Our experiments have shown that the extent of substitution indeed varies with the size and electronic structure of the dopant ions. Further, substitution on the Cu(1) site appears preferable to substitution on the Cu(2) site. Perhaps unsurprisingly, those ions closest in character to Cu^{2+} ions appear to substitute with the greatest ease. Strongly charged ions such as Si^{4+} are reluctant to substitute for copper, and prefer to react with the BaO formed at the surface of the grains by the decomposition of the 1:2:3 phase. Samples prepared with a nominal 7% of the Cu replaced by Si still have a transition temperature of 85K. In contrast, more weakly charged ions such as Mg^{2+} substitute more readily for copper and samples prepared with a nominal 7% of the Cu replaced by Mg have a transition temperature of 40K.

One element which appears to have a particularly strong effect on the transition temperature is zinc, and there is an interesting comparison here between Mg^{2+} and Zn^{2+} doping. Both ions would reduce the Cu-O chain length by substitution on the Cu(1) sites. In addition, Zn^{2+} has a d^{10} electronic configuration compared with Cu^{2+} which has a d^7 configuration, and the extra electron in the Cu-O conduction band could also reduce the transition temperature by compensating for the hole charge carriers. In order to locate the zinc ions and to determine the extent of substitution, we have performed powder neutron diffraction measurements to refine the site occupancies. These measurements are described in this paper, together with X-ray diffraction measurements and magnetic inductance measurements to determine the superconducting transition temperature.

Many of the questions about the superconducting mechanism would be resolved if more superconducting phases were synthesised. Following the discovery of the 1:2:3 phase, the previously unknown Y_2O_3-BaO-CuO phase diagram has been studied in some detail, both in an attempt to discover new superconducting phases and also to optimise the preparation of the 1:2:3 phase for technological applications. These experiments have also been encouraged by the persistent reports of much higher transition temperatures observed in trace amounts in some samples, although no other superconducting phases have been found in this pseudoternary system to date.

The reports of superconductivity above room temperature in strontium substituted samples (refs. 15,16) encouraged us to investigate the corresponding

Y_2O_3-SrO-CuO phase diagram and our results to date are described in this paper. The formation of the phases, their structure and their temperature stability have been studied using a combination of X-ray powder diffraction and electron spin resonance (ESR) spectroscopy. The latter is capable of detecting very small amounts of copper-containing phases embedded in other phases which would not be distinguished by conventional X-ray diffraction measurements.

SUBSTITUTIONAL CHEMISTRY OF THE 1:2:3 PHASE

Y_2O_3, La_2O_3, CuO, ZnO and $BaCO_3$ were used as the starting materials (BDH or Aldrich). The nominal compositions of the substituted 1:2:3 phase were chosen to be $YBa_2(Cu_{1-x}Zn_x)_3O_{7-y}$ and $Y(Ba_{1-x}La_x)_2Cu_3O_{7-y}$, so that 100 x represents the percentage of ions replaced in each case. Suitable amounts of the powders were thoroughly ground in an agate pestle and mortar until a uniform coloured powder was formed. The powder was then pressed into pellets and placed in an alumina boat for firing. The pelletisation was done to maximise the speed of reaction, to minimise the loss of volatile constituents such as BaO, and to minimise the reaction of the material with the alumina boat.

Samples containing variable amounts of zinc were fired in flowing oxygen at 950°C for 24 hours, then cooled in the furnace to below 400°C to ensure full oxidation. A similar procedure was followed for the samples containing variable amounts of lanthanum except that 24 hours was found to be insufficient time for complete reaction of the pellets and a second series of samples were fired for 72 hours. The results described in this paper pertain to this second batch of samples.

The samples were then characterised by X-ray diffraction, neutron diffraction, electrical resistance and magnetic inductance measurements. Neutron powder diffraction studies were carried out at room temperature using the Spallation Neutron Source (ISIS) at the Rutherford Laboratory and the full experimental details can be found elsewhere (ref. 17). The oxygen content of the samples was determined by chemical analysis (ref. 18). Samples were tested for superconductivity in the temperature range 5-300K using both electrical resistance measurements and magnetic inductance measurements.

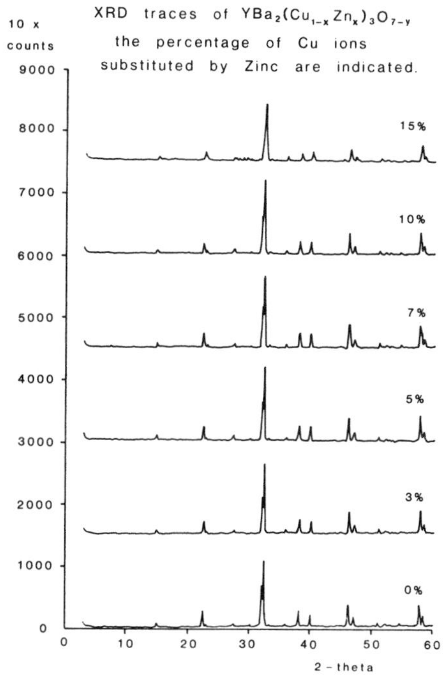

Fig. 3
X-ray diffraction patterns of a series of zinc-substituted samples. The curves have been translated along the vertical axis.

Figure 3 shows the X-ray diffraction patterns of a series of zinc-substituted samples. No change in the diffraction patterns with zinc content is evident, suggesting that the zinc ions substitute completely for the copper ions up to 10% replacement.

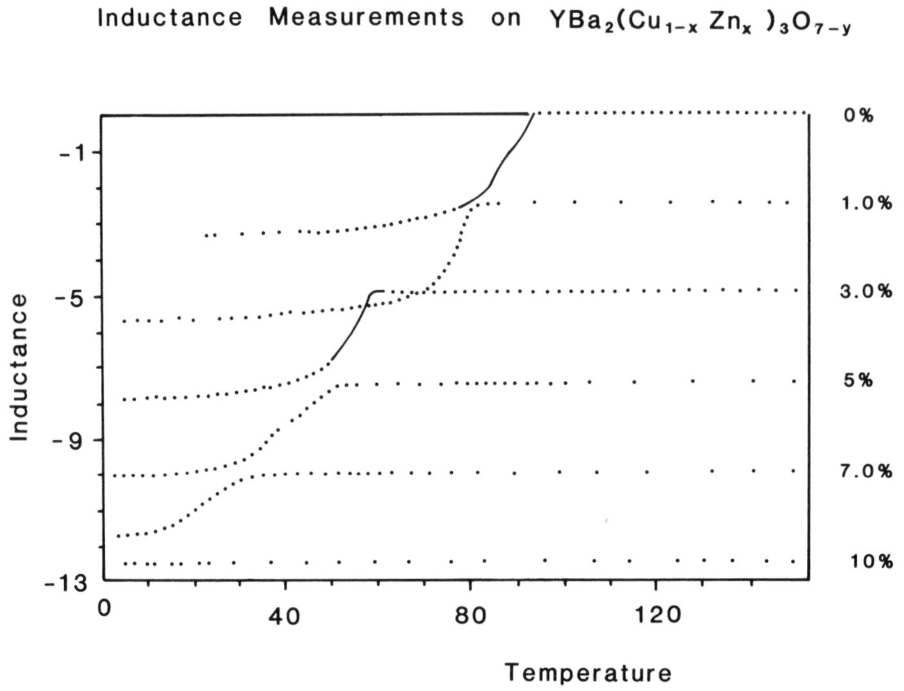

Fig. 4
Superconducting transitions for a series of powdered zinc-substituted samples as determined by magnetic inductance measurements. The curves have been translated along the vertical axis.

Figure 4 shows the corresponding magnetic inductance transitions for the same series of zinc-substituted samples. The superconducting transition temperature rapidly decreases with zinc content, and no transition occurs above 5K for 10% replacement.

In order to confirm these structural observations, and also to locate the zinc ions on the Cu(1) or Cu(2) sites, a powder neutron diffraction study of a (nominally) 7% substituted sample with a transition temperature of 30K was carried out (ref. 17). The refined structural parameters were then compared with those obtained from an undoped material. The refinements were satisfactory and gave broad agreement between most of the parameters common to both compounds. It is of interest to note that in the doped and undoped samples the O(1) site occupancy refines to 0.88(3) and 0.90(2) respectively, giving an overall oxygen stoichiometry of $O_{6.88(3)}$ and $O_{6.90(2)}$. This is in very good agreement with the results of our chemical analysis of the oxygen content, which gave 6.88(2) and 6.89(2) respectively (ref. 7). Thus the oxygen content of the 1:2:3 phase does not seem to be affected by the presence of the zinc ions, and so the change in transition temperature cannot arise from oxygen deficiency.

The most interesting feature of the refinements, however, concerns the amount and location of the substituted zinc ions. The copper site occupancies clearly indicate that the zinc occupies the Cu(1) site since the Cu(1) occupancy by copper drops to 0.83(6) to give a Cu(1) site occupancy by zinc of 0.17(6). Cu(2) remains fully occupied by copper with an occupancy of 1.00(4). The stoichiometry of the zinc-substituted material can therefore be written as $YBa_2(Cu_{0.943}Zn_{0.057})_3O_{6.88}$. This means that 5.7(2.0%), rather than the nominal 7% Zn, has been incorporated into the structure, and that 17.1(6.0)% of the Cu(1) atoms have been replaced by zinc atoms.

In other words, the average chain length in this composition is approximately four.

From these results there is a possible simple explanation for the observed transition temperatures in these zinc-substituted materials. To a first approximation there is a roughly linear relationship between the zinc content and the drop in transition temperature. The fact that this pattern emerges when essentially all of the substitution is taking place at the Cu(1) site, suggests that full occupancy by copper of the chains along the b axis is crucial to the retention of a high superconducting transition temperature and that these chains are either entirely responsible for the superconducting effect or significantly enhance this effect.

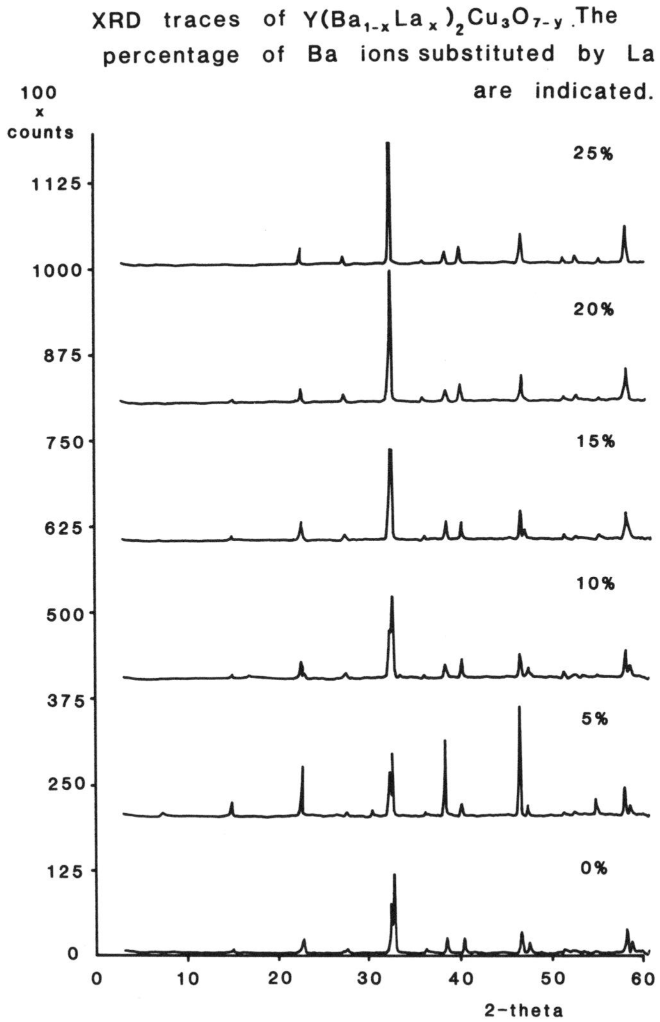

Fig. 5
X-ray diffraction patterns of a series of lanthanum-substituted samples. The curves have been translated along the vertical axis.

Figure 5 shows the X-ray diffraction pattern of a series of lanthanum-substituted samples. No other phases are observed in the diffraction patterns suggesting that the lanthanum ions substitute completely for the barium ions at least up to 25% replacement. However, the orthorhombic distortion of the unit cell (b-a) initially increases then decreases with lanthanum content as previously observed until the phase becomes tetragonal (ref. 8). Magnetic inductance measurements in Fig. 6 and electrical resistance measurements in Figs. 7 and 8 both show that the transition

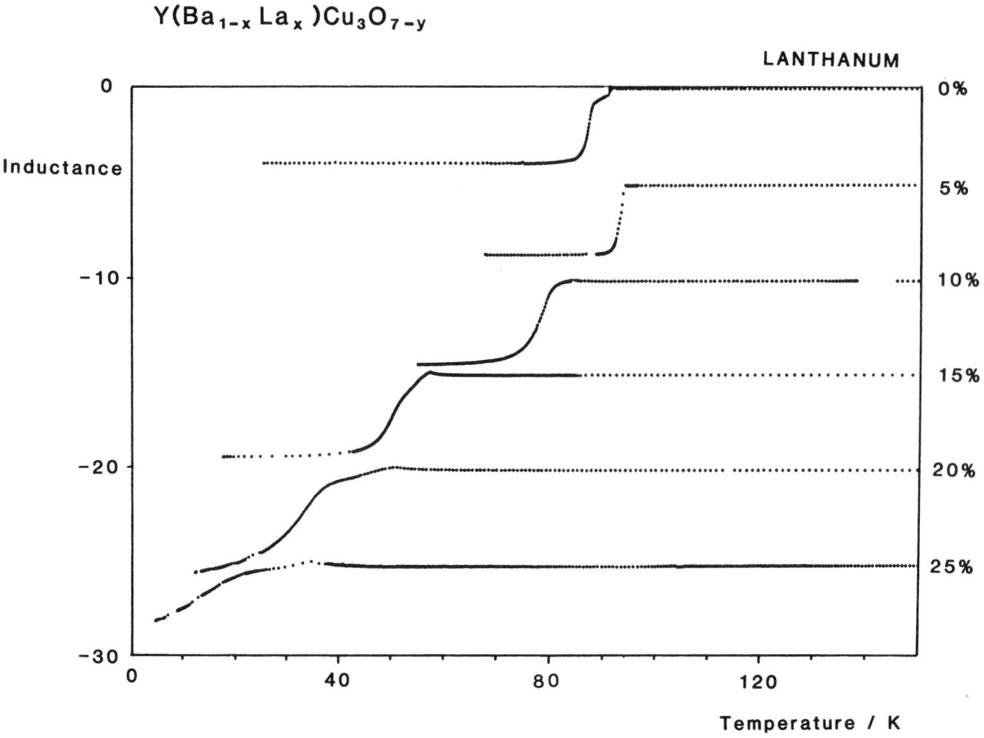

Fig. 6
Superconducting transitions for a series of lanthanum-substituted samples as determined by magnetic inductance measurements.

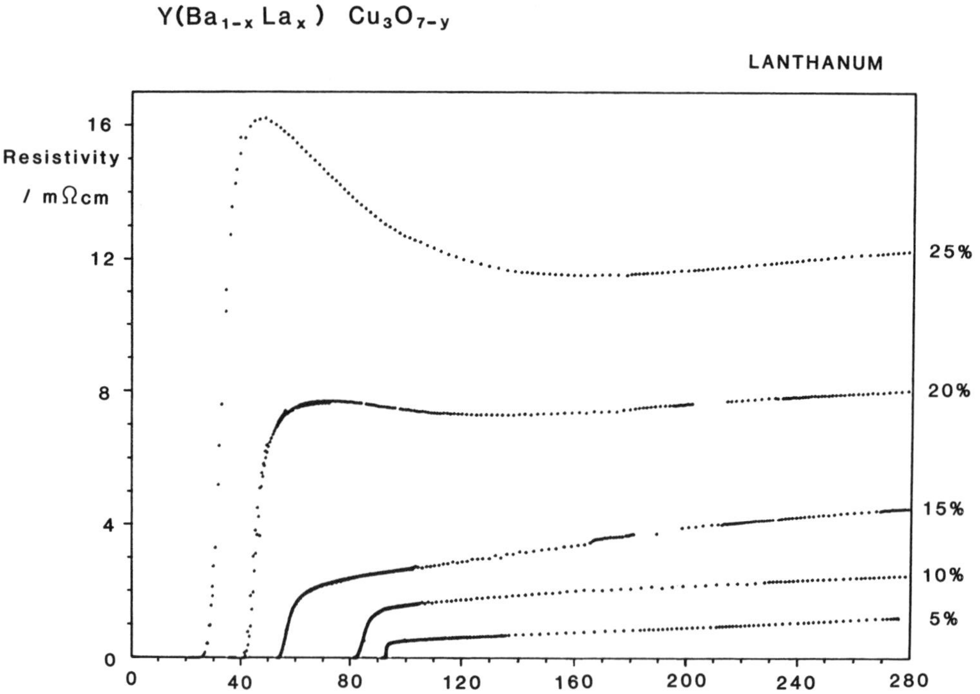

Fig. 7
Superconducting transitions for a series of lanthanum-substituted samples as determined by electrical resistance measurements.

temperature initially increases, then decreases, with lanthanum content. The magnetic inductance measurements also show that the samples are bulk superconductors for all compositions.

These results can be simply explained if the increased charge of the La^{3+} ions compared to the Ba^{2+} ions increases the oxygen content of the 1:2:3 phase. It is postulated that the extra oxygen ions initially occupy the vacant sites between the Cu-O chains in the basal plane, then occupy the vacant sites between the Cu-O chains as shown in Fig. 2. The 5% sample would be closest to the presumed optimum content of 7.0 for the 1:2:3 phase.

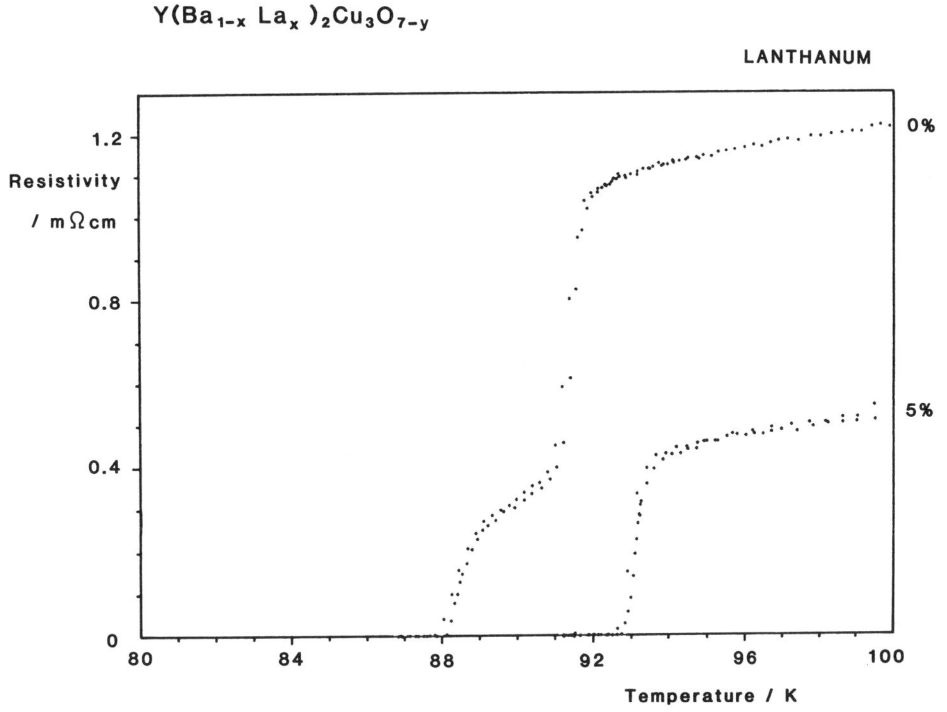

Fig. 8
Superconducting transitions for an undoped and a 5% lanthanum-substituted sample showing the increase in the transition temperature and the reduction in the normal-state electrical resistivity produced by the substitution.

STUDIES OF THE Y_2O_3-SrO-CuO PHASE DIAGRAM

We now turn to our studies of the Y_2O_3-SrO-CuO phase diagram. Approximately thirty samples were made with the nominal compositions indicated by the positions of the open squares in Fig. 9.

For comparison, Fig. 10 shows part of the corresponding Y_2O_3-BaO-CuO phase diagram (refs. 19,20). All these samples were then characterised by powder X-ray diffraction measurements and the samples containing copper ions were also examined by ESR spectroscopy at X-band frequencies (9.3GHz) at room temperature.

Figure 11 shows the X-ray diffraction patterns of some of the phases in the strontium system, and Fig. 12 shows some of the corresponding X-band ESR spectra.

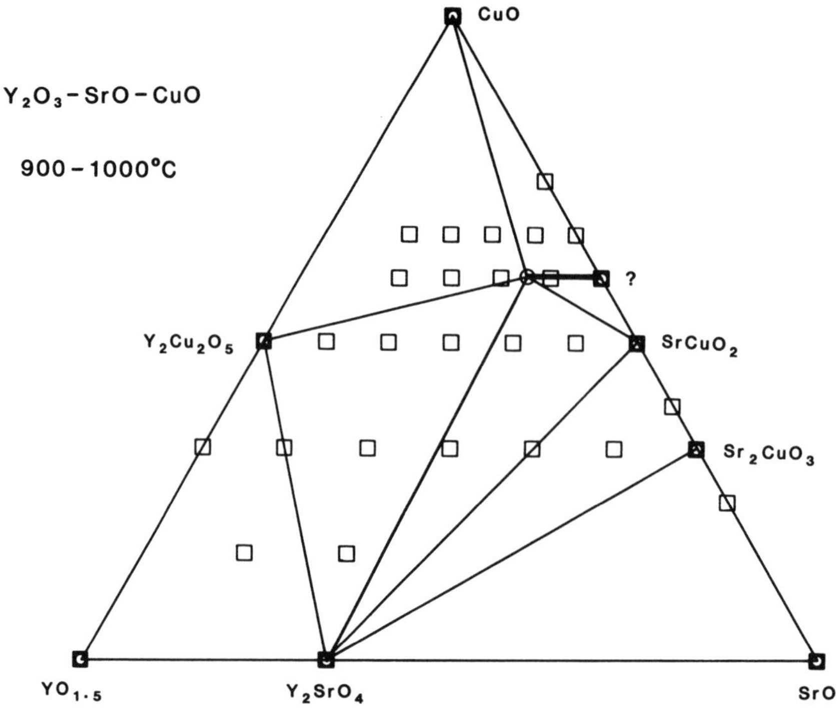

Fig. 9
Compatability regions in the Y_2O_3-SrO-CuO pseudoternary system in the temperature range 900°-1000°C as determined by the present work. The squares indicate the nominal compositions of the prepared samples. The heavy line indicates the apparent range of solid solution for the phase with the unknown structure.

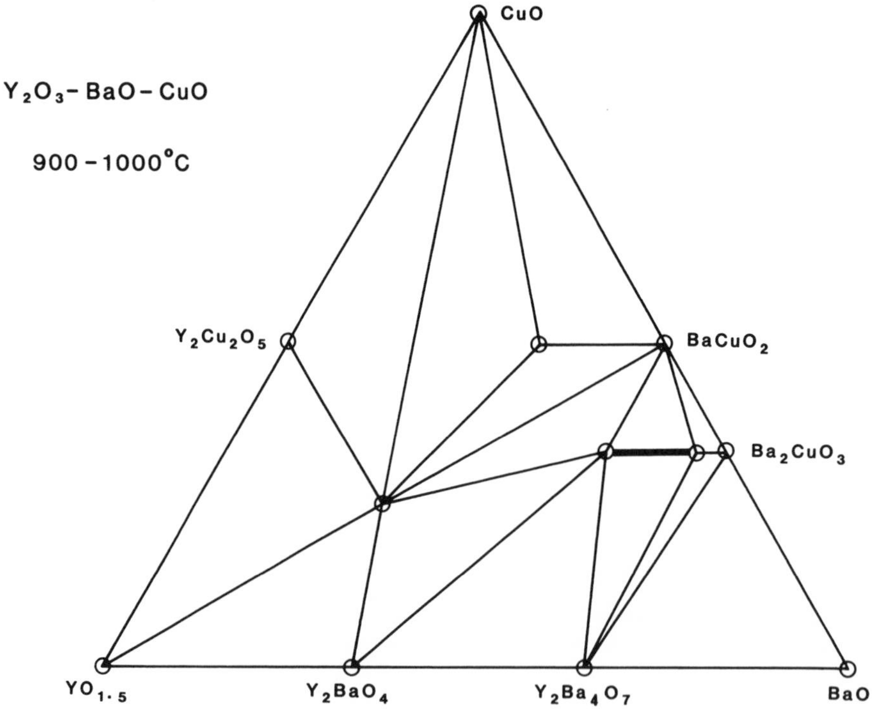

Fig. 10
Compatability regions in the Y_2O_3-BaO-CuO pseudoternary system in the temperature range 900°-1000°C. The points 2:1:1, 1:2:3 and 1:3:2 refer to the phases Y_2BaCuO_5, $YBa_2Cu_3O_{7-y}$ and $YBa_3Cu_2O_{4.5}$ respectively. The heavy lines indicate the apparent range of solid solution for the 1:3:2 phase.

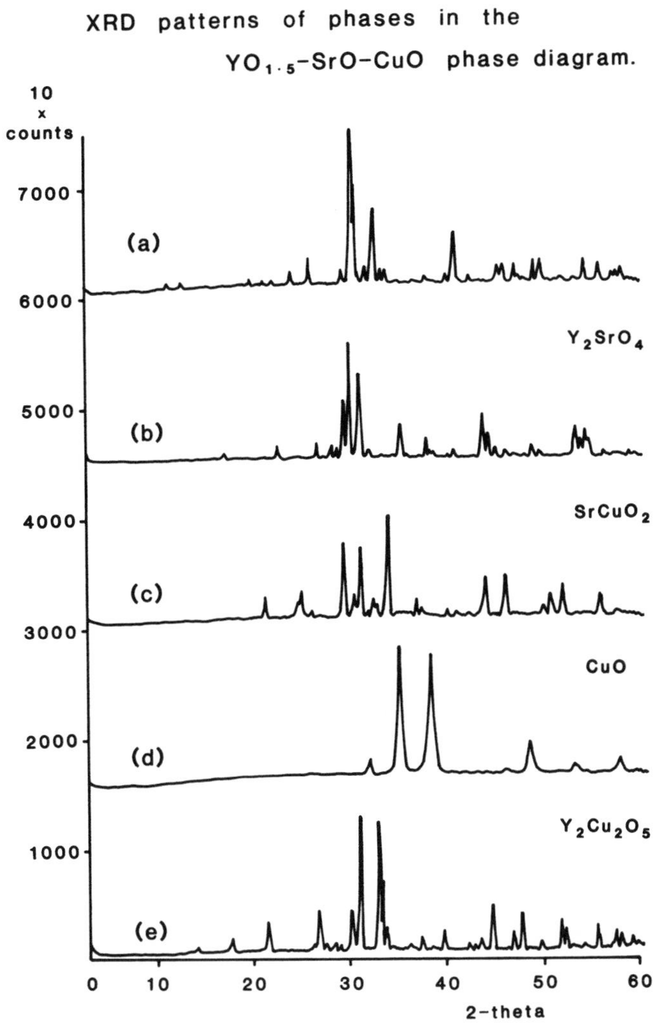

Fig. 11
X-ray diffraction patterns of some of the phases found in the Y_2O_3-SrO-CuO pseudoternary system.

Y_2O_3, CuO and $SrCO_3$ were used as the starting materials (BDH or Aldrich). Suitable amounts of the powders were ground in an agate pestle and mortar pressed into pellets and placed in an alumina boat for firing. The samples were generally fired at 1000°C for 72 hours in air, then allowed to cool to room temperature in the furnace. Some of the copper-rich samples were fired at 900°C for 72 hours in air since they were observed to melt at the higher temperature.

Initial efforts were concentrated upon the investigation of compounds in the pseudobinary systems along the three edges of the triangle in Fig. 9. The Y_2O_3-SrO system contains several phases stable at 1000°C, but the only phase of interest for this paper is Y_2SrO_4 which has an orthorhombic structure with $\underline{a}$ = 10.77Å, $\underline{b}$ = 11.914Å and $\underline{c}$ = 3.410Å (ref. 21). The X-ray powder diffraction pattern of this insulating phase is shown in Figure 11(b).

CuO shows some very interesting magnetic properties which are not fully understood. The material is monoclinic with $\underline{a}$ = 4.684Å, $\underline{b}$ = 3.425Å, $\underline{c}$ = 5.129Å and $\underline{\beta}$ = 99°28' (ref. 22) and the X-ray diffraction pattern of this phase is shown in Fig. 11(d). The Cu^{2+} ions are in square planar coordination and two plausible ground states for the spins in this phase are: (a) long range antiferromagnetic ordering of pairs of spins to form dimers. Magnetic susceptibility measurements provide evidence for both of these (ref. 23) and it is possible that both ground states can occur depending upon the degree of oxygen deficiency. The ESR spectrum of CuO at room temperature is shown in Figure 12(d) and consists of a broad symmetrical resonance with a mean g value close to free spin and a peak-to-peak width of around a kG. ESR measurements at different temperatures and frequencies should help to resolve this question.

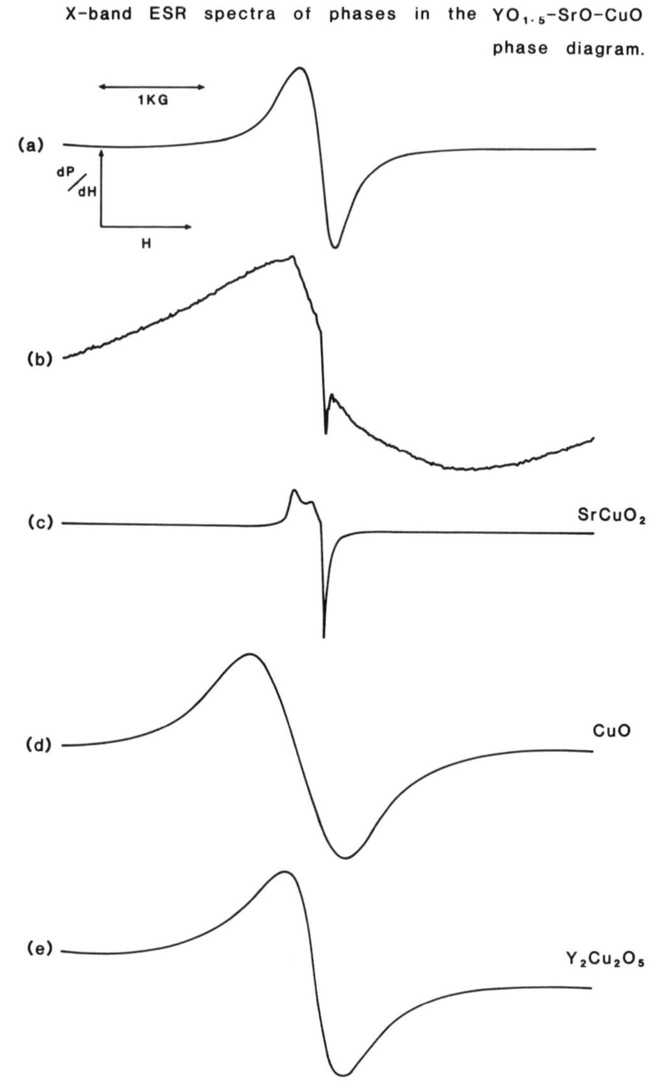

Fig. 12
X-band ESR spectra of some of the copper-containing phases found in the Y_2O_3-SrO-CuO pseudoternary system.

In the Y_2O_3-CuO system, only one pseudobinary compound is known to be stable at 1000°C. $Y_2Cu_3O_5$ has an orthorhombic structure with a = 3.527Å, b = 5.417Å and c = 12.488Å) (ref. 24) and the X-ray powder diffraction pattern of this phase is shown in Fig. 11(e). The Cu^{2+} ions in this structure are in very distorted tetrahedral coordination, and the X-band ESR spectrum shown in Fig. 12(e) consists of a broad symmetrical resonance with a mean g value close to free spin and a peak-to-peak width of ca. 500G. Magnetic susceptibility measurements on this material have shown that the Cu^{2+} ions order antiferromagnetically at 13K (ref. 25). It was noticed that an incomplete reaction occurred in some of the yttrium-containing compositions if the the samples were fired for a shorter period than 72 hours. Under these circumstances, the Y_2O_3 and the CuO react rapidly to form $Y_2Cu_2O_5$ as an intermediate phase.

The SrO-CuO pseudobinary system is even more complicated, and also contains several stable phases. The investigation of this pseudobinary system is made difficult by the low melting point of the copper-rich compositions, the volatility of strontium oxide and the reactivity of the strontium-rich compositions with atmospheric moisture. Similar problems are found in the corresponding BaO-CuO pseudobinary system (refs. 19,20).

Sr_2CuO_3 has an orthorhombic structure with a = 12.684Å, b = 3.909Å and c = 3.496Å (ref. 26). Traces of $SrCuO_2$ and $SrO.2H_2O$ (ref. 27) were also visible in the diffraction pattern of this sample, and the presence of a small amount of these phases is understandable since the SrO is volatile and Sr_2CuO_3 is likely to be hygroscopic. Similar phenomena are reported in the case of Ba_2CuO_3 (refs. 19,20). Rather surprisingly, we could not detect any ESR signal from the material apart from a weak signal due to the $SrCuO_2$ as discussed below. The reason for the absence of the ESR signal is not understood since the Cu ions are in square-planar coordination.

$SrCuO_2$ is reported to have an orthorhombic structure with a = 3.565Å, b = 16.326Å and c = 3.921Å and the X-ray powder diffraction pattern of this phase is shown in Fig. 11(c). The Cu^{2+} ions in this structure are in square planar coordination, and the X-band ESR spectrum shown in Fig. 12(c) consists of a narrow resonance with strong anisotropic features. Experiments at Q-band frequency are underway to accurately determine the g factors for this material.

In addition to these documented phases, we detected another SrO-CuO phase with an approximate Sr:Cu ratio of 2:3 and the X-ray powder diffraction pattern shown in Fig. 11(a). The X-band ESR spectrum of this phase shown in Fig. 12(a) consists of a narrow symmetrical resonance with a mean g value close to free spin and a peak-to-peak width of ca. 400 G. This phase appears to have a range of solid solution as discussed below.

After investigation of the three pseudobinary systems, the complete pseudo-ternary system was investigated to see whether any new phases could be formed. By means of the X-ray diffraction and ESR measurements, the tie lines shown in Fig. 9 could be constructed. ESR proved a powerful technique in this regard since the presence of the samples of trace amounts of $Y_2Cu_2O_5$ and $SrCuO_2$ was readily detectable.

The great majority of the prepared samples were poor electrical conductors as judged by their weak perturbation of the ESR spectrometer resonant cavity. Only those samples with compositions inside the triangle shown in Fig. 9 with vertices CuO, $SrCuO_2$ and the unknown phase showed appreciable electrical conductivity and it was necessary to finely grind these samples before recording the ESR spectra. Pellets of these samples were therefore examined for superconductivity in the temperature range 5-300K, but none of the samples proved to be superconducting.

The unknown phase mentioned above appears to have a solubility range extending approximately over the compositions shown in Fig. 9. As the yttrium content of this phase increased, both the electrical conductivity of the material and the width of the ESR signal increased. The 1:2:3 phase containing strontium instead of barium does not form under these conditions; a mixture of Y_2SrO_4, $SrCuO_2$ and the unknown phase is formed instead. Figure 12(b) shows the ESR signal obtained from a sample with the nominal composition of the 1:2:3 phase; the narrow signal arises from the $SrCuO_2$ and the broad signal is tentatively attributed to the unknown phase with the composition at the yttrium-rich end of the solid solution range.

In order to further increase the electrical conductivity of the unknown phase, an attempt was made to replace some of the strontium with barium. However, X-ray diffraction measurements showed that the barium preferred to form the 1:2:3 phase rather than form a solid solution with the unknown phase; in samples where over half of the strontium had been substituted a superconducting transition at 90K was observed.

It is of interest to compare the Y_2O_3-SrO-CuO phase diagram with the corresponding Y_2O_3-BaO-CuO phase diagram. One major difference is that the strontium analogues of the main pseudoternary phases Y_2BaCuO_5, $YBa_2Cu_3O_{7-y}$ and $YBa_3Cu_2O_{6.5}$ do not form under the preparation conditions employed in this work. In partial compensation, an unknown phase appears for copper-rich compositions which does not appear to have a barium analogue. It is plausible, however, that these phases could be made under preparation conditions involving different temperatures and atmospheres; indeed the preparation of $YSr_2Cu_3O_{7-y}$ has already been claimed (ref. 4).

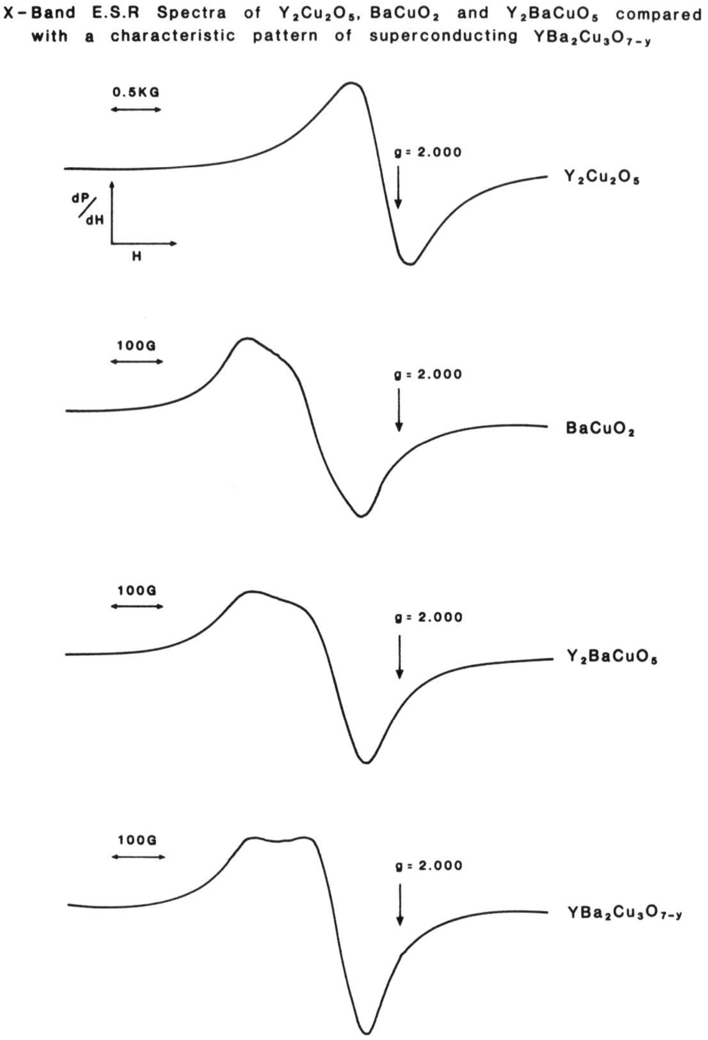

Fig. 13
X-band ESR spectra of some of the copper-containing phases found in the Y_2O_3-BaO-CuO pseudoternary system.

Figures 13 and 14 show the X-band and Q-band ESR signals of some of the phases in the Y_2O_3-BaO-CuO phase diagram. The ESR signal in the 1:2:3 phase is attributed to the presence of small amounts of Y_2BaCuO_5 and $BaCuO_2$ in the samples (ref. 28). The Cu^{2+} ions of Y_2BaCuO_5 are in distorted square-pyramidal coordination (ref. 29) and the Cu^{2+} ions in $BaCuO_2$ are apparently in a combination of square-pyramidal and square-planar coordination (ref. 30). The ESR spectra of these two phases are very similar to each other and to $SrCuO_2$. It is clear that the magnetic and magnetic resonance properties of the copper-containing phases in these pseudoternary phase diagrams are extremely interesting and more detailed measurements are currently underway.

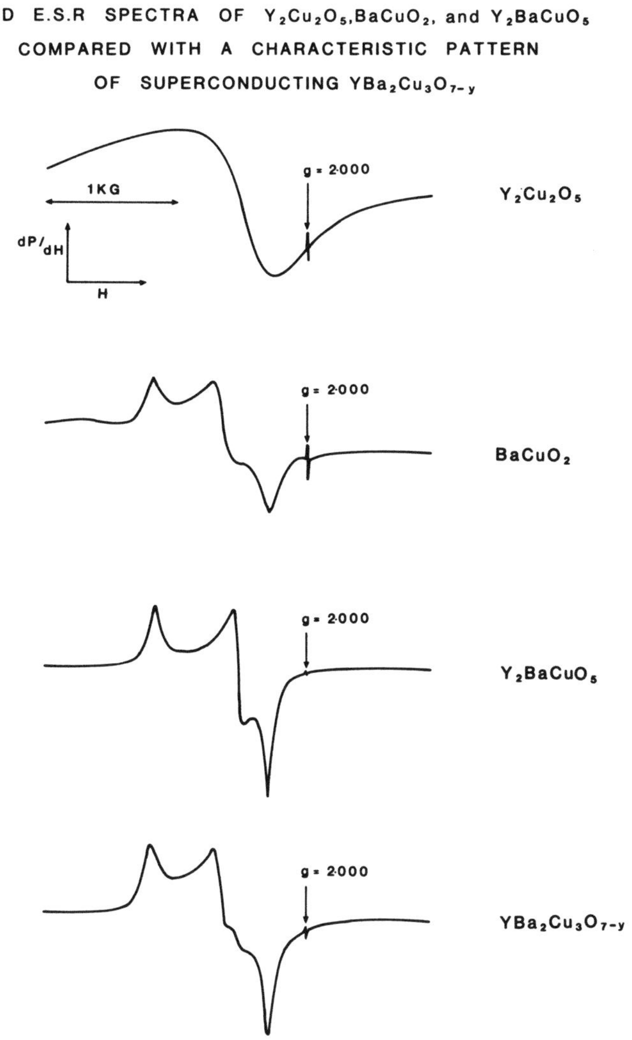

Fig. 14
Q-band ESR spectra of some of the copper-containing phases found in the Y_2O_3-BaO-CuO pseudoternary system.

ACKNOWLEDGEMENTS

Support from the S.E.R.C. is gratefully acknowledged. We would like to thank C. C. Welch and P. Miller for the measurements of the superconducting transitions.

REFERENCES

1. W. I. F. David, W. T. A. Harrison, J. M. F. Gunn, O. Moze, A. K. Soper, P. Day, J. D. Jorgensen, D. G. Hinks, M. A. Beno, L. Soderholm, D. W. Capone, I. K. Schuller, C. U. Segre, K. Zhang and J. D. Grace, Nature, 327, 310-312 (1987).
2. T. Wada, S. Adachi, T. Mihara and R. Inaba, Japan. J. Appl. Phys., 26, L706-L708 (1987).
3. B. W. Veal, W. K. Kwok, A. Umezawa, G. W. Crabtree, J. D. Jorgensen, J W Downey, L. J. Nowicki, A. W. Michell, A P. Paulikas and C. H. Somers, Appl. Phys. Lett., 51, 279-281 (1987).
4. M. Oda, T. Murakami, Y. Enomoto and M. Suzuki, Japan. J. Appl. Phys., 26, L804-L806 (1987).

5. Z. Qi-Rui, C. Lie-Zhao, Q. Yi-Tai, C. Zu-Yao, G. Wei-Yan, Z. Yong, P. Guo-Qang, Z. Han, X. Jaian-Sheng, Z. Ming-Jian, Y. Dao-Qi, H. Zheng-Hui, S. Shi-Fang, F. Ming-Hu and Z. Tao, Solid State Comm., 63, 535-536 (1987).
6. T. Wada, S. Adachi, O. Inoue, S. Kawashima and T. Mihara, Japan. J. Appl. Phys., 26, L1475-L1477 (1987).
7. L. Soderholm, K. Zhang, D. G. Hinks, M. A. Beno, J. D. Jorgensen, C. U. Segre and I. K. Schuller, Nature, 328, 604-605 (1987).
8. R. Liang, Y. Inaguma, Y. Takagi and T. Nakamura, Japan. J. Appl. Phys., 26, L1150-L1152 (1987).
9. W. I. F. David, W. T. A. Harrison, R. M. Ibberson, M. T. Weller, J. R. Grasmeder and P. Lancaster, Nature, 328, 328-329 (1987).
10. E. Takayama-Muromachi, Y. Uchida, A. Fujimori and K. Kato, Japan. J. Appl. Phys., 26, L1546-L1549 (1987).
11. W. W. Warren, R. E. Walstedt, G. F. Brennert, G. P. Espinosa and J. P. Remeika, Phys. Rev. Lett., 59, 1860-1863 (1987).
12. H. Riesemeier, C. Grabow, E. W. Scheidt, V. Muller, K. Luders and D. Riegel, Sol. State Comm., 64, 309-312 (1987).
13. Y. Maeno, T. Tomita, M. Kyogoku, S. Awaji, Y. Aoki, K. Hoshino, A. Minami and T. Fujita, Nature, 328, 512-514 (1987).
14. G. Xiao, F. H. Streitz, A. Gavrin, Y. W. Du and C. L. Chien, Phys. Rev., B25, 8782-8784 (1987).
15. H. Ihara, N. Terada, M. Jo, M. Hirabayashi, M. Tokumoto, Y. Kimura, T. Matsurbara and R. Sugise, Japan. J. Appl. Phys., 26, L169-L171 (1987).
16. B. Jayaram, S. K. Agaewal, A. Gupta and A. V. Narlikar, Solid State Comm., 63, 713-716 (1987).
17. R. Jones, P. P. Edwards, M. R. Harrison, W. I. F. David, C. C. Wilson and W. S. Hewells, submitted to Nature, (1988).
18. D. C. Harris and T. A. Hewston, J. Solid State Chem., 69, 182-185 (1987).
19. K. G. Frase, E. G. Liniger and D. R. Clarke, submitted to Comm. Amer. Ceram. Soc.
20. K. G. Frase and D. R. Clarke, submitted to Mater. Sci.
21. JCPDS Powder Diffraction File, "Y_2SrO_4", 32-1272.
22. JCPDS Powder Diffraction File, "CuO", 5-661.
23. B. Roden, E. Braun and A. Freimuth, Solid State Comm., 64, 1051-1052 (1987).
24. JCPDS Powder Diffraction File, "$Y_2Cu_2O_5$", 33-511.
25. R. Troc. Z. Bukowski, R. Horyn and J. Klamut, Phys. Lett., A125, 222-224 (1987).
26. JCPDS Powder Diffraction File, "Sr_2CuO_3", 34-283.
27. JCPDS Powder Diffraction File, "$SrO.2H_2O$", 28, 1222.
28. G. J. Bowden, P. R. Elliston, K. T. Want, S. X. Dou, K. E. Esterling, A. Bourdillon, C. C. Sorrell, B. A. Cornell and F. Separovic, J. Phys., C20, L545-L552 (1987).
29. C. Michel and B. Raveau, J. Solid State Chem., 43, 73-80 (1982).
30. H. N. Migeon, F. Jeannot, M. Zanne and J. Aubry, Revue de Chimie Minerale, 13, 440-445 (1976).

Theory of superconductivity in systems with strong Coulomb correlations, with applications to the organic and oxide superconductors

S. Mazumdar

Physical Chemistry Division, National Chemical Laboratory, Pune-411 008, India.

Abstract - A general theory of superconductivity (SC) is developed for systems with repulsive interactions between electrons. Two dimensionality and band filling close to one-quarter are essential requirements for SC. Pairing results from the combined effect of the on-site and nearest neighbor Coulomb repulsion driving a charge density wave (CDW) with periodicity $4k_F$ and antiferromagnetic correlation between nearest occupied sites. Superconducting pairs are polarons moving in the background CDW. Application to the organic superconductors all of which have quarter-filled bands is straightforward. In $La_{2-x}M_xCuO_{4-y}$ where M=Sr, Ba, the dopant-induced insulator-to-metal transition is accompanied by a quasiionic-to-quasicovalent transition in which all the holes are transferred from the copper to the oxygen at $x \sim 0.15$. With twice as many oxygens as coppers, doped systems are therefore nearly quarter-filled. The metal-oxygen layers in Y-Ba-Cu-O and Ba-Pb-Bi-O similarly consist of quarter-filled oxygen bands. The list of experiments explained within the present model includes such diverse phenomena as the limitation of SC to the orthorhombic structure, the perplexing role of electron-phonon interactions in these Coulomb correlated systems, loss of SC with loss of the chain oxygen in Y-Ba-Cu-O etc. A very wide variety of experiments are explained within the present model, proving its validity.

INTRODUCTION

The discovery of high temperature SC has led to intense theoretical effort to understand the mechanism of this exotic phenomenon (see reference 1 for a recent review). However, too much emphasis has been placed on the high temperature itself and the very large number of other unusual properties are only beginning to be recognized. As pointed out by Anderson et al (ref. 2), almost none of the properties of the copper oxide based superconductors correspond to those of the "normal" superconductors, while the behavior of these and several other classes of narrow-band materials are similar. It should in principle be possible to explain the SC in all these exotic materials within one general theory, while the high T_c itself is a consequence of some special feature of the copper oxides within the general theoretical framework.

We develop a general theoretical framework here, focussing on the organic superconductors (ref. 3), the "low" T_c oxide $BaPb_{1-x}Bi_xO_3$ and the high T_c oxides $La_{2-x}M_xCuO_4$ where M=Sr, Ba and $YBa_2Cu_3O_{7-y}$. For convenience, these will be mostly written as Ba-Pb-Bi- O, La-M-Cu-O and Y-Ba-Cu-O respectively. The emphasis will be on understanding the complete emperimental behavior on a qualitative level rather than calculating T_c. The similarities between the low T_c and high T_c oxides have been emphasized by several authors, while the large number of similar characteristics between the latter and the organics (ref. 3) have been pointed out recently (ref. 4). The common themes are quasi-two-dimensionality and electron correlation, but we go beyond this in the present work. We shall theoretically argue that while semiconducting La_2CuO_{4-y} can be considered as half-filled band of Cu-holes, the doped material beyond a small critical dopant concentration has all the holes on the O-atoms, so that the system behaves like a nearly quarter-filled band (there being twice as many oxygens as coppers in the CuO_2 layer), exactly as the organics. Recognition of this simultaneously eliminates several theoretical problems and explains nearly all the experimental anomalies. Besides explaining the currently available data, we also make several predictions. We are unaware of any theoretical model that explains all the experiments discussed in the present paper.

Antiferomagnetism (AF) has been recognized to hold the key (ref. 5) to the theory of SC in the high T_c oxides. However, as discussed here (and by other investigators, see below) AF by itself cannot drive the SC. Pairing is shown to be a result of the combined effect of strong on-site Coulomb repulsion which drives the AF as well as intersite Coulomb repulsion which gives a background charge density wave (CDW) near the quarter-filled band sector. Superconducting pairs are polarons moving within this background CDW. In the oxides these polarons are nearest neighbor O^{1-}-O^{1-} species.

While we have focussed only on the organics and the oxides here, we believe that the present theory is flexible enough that it can be extended to explain the exotic SC in other strongly correlated systems like the heavy-fermions (ref. 6).

BRIEF REVIEW OF THE THEORETICAL SITUATION

We present a brief review of the various theoretical models that have been suggested for the copper oxides. An early review (ref. 1) has already appeared. As discussed there, and as is indicated by the absence of the isotope effect in Y-Ba-Cu-O (ref. 7), the BCS mechanism cannot be the origin of SC in this system. A weak isotope dependence of T_c has recently been observed (ref. 8) in La-M-Cu-O. The weakness of the mass dependence, along with the confirmed AF in the undoped system (ref. 9) argue against phonon-based mechanisms (see later for further discussions of this isotope effect). We are unaware of similar investigations in Ba-Pb-Bi-O, but we believe that the mechanism of SC in this material is the same. Finally, absence of isotope effect in the organics has been noted earlier (ref. 10).

In addition to the absence of the isotope effect, SC in these systems is reminiscent (ref. 2) of superfluidity in He II, suggesting Bose type condensation (refs. 2,11). This is supported by "metallic" specific heat coefficient γ as $T \rightarrow O$ in Ba-Pb-Bi-O (ref. 12), the high T_c oxides (ref. 13) and the organic superconductors (ref. 14). Similar anomalies exist with large tunneling gaps and the exceptionally large upper critical fields. Because of the above we also neglect theories which are "BCS-isomorphs", involving BCS-type couplings of band electrons with bosons other than phonons (plasmons, excitons etc.).

Finally we come to the theories emphasizing electron correlation, which can be broadly classified into single-band (refs. 2,15-20) and two-band (refs. 21-23) models. The single band models are largely based on the two dimensional (2D) simple Hubbard Hamiltonian with strong on-site electron repulsion U,

$$H_{Hub} = U \sum_i n_{i\uparrow} n_{i\downarrow} + \sum_{\langle ij \rangle \sigma} t_{ij} (c^+_{i\sigma} c_{j\sigma} + c^+_{j\sigma} c_{i\sigma}) \qquad (1)$$

Here $c^+_{i\sigma}$ creates a fermion with spin σ at site i, $\langle ij \rangle$ are nearest neighbors in a 2D lattice, t_{ij} a fermion hopping integral and $n_{i\sigma}$ is the number of fermions with spin σ at site i. Theoretical work with the high T_c oxides have been largely confined to La-M-Cu-O, and the active sites are presumed to be the copper-atoms in the CuO_2 layer. The vacuum consists of Cu^{1+} and O^{2-}, so that La_2CuO_{4-y} with formally all the coppers in the Cu^{2+} state (one hole in the $3d_{x^2-y^2}$ orbital) has a half-filled band with density of holes per site 1. For large U the system is a Mott-Hubbard semiconductor. Both s-wave (refs. 15-18) and d-wave (ref. 19) pairing havebeen suggested in the weakly doped Hubbard model. Numerical simulations had earlier suggested the possibility (ref. 24) of SC in the Hubbard model, but this work has recently been criticized (ref. 25). More recent numerical work (ref. 26) do not find SC within the single band Hubbard model. We believe that these calculations, even though based on very small systems, predict the correct infinite system results because pairing in the above theories is <u>local.</u> Intuitively this absence of pairing may be understood: while the antiferromagnetic coupling that is supposed to lead to pair binding goes as $J_{ij} = 2t_{ij}^2/U$, delocalization goes as $t_{ij} >> J_{ij}$. This is independent of ρ and dimensionality.

An interesting variation of the rigid band Hubbard models have been suggested (ref. 20), within which the ground state of (1) exhibits enhanced 2D bond alternation (exactly as in 1D, ref. 27) for = 1, in the presence of coupling with intersite phonons. The weakly doped $\rho = 1$ system is supposed to maintain this short-bond long-bond structure, and pairing involves the holes on a short bond in this doped spin-Peierls phase. It has, however, been subsequently shown (ref. 28) that the on-site correlation destroys the bond-alternation in 2D, so that the spin-Peierls phase does not occur here. Finally, the observation that doping leads to holes on the oxygen-atoms (ref. 29) rather than the copper-atoms makes eqn. (1) irrelevant and the arguments about SC in the Hubbard model academic any way.

In the two-band models (refs. 21-23) a $\rho = 1$ Mott-Hubbard band corresponding to the Cu^{2+} ions is again assumed, and doping leads to a few holes on the oxygens. The oxygen holes are paired through being coupled to the AF of the Cu^{2+} background. These models are therefore highly specific to La-M-Cu-O, and furthermore, do not explain several experimental features. Similar very specific models have been proposed for the organics (triplet SC, exchange of spin fluctuations between neighboring chains, etc.) but these also do not explain important experimental facts (for instance, the limitation of organic SC to $\rho = 0.5$). We believe that nearly all the experimental peculiarities can be understood within an extended Hubbard Hamiltinian, which has been previously proposed (refs. 30-33) as a unified theoretical model for segregated stack quasi-one-dimensional charge-transfer solids. In the next section we discuss how SC is obtained within the 2D extended Hubbard model near $\rho = 0.5$, and in the following sections discuss the applicability of the same to the organaics and the oxides. Nearly all the anomalies and contradictions are explained.

THE EXTENDED HUBBARD MODEL FOR SUPERCONDUCTIVITY

We go beyond the simple Hubbard model of eqn. (1) and add a nearest neighbor Coulomb repulsion to get the extended Hubbard Hamiltonian,

$$H_{EH} = H_{Hub} + \sum_{\langle ij \rangle} V_{ij} n_i n_j \qquad (2)$$

Here $n_i = \sum_{\sigma} n_{i\sigma}$ is the total number of electrons at site i. Within a square lattice $t_{ij} = t$, $V_{ij} = V$, but for application to the organics we shall choose $t_{ij} = t_{\parallel}, t_{\perp}$, $V_{ij} = V_{\parallel}, V_{\perp}$, where $t_{\parallel}$, $V_{\parallel}$ correspond to intra-stack interactions and $t_{\perp}$, $V_{\perp}$ interstack interactions. We shall consider eqn. (2) near the quarter-filled band sector with $\rho = 0.5$.

The geometric broken symmetries within eqn. (2) for $\rho = 0.5$ have been thoroughly investigated (ref. 34) within the $t_{\perp} = V_{\perp} = 0$ 1D limit. The $4k_F$ CDW and $2k_F$ SDW ($k_F = \pi/4a$ is the Fermi wavevector, a = lattice spacing) coexist here, both requiring the same site occupancy modulation, with alternate sites occupied. CDW is also obtained in a related 2D spinless fermion system (ref. 35) with nearest neighbor Coulomb interactiuon ($n_i = 0$, 1 only). Therefore, for large U and sufficiently large V_{ij} ($V_{\parallel} > 2t_{\parallel}$ is required in 1D limit for long range CDW order) a 2D CDW is always obtained within eqn. (2). For slightly smaller V_{ij} one still expects short range ordered CDW. For $t_{\parallel}$ much large than nonzero $t_{\perp}$ a $2k_F$ SDW will accompany the CDW with the spins on the occupied sites alternating along a $\rho = 0.5$ stack. For $t_{\parallel} \sim t_{\perp}$, fraustration can destroy long range SDW even though the 2D CDW is stronger, - spin coupling between nearest neighbor occupied sites on the same stack and on different stacks are comparable. The $U \rightarrow \infty$ approximation (no double occupancies allowed) must be valid for SC to occur within eqn. (2). This condition is met at $\rho = 0.5$ since both V_{ij} and U reduce $\langle n_{i\uparrow} n_{i\downarrow} \rangle$ here (refs. **30-33**). Equation (2) can be written in this limit as,

$$H_1 = \sum_{\langle ij \rangle} V_{ij} n_i n_j + \sum_{ij} t_{ij} \left[(1-n_{i,\sigma})\, c^{+}_{i\sigma} c_{j\sigma} (1-n_{j,\sigma}) + h.c.\right] \qquad (3)$$

where fermion hops occur only to unoccupied sites, $n_i = 0$ or 1 and magnetic terms of the order (t^4/U^3) have been neglected. To understand the SC let us consider the weakly incommensurate case $\rho = 0.5 + \epsilon$, $\epsilon \rightarrow 0^+$ near the $t_{ij} \rightarrow 0$ limit. In Fig. 1(a) we show the situation in 1D, where we have added a single fermion to the static CDW-SDW configuration, creating two "Hubbard dimers" (ref. 36), each dimer being a nearest neighbor occupied pair. Single particle hop from the configuration ... 010111010 ... to the configuration ... 010110110 ... has zero Coulomb barrier, where the numbers denote site occupancies. A series of such single particle hops, leading to infinite separation of the dimer pair (ref. 36), is possible in 1D (see Fig. 1(a)). As more fermions are added these occupy the sites that were unoccupied in the commensurate $\rho = 0.5$ CDW configuration and once again low energy correlated motion as well as incommensurate correlated CDW (ref. 32) is possible.

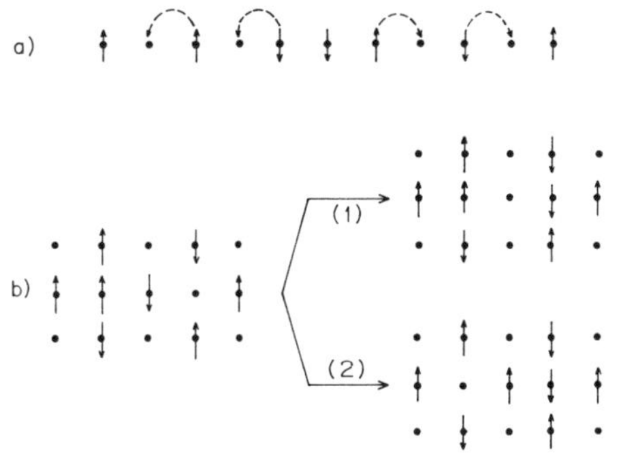

Fig. 1. Charge transfers in the $\rho = 0.5 + \epsilon$ CDW in (a) 1D, (b) 2D. Solid and dashed arrows represent electrons and nearest neighbor hops respectively. Dimers separate freely in 1D by consecutive single particle hops, but there exists strong Coulomb barrier to similar process in 2D in path (1). The pair hopping in path (2) has no such barrier.

The above are not true in 2D, as shown in Fig. 1(b), where we have added the extra fermion to the center of a cage formed by the CDW. Four nearest neighbor dimers are formed now, and as shown in path 1 a single particle hop now has a large Coulomb barrier $2V_{\perp}$. What is more important is that for consecutive hops as in Fig. 1(a), there is a strong Coulomb barrier at each step. Separation of the dimers is therefore much more difficult in 2D. Note, however, that a two particle hop, as shown in path 2, has zero Coulomb barrier. Furthermore, independent of whether there is long range 1D SDW or fraustration due to comparable $t_{\parallel}$ and $t_{\perp}$, at least two out of the four dimers within a cage are always singlet coupled. Consecutive two particle hops can now occur within the background CDW, where each pair is very similar to a polaron.

We have demonstrated here pairing due to purely repulsive interactions. Pairing in the incommensurate case occurs for the same reason as CDW in the commensurate case, - Coulomb repulsion is minimized this way. The pair binding may also thought to be a consequence of coupling to a different degree of freedom (background CDW), regarded essential by Emery (ref. 21) in correlated systems. What are to be noted are, (a) binding is occurring within a single band here, and (b) the binding energy is larger than the delocalization energy.

It is emphasized that this binding is not a special case in Fig. 1, it is a consequence only of two dimensionality and nonzero V_{ij}, and persists for $1/2 < \rho < 2/3$, assuming that the $U \rightarrow \infty$ approximation remains valid for large ρ. Between these two commensurate points minimum energy configurations are obtained by simply adding fermions to the empty sites in the $\rho = 0.5$ CDW configuration, so that the local site occupancy scheme remains the same as in Fig. 1(b). This is shown in Fig. 2 for $\rho = 0.6$. We have chosen an arbitrary configuration which minimizes the Coulomb repulsion, and find that while all pair motions are energetically feasible, single particle hops (in particular a series of hops) cost too much Coulomb energy not only within the given configuration, but within any configuration reached by any number of pair hoppings from the former. However, a series of pair hoppings connect all possible minimum energy configurations.

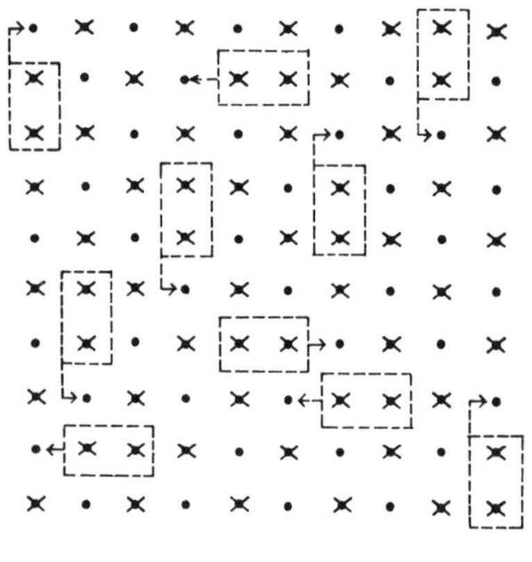

Fig. 2. Free particle like pair motion in $\rho = 0.6$. The crosses are electrons, and nearest neighbor spin pairing due to large U has been assumed.

The reduced Hamiltonian within the paired subspace is,

$$H_{red} = -(\bar{t}^2/2\bar{V}) \sum_{\langle ij\rangle \langle jk\rangle \sigma} [1-n_{i,-\sigma}) c^{+}_{i\sigma} c^{+}_{j,-\sigma} c_{j\sigma} c_{k,-\sigma} (1-n_{k\sigma}) + \text{h.c.}] \tag{4}$$

where the large U is responsible for singlet pairing and $\bar{t}$ and $\bar{V}$ are weighted averages of $t_{\parallel}$, $t_{\perp}$ and $V_{\parallel}$, $V_{\perp}$. The term neglected in eqn. (4) corresponds to the superconducting state is intermediate between the commensurate CDW and the incommensurate correlated conducting state.

From Figs. 1 and 2 the paired state is a true liquid state as in Anderson's resonating valence bond (RVB) theory, except here the binding is due to V_{ij} and occurs near $\rho = 0.5$ and not $\rho = 1$. Each many-electron configuration is exactly equally probable and the wavefunction is of the form $\Psi = \sum_m e^{iKa} \Phi_m$ where Φ_m is any configuration and K a total wavevector for the entire system. The ground state (K=0) is a coherent charged superfluid with Bose type pair condensation. The nearest neighbor pairing here is very similar to that proposed in the bipolaron model of superconductivity (ref. 37) involving also local pairing. The theory is also conceptually very close to a recent model of bipolaron condensation in doped CDW systems (ref. 38), the difference being that the background CDW here is due to electron correlation. Similarities exist also with phenomenological models (ref. 39) of coexisting CDW or SDW and SC.

We do not present any calculations of the thermodynamics but point out some general theoretical and experimental features. The critical temperature $T_c \sim (\bar{t}^2/2\bar{V}).\ n_c$ where n_c is the number of carriers. We defer the "calculations" of T_c until later, where we will show that the predicted critical temperatures for the organics, Ba-Pb-Bi-O and the copper oxide based superconductors are all reasonable. Independent of T_c and the other parameters the following are expected in general:

(i) A "gapless" specific heat behavior, the lowest excitations being Bose like. This is observed experimentally (ref. 12-14).
(ii) A large ($>$ $3.5T_c$ in most cases) tunneling gap. Single-particle excitations are due to the higher order term in eqn. (4) and are like those of an incommensurate correlated CDW state to the corresponding conducting state. The actual magnitude of the gap will vary, but will be strong function of Coulomb repulsion and the degree of two dimensionality in case of the organics. Large tunneling gaps have been experimentally observed both in the organics (ref. 40) and the oxides (ref. 41).
(iii) A very large upper critical field, - spin pairing is preexisting due to the large U and very strong. Again this is experimentally observed in the organics (ref. 42) and the oxides (ref. 43).
(iv) Unusually strong nonmagnetic impurity effects. The irradiation effects in the organics (which are otherwise large clean single crystals) are particularly strong (ref. 3). Such strong impurity effects are consequence of very short coherence lengths and will occur in any local-pairing theory.

Since the above features would be predicted within several of the proposed theories we now discuss more specific experiments in the organics and the oxides, to show the applicability of the extended Hubbard model to these systems. In particular for the copper oxides we develop a theory of insulator-to-metal transition that is very different from what has been published until now. Agreement between the present theory and experiments is then pointed out.

APPLICATION TO THE ORGANIC SUPERCONDUCTORS

Application of the present theory to the organics is straightforward (ref. 4). The active sites are the lowest unoccupied molecular orbitals of the organic molecules TMTSF (tetramethyltetra-selena fulvalene) and BEDT-TTF (bisethylenedithiotetrathiafulvalene, hereafter ET). The chemical formulas of the materials are $(TMTSF)_2X$ and ET_2X, where X is a monovalent inorganic ion, viz., PF_6, ClO_4, I_3 etc., so that stoichiometry requires one hole per two organic molecules ($\rho = 0.5$). The donor molecules form quasi-two-dimensional layers separated by the inorganic anions. The anisotropy within the layers is quite strong for TMTSF ($t_{\parallel} \sim 10t_{\perp}$), within the ET-layer the anisotropy is weaker ($t_{\parallel} \sim 2\text{-}3t_{\perp}$) (ref. 3).

As the temperature is lowered a typical TMTSF-based system first goes into a SDW state. Application of pressure is necessary for SC in all cases except for $(TMTSF)_2ClO_4$, which is an ambient pressure superconductor (provided cooling is slow enough). The ET_2X materials do not exhibit a SDW but CDW has been found in some materials (ref. 3). This is in conformity with that we have said about spin fraustrations and the resultant absence of SDW in strongly 2D $\rho = 0.5$ systems. Ambient pressure SC has been observed in β-ET_2I_3, β-ET_2AuI_2, and more recently in pressure-recycled (refs. 3,44) β-ET_2I_3. What is interesting in this last case is that T_c jumps from about 1K to nearly 8K after pressure treatment.

The SC in the above materials is understood within the present theory. Strong correlations drive the transition to the CDW or SDW. In the case of $(TMTSF)_2X$, because of the wider 1D bandwidth than usual $V_{\parallel} < 2t_{\parallel}$ and therefore CDW with long-range order is not obtained. This is irrelevant in the present context, since SDW and CDW require the same site occupancy modulation at $\rho = 0.5$ (unlike $\rho = 1$). The effect of pressure (increasing $t_{\parallel}, t_{\perp}$) is to melt the CDW/SDW, but before the metallic state with arbitrary occupancy is reached, an intermediate state with a few pairs as in Fig. 1 is formed. It is this state which is superconducting. Assuming the critical density of pairs required for Bose condensation (ref. 45) to 2-5% we get a $T_c < 10K$ for the organics from the estimated values of $\bar{t}$ and $\bar{V}$.

The enhancement of T_c following the pressure treratment in β-ET_2I_3 can be understood by assuming incomplete charge-transfer in the "low T_c" phase before the pressure treatment. Note that within our theory the incommensurability effects for $\rho < 0.5$ and $\rho > 0.5$ are not the same. Some single particle motion without surmounting the Coulomb barrier is always allowed for $\rho < 0.5$. This lowers T_c. We speculate that application of pressure induces complete charge-transfer in β-ET_2I_3 (increase of charge-transfer with pressure occurs, for instance, in TTF-TCNQ). It is tempting to speculate the same about the other ambient pressure superconductors, viz., imperfect nesting due to incomplete charge-transfer. The important role of imperfect nesting (from a different perspective) has been emphasized by Jerome and Creuzet (ref. 3). Detailed comparisons to experiments investigating the superconducting state may be found in reference 4.

We complete this section by pointing out that the understanding of the normal state behavior of charge-transfer solids requires both large U and a substantial $V_{\parallel}$. This is most easily seen from:

(a) Optical properties: An absorption due to $V_{\parallel}$ in all $\rho < 1$ solids. In reference 33 we have discussed in detail why this absorption cannot be simply due to electron-phonon interactions within a large U model. This absorption is seen in all the organic superconductors.
(b) Magnetic susceptibility: While the magnetic susceptibility is strongly enhanced in some charge transfer solids, the enhancement factor is close to 1 in others. This is due to the strong ρ-dependence of $\langle n_{i\uparrow} n_{i\downarrow} \rangle$ within eqn. (2). Within the extended Hubbard model, susceptibilities are strongly enhanced for $0.5 < \rho < 0.6$ and weakly enhanced for $0.67 < \rho < 0.8$.
(c) Similarly the $4k_F$ instability in the organic systems is limited to $0.5 < \rho < 0.6$. This is also an effect of substantial $V_{\parallel}$ and the consequent ρ-dependent $\langle n_i n_j \rangle$.
(d) Finally, the limitation of organic SC to $\rho = 0.5$ is the strongest proof of an important role of V. The present SC mechanism is the only one which emphasaizes the role of band-filling.

APPLICATION TO THE OXIDE SUPERCONDUCTORS

Finally we come to the chief theme of the present paper, viz., a theory of the oxide superconductors. As discussed in the second section, in spite of the feverish experimental research, no existing theoretical model explains even a small fraction of the seemingly very perplexing data. Partly this is because of the large number of apparently contradictory experimental results. For e.g., undoped La_2CuO_{4-y} is antiferromagnetic, but the weakly doped $La_{2-x}M_xCuO_{4-y}$ exhibit several features that can be, and have been, associated with CDW, - even though AF and CDW normally do not coexist. It is in any event not clear how such CDW like behavior can persist for the incommensurate systems with moderately large x. A second source of the theoretical confusion is perhaps the tendency to ascribe the origin of the SC to every new experimentally observed property. Thus the proximity of the SC to the antiferromagnetic state has led to the widely accepted hypothesis that the superconducting pairing is due to the AF, even though this does not follow automatically. Similarly, the observation that the missing oxygens in $YBa_2Cu_3O_6$ are the ones on the quasi-one-dimensional chains in the crystal has led to the speculation that SC is one dimensional in this system. A very large number of such conjectures exist by now.

The theory presented here explains all the following in the oxide superconductors:

(i) AF in undoped La_2CuO_{4-y} (ref. 9),
(ii) CDW like features that are seen in La-M-Cu-O, superconducting Y-Ba-Cu-O and Ba-Pb-Bi-O, with soft phonons as well as vibrational doublets characteristic of systems in which there is an underlying CDW and the electrons are coupled with vibrations (refs. 46,47),
(iii) the very important role of the orthorhombic phase (ref. 48), and other such indications (refs. 49,50) of the role of phonons in this system with strong Coulomb correlation,
(iv) Photoemission and Auger data showing unusually large amounts of monovalent Cu^{1+} and O^{1-} (refs. 29, 51-53),
(v) epr and nmr data showing near total absence of Cu^{2+} in superconducting Y-Ba-Cu-O (ref. 54); Curie like magnetic susceptibility in spite of this,
(vi) an absorption at 0.4 - 0.5 ev seen only in superconducting La-M-Cu-O and Y-Ba-Cu-O (refs. 55,56),
(vii) magnetic susceptibility data with highly unusual dopant induced behavior in La-M-Cu-O (ref. 57),
(viii) Hall effect data showing a very large and sudden drop in the Hall coefficient beyond $x = 0.15$ in the above material (ref. 58), and finally,
(ix) the role of the chain oxygens in $YBa_2Cu_3O_{7-y}$ (refs. 59,60). This last feature is the most difficult to understand. Clearly if the assumption is made that SC in the La-M-Cu-O and Y-Ba-Cu-O has the same mechanism, then the SC must be confined to the copper-oxygen layers. In such a case there is no reason to lose SC when the chain oxygens are lost ! Not only do we explain the above phenomenon, but we shall even explain why the various Cu-O bond lengths are so different (ref. 61) in this material, and why the chain oxygens are the most weakly bonded, as evidenced from their loss on heating.

In addition to the above, the present model also explains the other unusual set of observations that we have already discussed, viz., "metallic" γ , large tunneling gap, very large upper critical field etc. all associated with local pairing and Bose condensation. Indeed we are unaware of any reproducible experimental observation that is not explained within the theory. Explanation of all the above normal state properties as well as SC requires only the recognition that all the conducting holes are on the oxygens even though in the semiconducting La_2CuO_{4-y} the holes are on the coppers. The dopant-induced insulator-to-metal transiton in La-M-Cu-O is accompanied by an ionic-to-covalent transition. The metal-oxygen layers in this material, as well as in Y-Ba-Cu-O and Ba-Pb-Bi-O consist of quarter-filled moderately strongly correlated oxygen bands, with the coppers mostly in the d^{10} configuration. We present here the theory of this ionic insulator-to-covalent conductor transition and follow it up with detailed explanations of all of the above experiments, along with several others. This will also be accompanied by several experimental predictions.

We begin by writing down the standard two band model for the CuO_2 layer in La_2CuO_{4-y},

$$H = \sum_{\substack{ij\\ \sigma}} \epsilon_{ij} c^+_{i\sigma} c_{j\sigma} + 1/2 \sum_{\substack{i,j\\ \sigma,\sigma'}} U_{ij} n_{i\sigma} n_{j\sigma'} \tag{5}$$

where $c^+_{i\sigma}$ creates holes, and the i,j run over all the copper and oxygen sites in the second term, i.e., we include the long range Madelung energy. The vacuum consists of Cu^{1+} and O^{2-}. Here $\epsilon_{ii} = \epsilon_d, \epsilon_p$, where ϵ_d, ϵ_p are site-energies that determine the hole occupancy. Experiments indicate that $\epsilon_d < \epsilon_p$. For $i \neq j$ we restrict the first term to i,j nearest neighbors so that $t_{ij} = t$ is the hybridization between copper and oxygen.

Within each CuO_2 layer in La_2CuO_{4-y} at least half the oxygens must be O^{2-} for charge balance, since lanthanum is certainly in the La^{3+} state. Because of the long range nature of the Madelung energy, it is also to be expected that these O^{2-} species are evenly distributed so that each CuO_2 unit has on the average one O^{2-}. This then leaves lthe lowest energy state of the remaining oxygen and the copper to be determined. In the limit of $t = 0$, two distinct states, - covalent Cu^+O^- and ionic $Cu^{2+}O^{2-}$ are possible. Since in the literature (ref. 62) this situation has normally been discussed in the electron language, we will do so here, temporarily forsaking the symbols describing holes in eqn. (5). We define I_d as the second ionization energy of copper ($Cu^{1+} \longrightarrow Cu^{2+}$), A_p as the second electron affinity of oxygen ($O^{1-} \longrightarrow O^{2-}$) and E_M the Madelung energy difference between the two configurations per unit (CuO). Then the inequality

$$I_d - A_p - E_M \lessgtr 0 \tag{6}$$

determines whether the covalent or the ionic configuration is the lower state. For a smaller left hand side, the CuO_2 layer contains only Cu^{2+}, while if this quantity is greater than zero Cu^{1+} is preferred.

For finite t the ionicity of copper is not restricted to only 1 or 2. However, it is still possible to describe the system in terms of quasicovalent and quasiionic and the transition between these two states remains first order type as long as $t/\alpha V_M$ is less than a critical quantity, where $-\alpha V_M = E_M$, α the Madelung constant and V_M the nearest neighbor Coulomb attraction (ref. 62). This has shown numerically in one dimension, while experimentally quasi-one-dimensional mixed-stack charge-transfer solids are known which are in the borderline defined by eqn. (6) and in which such first order like neutral-to-ionic transitions are observed as a function of temperature and pressure (ref. 63). In the present case t is small ($\sim$ 1 ev), α is larger than in one dimension and V_M corresponds to the attraction between doubly charged (Cu^{2+} and O^{2-}) species, so that we believe this condition is met. Then in the semiconducting material, $I_d - A_p - E_M < 0$ and we have formally the quasiionic state described by a Hubbard model on only the copper lattice

$$H_{Cu} = U_d \sum_i n_{i\uparrow} n_{i\downarrow} + t_d \sum_{\substack{\langle ij\rangle\\ \sigma}} (d^+_{i\sigma} d_{j\sigma} + d^+_{j\sigma} d_{i\sigma}) \tag{7}$$

Here H_{Cu} is the effective single-band Hamiltonian obtained from eqn. (5), and $d^+_{i\sigma}$ creates holes on the d-orbitals of copper. We have not included any long range Coulomb interaction since these are irrelevant (qualitatively speaking) for the description of the half-filled band Mott insulator. The optical gap is the charge-transfer gap ($Cu^{2+}\ O^{2-} \longrightarrow Cu^{1+}\ O^{1-}$) within the strongly correlated model (ref. 64) and

$$t_d = \frac{t^2}{\Delta E_{CT}} \tag{8}$$

where ΔE_{CT} is the charge-transfer energy.

We believe that the description of the conducting doped $La_{2-x}M_xCuO_{4-y}$ (and Y-Ba-Cu-O and Ba-Pb-Bi-O) is quite different. For x = 0 once again there are two possibilities even with the holes going to oxygen. One is the situation where the original N holes remain on the coppers, and xN holes (O^{1-} species) move through the oxygen network within the background AF of the Cu^{2+} ions (ref. 21). A second possibility, not discussed until now, is that even the original copper holes move over to the oxygens, and all the holes are now involved in the delocalization. Note that nearly half the oxygens remain as O^{2-} and this does not violate charge balance. This second possibility occurs because La_2CuO_{4-y} is very close to the borderline of eqn. (6), as seen for e.g., in the loss of AF for y = 0. Thus the insulator-to-metal transition would be accompanied by a quasiionic-to-quasicovalent transition. Once again, of course, complete delocalization over both the copper and the oxygen is possible, wiping out the distinction between the two states considered here. But we believe that this does not happen, for the same reason that it does not occur in the semiconducting CuO_2 layer. Of course in this quasicovalent conducting state a few of the coppers will remain in the Cu^{2+} configuration, and therefore there is a small critical x_c only beyond which the transition takes place. To understand this situation let us consider the energetics within a CuO_2 layer.

Let us define E'_M as the Madelung energy per CuO unit as before, but now a fraction x of the oxygens which used to be O^{2-} are O^{1-} (i.e., $E'_M < E_M$). We define $\epsilon(\bar{t}_p)$ as the delocalization energy of the dopant induced holes in the standard two-band state, where the coppers are in the Cu^{2+} configuration. Similarly $\epsilon(t_p)$ is the delocalization energy with all the holes on the oxygens. The hopping integral $\bar{t}_p$ is a weighted average of

$$t_1 = \frac{t^2}{\Delta E_{CT}}, \quad t_2 = \frac{t^2}{U - \Delta E_{CT}} \tag{9}$$

where in the two processes the two intermediate configurations for hopping between O^{1-} and O^{2-} are $O^{1-}Cu^{1+}O^{1-}$ and $O^{2-}Cu^{3+}O^{2-}$ respectively. The integral t_p has the intermediate configuration $O^{2-}Cu^{2+}O^{2-}$ and is given by

$$t_p = \frac{t^2}{\Delta E'_{CT}} \tag{10}$$

where $\Delta E'_{CT}$ corresponds to the reverse charge-transfer process $Cu^{1+}O^{1-} \longrightarrow Cu^{2+}O^{2-}$. In the absence of correlations $\epsilon(t_p)$ would have been given by

$$\epsilon(t_p) = N^{-1} \sum_{occ} [2t_p(\cos k_x + \cos k_y) + 4t_p \cos k_x \cos k_y] \tag{11}$$

where the second term occurs because of next neighbor hopping in the oxygen network, the magnitude of which is the same as the nearest neighbor hopping in the uncorrelated limit. A similar expression is valid for $\epsilon(\bar{t}_p)$, but both the band energies are considerably reduced in presence of electron correlation, which as we discuss later, is relevant even for the oxygen band.

Which particular conducting state, the quasiionic with a few O^{1-} holes vs. the quasicovalent with all the holes on the oxygens is energetically more stable is now determined by

$$I_d - A_p - E_M \lessgtr -[|\epsilon(t_p)| - |\epsilon(\bar{t}_p)|] \tag{12}$$

The following are now to be taken into consideration. Firstly, the number of occupied states corresponding to $\epsilon(t_p)$ is much larger than the number corresponding to $\epsilon(\bar{t}_p)$, so that even with correlated holes we expect $|\epsilon(t_p)| \gg |\epsilon(\bar{t}_p)|$. This condition is further strengthened by the fact that $t_p > \bar{t}_p$ since t_2 is very small. Secondly, both from theoretical estimates and spectroscopic evidence (see below) even the semiconducting La_2CuO_{4-y} is very close to the borderline defined in eqn. (6). Here $E'_M < E_M$ because of two reasons: (a) at least a fraction x of the oxygens, which are O^{2-} in La_2CuO_{4-y} are O^{1-}, and (b) La^{3+} is substituted with Sr^{2+}. Both of these make E'_M substantially smaller than E_M. Increased delocalization along with smaller Madelung energy then makes the left hand side in eqn. (11) larger, so that for $x > x_c$ in $La_{2-x}M_xCuO_{4-y}$ the active sites are the oxygens, described by the extended Hubbard Hamiltonian

$$H_o = U_p \sum_i n_{i\uparrow} n_{i\downarrow} + V_p \sum_{\langle ij \rangle} n_i n_j + t_p \sum_{\substack{\langle ij \rangle \\ \sigma}} (p^+_{i\sigma} p_{j\sigma} + p^+_{j\sigma} p_{i\sigma}) \quad (13)$$

where the sums over sites are restricted to the oxygen sites, $p^+_{i\sigma}$ creates holes on the oxygen p-orbitals, and V_p is the nearest neighbor repulsion between these holes. This repulsion is important now since the system is nearly quarter-filled and not half-filled. It is of course understood that V_p in this quarter-filled case incorporates all the long range Coulomb effects exactly as U does the same in the half-filled case. Since such repulsion would tend to give an oxygen based CDW, we have not included the next nearest hopping which would be substantially reduced by V_p. Conversely, the magnitude of V_p as observed from experiments would be smaller than its true value. Note that the coppers are now predominantly Cu^{1+}.

Eqn. (12) now gives SC exactly as described before. The O^{1-} holes form a 2D CDW (see Fig. 3), and the superconducting pairs are O^{1-}-O^{1-} species corresponding to the hole pairs in Fig. 1. We shall proceed through most of the rest of the section with assumption of a similar quarter-filled 2D O-band in Y-Ba-Cu-O and Ba-Pb-Bi-O. This explains all the observed phenomena. Only in the end do we discuss the role of the CuO chains in $YBa_2Cu_3O_{7-y}$. From this discussion it will be seen that this assumption is completely justified.

We have, of course, not theoretically proved the quasiionic to quasicovalent transition but have only attempted to justify such a scenario. But such a transition cannot be proved ab initio. This is because band calculations would miss the correlation induced features, while numerical simulations in 1D (ref. 62), where a similar transition is well understood, are dependent on empirical parametrization. Such simulations can be done here, but the validity of the model is proved only by comparing to experiments. We discuss below all the unusual features of the oxide superconductors in the light of the present model.

Let us then summarize the hypothesis made here. It is being claimed that the 2D conducting metal-oxygen layers in these superconductors have correlated quarter-filled bands in which the holes are almost entirely on the oxygens. As discussed below, we believe that the quasiionic-to-quasicovalent transition occurs at x very close to 0.15. In Y-Ba-Cu-O the O-based CDW already exists, while the required ratio of O^{1-} to O^{2-} in the oxygen layer in $BaPb_{1-x}Bi_xO_3$ probably is reached at x = 0.25. This requires a considerable fraction of the Pb in the mixed valent state. As shown below, nearly every reproducible peculiarity in the oxides can be explained with this scenario.

Antiferromagnetism and semiconduction in La_2CuO_{4-y}

This particular feature is understood within all theories emphasizing electron correlation. As discussed here, eqn. (7) is the effective ground state Hamiltonian, and the semiconducting and the optical gaps are the charge-transfer gap. On the other hand, the loss of AF for y = 0 simply proves that even the semiconducting material is very close to the borderline defined in eqn. (6).

CDW like behavior in La-M-Cu-O, Y-Ba-Cu-O and Ba-Pb-Bi-O

CDW like behavior are seen from a number of experiments in all these materials. The optical phonon spectra (refs. 46,47), the doublets in the vibrational features of the superconducting species (ref. 46), i.r. active Raman modes in Ba-Pb-Bi-O and possibly in the copper oxides etc. A very large number of investigators have pointed out these and have speculated about breathing mode displacements of the oxygens around each copper (or bismuth, as the case may be) atom. For a copper-based CDW to occur such displacements should have oxygens moving in towards a copper and moving away from the neighboring coppers, so that there is tendency to disproportionation into Cu^{1+} and Cu^{3+}. Such a CDW is not theoretically possible for the large U_d driving the observed AF. Similarly there is evidence for two different Bi-O bond lengths (ref. 47) but also spectroscopic evidence of the equivalence of the bismuths (ref. 65).

All of the above are explained with a $\rho = 0.5$ oxygen-based CDW driven by V_p. The conduction electrons are still coupled to the Cu-O vibrations but the modes are different (see the following paragraph or Fig. 3). Note that the nature of this CDW is such that it is equivalent to what has been called the $4k_F$ CDW in one-dimensional systems (refs. 32,34), thus proving strong correlation even in the oxygen band.

The role of the orthorhombic crystal structure

A concensus has now been reached that SC is obtained only in the orthorhombic phase (ref. 48) of La-M-Cu-O, while due to the Cu-O chain in Y-Ba-Cu-O the superconducting sample is orthorhombic anyway. The situation is less clear in Ba-Pb-Bi-O because of the greater role of disorder. All existing theories neglect this phenomenon and assume that the tetragonal-to-orthorhombic transition is not related to SC. This is not so: the oxygen-based $4k_F$ CDW requires the orthorhombic crystal structure, as seen in Fig. 3. The metal-oxygen bond lengths are now different along the x and y-axes because the oxygens are inequivalent in the 2D CDW, each O^{1-} preferring neighboring O^{2-} and vice versa. The conduction electrons are coupled to the breathing mode, but the nature of the mode is very different from what has been assumed so far.

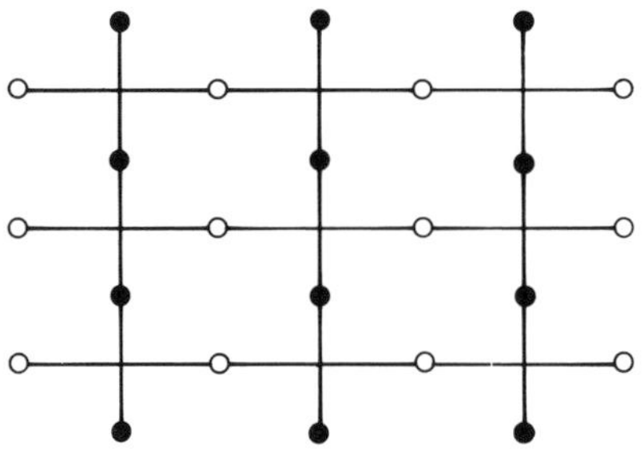

Fig. 3. The relationship between the orthorhombic crystal structure and the oxygen-based CDW. Here all copper atoms are where the lines cross, the black and white circles are O^{1-} and O^{2-} respectively. The covalent bonds between Cu^{1+} and O^{1-} are stronger and the bond lengths are shorter.

For $x > 0.2$ the tetragonal phase is stabilized in La-M-Cu-O. This is expected. We argue in the next paragraph that $x = 0.15$ corresponds to the $\rho = 0.5$ commensurate point. For ρ appreciably larger than this incommensurability sets in. One final comment is that, the temperature at which the tetragonal-to-orhombic transition takes place sets the upper limit for T_c.

Elasticity studies and neutron experiments emphasizing the role of phonons

We refer to the elasticity studies by Bourne et al (ref. 49), the neutron inelastic scattering experiments by Balakrishnan et al (ref. 50a) and the neutron powder diffraction by Paul et al (ref. 50b). All these studies are for La-Sr-Cu-O. It is found that the conduction electrons and vibrations are very strongly coupled. In particular $x = 0.15$, the Young's modulus first increases predictably as temperature is decreased, then at about 200K there is a sharp drop showing considerable lattice softening. For $x = 0.3$ the anomaly in the Young's modulus is very weak.

This is expected in our theory. Since the maximum in T_c occurs at ρ closest to 0.5 in the theory and at $x = 0.15$ experimentally, we associate $x = 0.15$ with $\rho = 0.5$. Even though the CDW is driven by V_p, coupling with the lattice gives the resultant phonon softening. The commensurability makes the CDW strongest at $\rho = 0.5$, while $x = 0.3$ is highly incommensurate. The anomaly in the Young's modulus at $x = 0.15$ is then due to the tetragonal-to-orthorhombic transition, and it is not a coincidence that the temperature at which the anomaly occurs is the same as the temperature at which the orthorhombic phase is reached at $x = 0.15$. No anomaly at $x = 0.3$ exists in reality. We predict strong anomalies in the Young's modulus at higher temperature for $x < 0.15$ from the work on the tetragonal-orthorhombic transition (ref. 48a).

Spectroscopic evidence for O^{1-}

Unusually large amount of O^{1-}, and consequently Cu^{1+} species have been reported by several research groups from ESCA studies (refs. 29,51-53). Complete absence of Cu^{3+} is also found.

A d-count of about 9.5 for copper occupancy has been given by Fujimori et al (ref. 51) while an even larger value (9.7) is given by Sarma and Rao (ref. 29). The amount of O^{1-} and O^{2-} seem to be comparable, which would agree with the copper valency. Indeed Rao et al (ref. 52) find evidence for "oxygen dimerization" and have even suggested hole-hole pairing from O-O bonds. While this work does not specify which oxygens are involved in pairing, it certainly agrees with the present work.

Absence of Cu^{2+} in epr and nmr measurements (ref. 54) in $YBa_2Cu_3O_{7-y}$

This is understood. Indeed we will argue that all the three coppers in this material are in the Cu^{1+} state. A prediction is made here. While epr gives substantial Cu^{2+} in the semiconducting material Y_2BaCuO_5 and other impurity phases (ref. 54), we predict an absence of Cu^{2+} in $YBa_2Cu_3O_6$. Several groups have reported Curie like susceptibility in $YBa_2Cu_3O_7$ however. This is due to the large O^{1-} concentration in the 2D layers as well as other O^{1-} (see below).

Electronic absorption at 0.4 - 0.5 ev

An optical absorption at 0.4 - 0.5 ev is seen (ref. 55) in superconducting samples of $La_{2-x}Sr_xCuO_{4-y}$. Etemad et al (ref. 55) have found that the oscillator strength of this absorption is close to zero for x = 0, grows rapidly and exhibits a peak at $0.15 < x < 0.20$, following which the oscillator strength decays rapidly. The same absorption is seen (ref. 56) in $YBa_2Cu_3O_{7-y}$ with $y \sim 0.1$ (superconducting) but is not found in $y \sim 0.8$ (semiconducting). Etemad et al (ref. 55) and also Kamaras et al (ref. 56) conclude that there is a strong correlation between this absorption and SC.

In the present theory this absorption is due to V_p. Such an absorption is expected (at $2V_p$ in 2D for $t_p = 0$) within the extended Hubbard model (ref. 30); the oscillator strength is proportional to the fraction of nearest neighbor electron hops that cost an energy $2V_p$, and this fraction is 1 in the static $\rho = 0.5$ CDW. Since we correlate the x = 0.15 in $La_{2-x}Sr_xCuO_{4-y}$ as $\rho = 0.5$ and have claimed that the layers in $YBa_2Cu_3O_7$ have $\rho = 0.5$ it is expected that maximum oscillator strength occurs here. As discussed below, the absence of this absorption in $YBa_2Cu_3O_6$ not only supports our model but also explains the precise role of the chains. We would also add that within no other theoretical picture the observed x-dependence of the absorption can be explained; the charge-transfer excitation intensity (ref. 66) can either remain constant or vary monotonically.

The Hall coefficient in $La_{2-x}Sr_xCuO_{4-y}$

Ong et al (ref. 58) have measured the Hall coefficient R_H of $La_{2-x}Sr_xCuO_{4-y}$ as a function of x and have found that while for $0 < x < 0.15$ the number of carriers obtained from R_H goes as x, there is a sudden drop in R_H by a factor of 30 between x = 0.15 and 0.18, while R_H is undetectable beyond x = 0.2.

This is expected within the present model. For $x > x_c$ (which can very well be 0.15), all the holes are carriers in the quasicovalent conductor.

Magnetic susceptibility of $La_{2-x}Sr_xCuO_{4-y}$

Schneemeyer et al (ref. 57) have measured the static magnetic susceptibility above T_c of the above material for a large number of x. AF at x = 0 is quickly destroyed and the high temperature (150K) susceptibility drops enormously to become weakly diamagnetic near x = 0.05, beyond which the susceptibility rises again to values larger than the x = 0 susceptibility at x = 0.25 where it exhibits a peak. Similar behavior of the susceptibility vs. x is seen for a wide range of temperature. The initial drop in the susceptibility and rapid destruction of AF is expected within the simple two-band model, the dopant-induced oxygen holes can lead to fraustrations in spin alignment. This has been discussed by Emery (ref. 21). Also, such holes will increase the probability of double occupancy in the correlated copper band considerably, thus also reducing the susceptibility. However, the increase in susceptibility beyond x = 0.05 is unaccounted for within this simple picture.

Within the present model for $x > x_c$ the moments are on the oxygens, and this explains the qualitatively different behavior of the temperature dependence of the susceptibility for moderately large x, aside from the magnitude increase with x. The increase is expected within the extended Hubbard model for a moderately large V_p, - the magnitude of the susceptibility within the model is strongly ϱ-dependent and is largest at $\varrho = 0.5$ (ref. 33).

The role of the chain oxygens in Y-Ba-Cu-O

We finally come to what is perhaps the most perplexing feature, - the loss of SC on heating $YBa_2Cu_3O_7$ which leads to a loss of the chain oxygen with the end product being semiconducting tetragonal $YBa_2Cu_3O_6$. No single theory explains SC in both $La_2M_xCuO_{4-y}$ and $YBa_2Cu_3O_7$. Indeed, until now it has not even been clear whether the loss of SC is due to the orthorhombic-to-tetragonal structural change, or whether the oxygen loss changes valencies (refs. 62,63).

Within the present theory these two phenomena (changes in structure and valency) are related, and the orthorhombic structure corresponds to the mixed valency of oxygen in the layers. We explain here not only the loss of SC, but even the observed Cu-O bond lengths in this material, which vary from 1.834 A to 2.341 A in the superconducting material, the values for the semiconductor being reasonably close. We will even explain the fact that the chain oxygens are the most weakly bound ones, and that all these observations are interrelated.

We will use the nomenclature of references 61(a). The coppers in the chains and in the layers will be denoted by Cu(1) and Cu(2) respectively. The chain oxygen is labelled O(1) while the oxygen bonded to both Cu(1) and Cu(2) is O(2). The layer oxygens are inequivalent in the orthorhombic superconductor and are labelled O(3) and O(4). The various bond lengths are: Cu(1)-O(1) = 1.947 A, Cu(1)-O(2) = 1.834 A, Cu(2)-O(2) = 2.341 A, Cu(2)-O(3) = 1.929 A and Cu(2)-O(4) = 1.961 A. For SC, orthorhombocity, as well as the 0.4 ev absorption seen in $YBa_2Cu_3O_7$ (ref. 56) we need alternating O^{1-} and O^{2-} in the layer. Similarly we need copper in the layers in the Cu^{1+} state. This would make O(3) largely monovalent and O(4) largely bivalent. NMR and epr find no Cu^{2+} at all, and this indicates that even Cu(1) is monovalent. This would agree with all other features, especially its coordination number. Then since lanthanum and barium are La^{3+} and Ba^{2+} most certainly, the total positive charges add up to 10, which has to be balanced by 7 oxygens, out of which four in the two layers give a total negative charge of 6 in $YBa_2Cu_3O_7$. Since there are two O(2) and only one O(1) this leaves no choice but to assign a valency of -2 to O(1) and -1 to O(2).

Before discussing $YBa_2Cu_3O_6$ we discuss the observed bond lengths in the context of the above assigned valancies. The extremely short Cu(1)-O(2) bond length is in agreement with the strong covalent bonding between Cu^{1+} and O^{1-}. Within the layer there are two nominally covalent bonds per Cu^{1+} which are therefore weaker than the Cu(1)-O(2) bond. However, because Cu(2) is then already covalently bonded within the layer, while O(2) is also strongly bonded to Cu(1), the Cu(2)-O(2) bond is very weak, thus explaining the large bond length. Finally, because the Cu(1)-O(2) is so strong, the Cu(1)-O(1) linkage is the weakest in the whole structure, and therefore O(1) is the oxygen lost on heating.

Now let us discuss $YBa_2Cu_3O_6$. The loss of O(1) can lead to two possibilities: the valencies of both the cations and the anions change, or the valancies of the cations remain the same and the oxygen valency is redistributed. Since the bond lengths in $YBa_2Cu_3O_6$ remain nearly the same the second must be occurring. The extra valency of -2 can go to the two O(2), so that these become O^{2-}, or they can go to the two O(3) in the two layers, which then become O^{2-}. Shorter Cu(1)-O(2) in $YBa_2Cu_3O_6$ (a longer bond between Cu^{1+} and O^{2-} is expected), loss of SC, loss of the 0.4 ev absorption and the transition to a tetragonal structure all indicate that the layer oxygens become completely bivalent in $YBa_2Cu_3O_6$. Now there is no mixed valence and the system is a semiconductor. In superconducting $YBa_2Cu_3O_{6.6}$ our guess is that one layer remains mixed valent with = 0.5. This should be observable in optical studies; we predict the 0.4 ev absorption with smaller oscillator strength.

We predict absence of Cu^{2+} from epr and nmr studies in $YBa_2Cu_3O_6$ and absence of the 0.4 ev absorption in the semiconducting Y_2BaCuO_5, in which lalrge amounts of Cu^{2+} is seen from epr measurements (ref. 54).

The critical temperatures

Only in the end we come to the explanation of the high T_c, since as discussed in the introduction, only the explanation of the other experimental behavior will decide whether a given mechanism is correct. We have already seen that $T_c <$ 5-10K is predicted in the organics. For the copper oxides t is supposed to be 1-1.5 ev, so that eqn. (9) with ΔE_{CT} = 2-2.5 ev gives $t_p \sim$ 0.5-1 ev. The magnitude of V_p is harder to estimate, since the absorption at $\sim$ 0.5 ev is $2V_p$ only at $t_p = 0$. Also for SC to occur, it is clear that the local pairing requires $2V_p < t_p$ since the former quantity represents the binding energy and the latter measures the delocalization of a single particle. Thus assuming $t_p/2V_p \lesssim 1$, and the critical concentration n_c of pairs again as 2-5%, we get a $T_c >$ 100K. For Ba-Pb-Bi-O the problem is harder because here not only there is mixed valence of oxygen, there is also mixed valent Pb. Our guess is that t_p therefore involves hopping integrals like t_2 in eqn. (8). Random arrangements of Pb and Bi, along with a greater site energy (electronegativity) difference between the metals and oxygen also reduce t. All these contribute to give T_c an order of magnitude lower.

We have, in the above, discussed only those properties which can be definitively explained within the present scenario. Aside from this explanation for several more perplexing features may be guessed. For example, T_c is strongly pressure dependent in La-M-Cu-O while very weakly dependent on pressure in Y-Ba-Cu-O. The reason for this is probably enhanced hole transfer from copper to oxygen in the former by pressure thus going closer to ρ = 0.5. The latter already has all the coppers in the Cu^{1+} state, due to the orthorhombic structure imposed by the CuO chain as well as the oxygen deficiency, and therefore pressure has little effect on T_c. The weak oxygen isotope effect in La-M-Cu-O may be because the transition from the tetragonal to the orthorhombic phase is really a condensation of the oxygen-based CDW, and even though this is driven by V_p lattice coupling is involved. Again, this would conform with the total absence of the isotope effect in orthorhombic Y-Ba-Cu-O. It is to be hoped that such features become more understandable in the future.

NOTE ADDED IN PROOF

During the preparation of this manuscript I became aware of the recent paper by Shafer et al (Phys. Rev. B36, 4047,1987) in which the authors carefully study the concentration of the $[Cu^{2+} - O^{1-}]$ complex as a function of x in La-Sr-Cu-O. They conclude that an abrupt transition occurs at x = 0.15 beyond which the $[Cu\text{-}O]^+$ concentration decreases with x. The phase transition, and in particular, its abrupt nature would agree with the present work. The authors ascribe SC to the $[Cu\text{-}O]^+$ species and its reduction with x beyond x = 0.15 to creation of oxygen vacancies. Present work indicates that the decrease in $[Cu\text{-}O]^+$ is due to transition to the quasicovalent state with most coppers in the Cu^{1+} state. I believe that the correlation of T_c with x, as well as with other properties is better described within the present model.

The idea of mixed valence of oxygen in Ba-Pb-Bi-O was born through discussions with M.S. Hegde and the CDW like optical phonon spectra (ref. 47). I have now received a preprint by P. Ganguly and M.S. Hegde (to be published in Physical Review B) which concludes the same about all conducting lead oxides. The authors ascribe all the oxygens giving rise to a 532-533 ev peak in the O(1s) XPS to O_2^{2-} species, but a major amount must be O^{1-} (note, however, that each metal atom is bonded to two O^{1-}, the bond angle being 180°; only when the bond angle is 90°, it could be referred to as O_2^{2-} in the present nomenclature). What is most interesting that if one correlates the intensities of the XPS peaks to real concentrations, then $BaPbO_3$ with very large O^{1-} concentration is not superconducting, but Ba-Pb-Bi-O with nearly equal amounts of O^{1-} and O^{2-} is a superconductor. This agrees with the present theoretical work, indicating that mixed valence alone is not a sufficient condition for SC.

ACKNOWLEDGEMENTS

The present work would not have been possible without the encouragement of Mieke Calis. The author acknowledges stimulating discussions with Professors C.N.R. Rao and M.S. Hegde, and particularly Professor P.S. Ganguly, who convinced him of the oxygen mixed valency. Theoretical criticisms by Professors N. Kumar, T.V. Ramakrishnan and S. Ramasesha are gratefully acknowledged. Discussions of magnetic resonance and ESCA data with Drs. S.K. Date and S. Badrinarayanan are acknowledgewd. The author thanks A.K. Gangopadhyay for speedy typing of the manuscript. Finally, the author is grateful to Professors V.J. Emery, S. Etemad, J.E. Hirsch, D. Jerome, C.N.R. Rao and D.B. Tanner for sending him their manuscripts prior to publication.

REFERENCES

1. T.M. Rice, Z. Phys. B, (1987).
2. P.W. Anderson et al, Phys. Rev. Lett., 58, 2790 (1987).
3(a). D. Jerome and F. Creuzet, Novel Mechanisms of Superconductivity, Plenum, New York (1987).
(b). J.M. Williams et al, Acc. Chem. Res., 18, 261 (1985).
(c). A.I. Buzdin and L.N. Bulaevskii, Sov. Phys. Usp., 27, 830 (1984).
(d). D. Jerome and H.J. Schulz, Adv. in Phys., 31, 299 (1982).
4. S. Mazumdar, Phys. Rev. Lett., (submitted).
5. V.J. Emery, Nature, 328, 756 (1987).
6. Z. Fisk et al, Nonlinearity in Condensed Matter, p. 142, Springer-Verlag, Berlin (1987).
7(a). B. Batlogg et al, Phys. Rev. Lett., 58, 2333 (1987).
(b). L.C. Bourne et al, 58, 2337 (1987).
8(a). B. Batlogg et al, Phys. Rev. Lett., 59, 912 (1987).
(b). Tanya A. Faltens et al, Phys. Rev. Lett., 59, 915 (1987).
9(a). D. Vaknin et al, Phys. Rev. Lett., 58, 2802 (1987).
(b). R.L. Greene et al, Solid St. Commun., 63, 379 (1987).
10(a). H. Schwenk et al, Phys. Lett., 102A, 57 (1984).
(b). C.P. Heidmann et al, Physica, 143B, 357 (1987).
11. P.W. Anderson and E. Abrahams, Nature, 327, 363 (1987).
12. C.E. Methfessel et al, Proc. Natl. Acad. Sci. USA, 77, 6307 (1980).
13. L.E. Wenger et al, Phys. Rev. B, 35, 7213 (1987).
14. G.R. Stewart et al, Phys. Rev. B, 33, 2046 (1986) and references therein.
15(a). P.W. Anderson, Science, 235, 1196 (1987).
(b). G. Baskaran et al, Solid St. Commun., 63, 973 (1987).
16. A.E. Ruckenstein et al, Phys. Rev. B, July 1 (1987).
17. H. Fukuyama and K. Yosida, Jap. J. Appl. Phys., 26, L371 (1987).
18. M. Cyrot, Solid St. Commun., 63, 1015 (1987).
19(a). H.J. Schulz, Europhys. Lett. (1987).
(b). M.M. Mohan and N. Kumar, J. Phys. C, 20, L527 (1987).
20. J.E. Hirsch, Phys. Rev. B., 35, 8726 (1987).
21. V.J. Emery, Phys. Rev. Lett., 58, 2794 (1987).
22. J.E. Hirsch, Phys. Rev. Lett., (1987).
23. P. Prelovsek, preprint (1987).
24. J.E. Hirsch, Phys. Rev. Lett., 54, 1317 (1985).
25. S. Mazumdar, Phys. Rev. Lett., to be published (1987).
26(a). H.Q. Lin et al, preprint (1987).
(b). J.E. Hirsch and H.Q. Lin, preprint (1987).
27. S.N. Dixit and S. Mazumdar, Phys. Rev. B, 29, 1824 (1984).
28. S. Mazumdar, Phys. Rev. B, 36, November 1 (1987).
29. D.D. Sarma and C.N.R. Rao, J. Phys. C, 20, (1987).
30. S. Mazumdar and Z.G. Soos, Phys. Rev. B, 23, 2810 (1981).
31. S. Mazumdar and A.N. Bloch, Phys. Rev. Lett., 50, 207 (1983).
32. S. Mazumdar et al, Phys. Rev. B., 30, 4842 (1984).
33. S. Mazumdar and S.N. Dixit, Phys. Rev. B., 34, 3683 (1986).
34. J.E. Hirsch and D.J. Scalapino, Phys. Rev. B., 29, 5554 (1984).
35. J.E. Gubernatis et al, Phys. Rev. B, 32, 103 (1985).
36. J. Hubbard, Phys. Rev. B., 17, 494 (1978).
37. A.S. Alexandrov et al, Phys. Rev. B., 33, 4526 (1987).
38. P. Prelovsek et al, J. Phys. C., 20, L229 (1987).
39. K. Machida and M. Kato, Phys. Rev. B, 36, 854 (1987).
40. M.E. Hawley et al, Phys. Rev. Lett., 57, 619 (1986).
41(a). M.E. Hawley et al, Phys. Rev. B, 35, 7224 (1987).
(b). M. Naito et al, Phys. Rev. B, 35, 7228 (1987).
42. R. Brusetti et al, J. Phys., 43, 801 (1982).
43. T.P. Orlando et al, Phys. Rev. B, 35, 5347 and 7249 (1987).
44. W. Kang et al, J. Physique, 48, 1035 (1987).
45. A.K. Rajagopal, Statistical Physics, supplement to the Journal of the Indian Institute of Science, p. 85 (1975).
46(a). S.L. Herr et al, Phys. Rev. B, 36, 733 (1987).
(b). D.A. Bonn et al, Phys. Rev. Lett., 58, 2249 (1987).
47(a). S. Tajima et al, Phys. Rev. B., 32, 6302 (1985).
(b). S. Uchida et al, J. Phys. Soc. Jpn., 54, 4395 (1985).

48(a). R.M. Fleming et al, Phys. Rev. B, 35, 7191 (1987).
(b). Ivan K. Schuller et al, Solid St. Commun., 63, 385 (1987).
(c). G. Van Tendeloo, Solid St. Commun., 63, 389 (1987).
49. L.C. Bourne et al, Phys. Rev. B, 35, 8785 (1987).
50(a). G. Balakrishnan et al, Nature, 327, 45 (1987).
(b). D. Mck. Paul et al, Phys. Rev. Lett., 58, 1976 (1987).
51. A. Fujimori et al, Phys. Rev. B., 35, 8814 (1987).
52(a). D.D. Sarma et al, Phys. Rev. B, 36, 2371 (1987).
(b). C.N.R. Rao et al, Mat. Res. Bull., 22, 1159 (1987).
53. A. Bianconi et al, Solid St. Commun., 63, 435 (1987).
54. G.J. Bowden et al, J. Phys. C., 20, L545 (1987).
55. S. Etemad et al, preprint (1987).
56. K. Kamaras et al, Phys. Rev. Lett., 59, 919 (1987).
57. L.F. Schneemeyer et al, Phys. Rev. B., 35, 8421 (1987).
58. N.P. Ong et al, Phys. Rev. B, 35, 8807 (1987).
59(a). A. Santoro et al, Mat. Res. Bull., 22, 1007 (1987).
(b). C.U. Segre et al, Nature, 329, 227 (1987).
60. P.K. Gallagher et al, Mat. Res. Bull., 22, 995 (1987).
61(a). G. Calestani and C. Rizzoli, Nature, 328, 606 (1987).
(b). P. Bordet et al, Nature, 327, 687 (1987).
62. Z.G. Soos and S. Kuwajima, Chem. Phys. Lett., 122, 315 (1985).
63(a). Y. Tokura et al, Solid St. Commun., 43, 757 (1982).
(b). Y. Tokura et al, Mol. Cryst. Liq. Cryst., 125, 71 (1985).
64. J. Zaanen et al, Phys. Rev. Lett., 55, 418 (1985).
65. G.K. Wertheim et al, Phys. Rev. B., 26, 2120 (1982).
66. C.M. Verma et al, Solid St. Commun., 62, 681 (1987).

Raman scattering studies of the copper–oxygen vibrations and oxygen order in La, Y–Ba, Sr–Cu oxides

Zafar Iqbal
Corporate Technology, Allied-Signal, Inc., Morristown, NJ 07960 USA

Abstract - The extensive investigations undertaken to date on the Raman scattering spectra of the 40K and the 90K superconductors $La_{2-x}Sr_xCuO_4$ and $YBa_2Cu_3O_{7-\delta}$, respectively, are assessed and a general interpretation of the Raman-active Cu-O vibrations in these systems is proposed. These results are compared with the data on the related $YBa_2Cu_3O_6$, Y_2BaCuO_5, $Y(Ba_{2-x}Y_x)Cu_3O_{7+\delta}$ and $La_2CuO_{4-\delta}$ phases. The usefulness of Raman scattering in indicating the degree of oxygen order in these materials is pointed out. In addition, polarized Raman results on ~1 mm. size single crystal platelets of $YBa_2Cu_3O_{7-\delta}$ are presented and compared with earlier micro-Raman studies on micron-size single crystals.

INTRODUCTION

The discovery of superconductivity near 30K in La-Ba-Cu oxide (ref. 1), at ~36K in $La_{2-x}Sr_xCuO_4$ (ref. 2) and at 90K in the Y-Ba-Cu oxide system (refs. 3 and 4) triggered world-wide activity involving the physics and chemistry of these exciting materials (for a collection of the early work see ref. 5). In terms of the Bardeen-Cooper-Schrieffer (BCS) theory (ref. 6) electron pairing followed by Bose-Einstein condensation was also expected to be mediated, at least in the 40K materials, by the atomic vibrations or phonons. The expectation that the Cu-O vibrations would lie at much higher frequencies than the previously known high temperature superconductors like Nb_3Sn, suggested a possible explanation for the observed high transition temperatures in the new oxides, since according to BCS theory the superconducting critical temperature, T_c, scales with the Debye temperature θ_D. Raman spectroscopic measurements were therefore undertaken by many groups both in the normal and the superconducting states of these materials, in order to identify the frequencies, symmetries and temperature-dependence of the copper-oxygen vibrations (see earlier overview in ref. 7). It is now, however, becoming increasingly evident (refs. 8 and 9) that phonons alone are not the primary pairing intermediary, particularly for the 90K superconductors. Raman studies have, in addition, provided structural information regarding the oxygen vacancies and oxygen order in these materials (ref. 10 and 11).

In this paper I will present an evaluation of the Raman data on well-characterized samples of the superconducting phases and compare these results with measurements on related but non-superconducting phases to provide a general interpretation of the observed spectra. Finally, questions regarding the short and medium-range order in these materials as obtained by Raman scattering studies will be discussed.

SAMPLE PREPARATION, CHARACTERIZATION AND SPECTROSCOPIC PARAMETERS

Superconducting $YBa_2Cu_3O_{7-\delta}$ (referred to as the Y 1-2-3 phase) discussed in this paper have been prepared by reacting well-mixed Y-oxide, Ba-carbonate and Cu-oxide weighed out in the nominal compositional ratio. The reactions are carried out initially in air in silica, alumina or gold boats and/or crucibles at 935°C for ~24 hrs. with a number of grindings. The sieved powder is then pelletized under ~10 Kbar pressure to about ~80% theoretical density and annealed in flowing O_2 typically for 12 hrs. at 900°C, followed by slow furnace-cooling. One variation of the Y 1-2-3 samples is obtained by using BaO_2 instead of $BaCO_3$ in the reaction. This provides excess oxygen and/or O_2^{2-} ions during the initial solid state reaction. A second variation is the use of evaporated Y, Ba and Cu citrates to give an atomically mixed precursor which is fired for a few hours to give extremely fine ($\lesssim 1\mu m$ size) powders of the Y 1-2-3 material.

Millimeter size single crystals of $YBa_2Cu_3O_{7-\delta}$ are grown from a CuO melt containing excess BaO at 970°C. Pt or Au crucibles are used for the crystal growth (ref. 12). The semiconducting Y_2BaCuO_5 phase is prepared via the firing of appropriate amounts of the oxides and Ba-carbonate in air at 1000°C for ~24 hrs. with a number of grindings. The oxygen depleted $YBa_2Cu_3O_6$ phase is prepared by annealing a sample of $YBa_2Cu_3O_{7-\delta}$ prepared via Ba-carbonate in 1 atm. Ar at 550°C for 9 hrs.

The metastable $Y(Ba_{2-x}Y_x)Cu_3O_{7+\delta}$ phase with x = 0.50 and 0.375 (referred to here for historical reasons as Y 3-3-6) is prepared via the firing of atomically mixed evaporated citrates of Y, Ba and Cu at 810°C under flowing O_2 for a relatively short period (typically ~2 hrs) (ref. 13). The samples are then pelletized and oxygen-annealed at 650°C and 520°C under 1 atm. and 3 atm. pressure respectively. The more stable Y:La (1:1) 3-3-6 phase is prepared via the ceramic route of firing the oxides and Ba-carbonate at 950°C for 24 hrs.

All samples were checked by X-ray diffraction and found to be greater than 95% by volume single phase. The transport and DC susceptibility of each of the samples were measured using four-probe AC techniques and a SHE 905 SQUID magnetometer.

The results on the La materials are from the literature. The synthetic conditions are given in the references noted in the text. The Raman data on thin films of Y 1-2-3 were obtained from deposits obtained by sputtering on $SrTiO_3$ (ref. 14).

All Raman data reported in this paper and those taken from the literature were obtained using conventional double monochromator Raman spectroscopy with a typical resolution of ~5 cm^{-1} and laser excitation either at 4880 or 5145Å. Except for micro-Raman results, the data discussed here have been recorded on pellets, single crystals or thin films using a line or defocussed back scattering geometry using 10-100 mW CW laser power at the sample. This consideration is very important since sample heating leading to laser-induced oxygen inhomogeneity can be introduced into the sample volumes being examined.

CRYSTAL STRUCTURES, NORMAL MODES AND SELECTION RULES

The K_2NiF_4 structure of $La_{1.8}Sr_{0.2}CuO_4$(La-Sr-Cu) and the oxygen deficient perovskite structure of $YBa_2Cu_3O_{7-\delta}$ are shown in Fig. 1. La-Sr-Cu consists of a layered tetragonal K_2NiF_4-type structure of space group I4/mmm composed of corner-sharing octahedra (ref.15). The ab plane O(1) atoms are at a distance of 1.893Å from the Cu atom and the two symmetry-related O(2) atoms along the c direction are at a distance of 2.412(6)Å. The Cu and O(1) layers are sandwiched between the layers formed by the La, Sr and O(2) atoms. The orthorhombic Y 1-2-3 phase of space group Pmmm is somewhat more complex (ref. 16). This tripled perovskite oxygen deficient structure has two copper sites. Cu(1) sites contain the chains of O(4) atoms running along the b direction with a Cu(1)-O(4) distance of 1.9429Å. The O(1) atoms along the c-direction are at a distance of 1.846Å from Cu(1) and thus represents the shortest Cu-O distance in the two superconducting structures. The Cu(2)-O(2) sites form the puckered layers in the ab plane which run orthogonal to the square planar chains. Within the layers the Cu(2)-O(2) distance is 1.9299Å and the Cu(2)-O(3) distance is 1.9607Å.

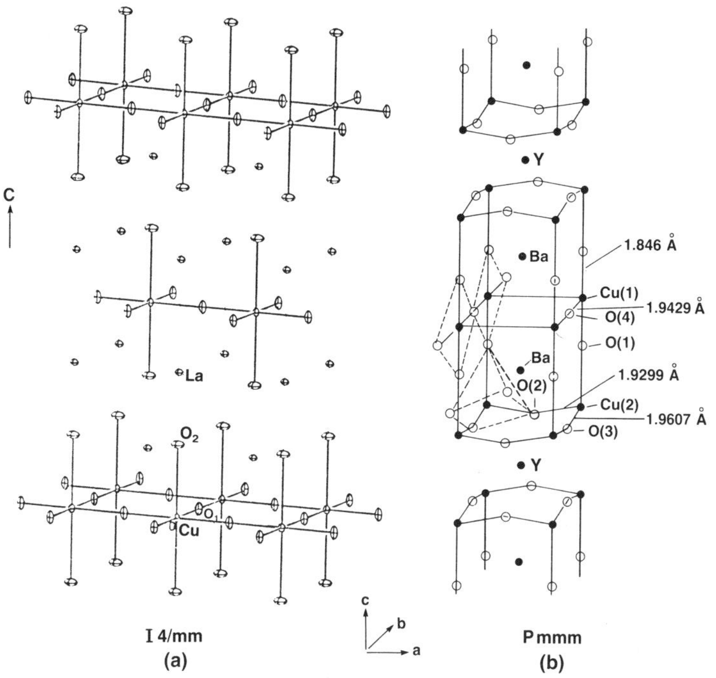

Fig. 1. Details of the crystal structure of $La_{1.8}Sr_{0.2}CuO_4$(a) and $YBa_2Cu_3O_{7-\delta}$(b).

Neutron results (ref. 17) show that the O(4) atoms are removed in the tetragonal and semiconducting $YBa_2Cu_3O_6$ material. Interestingly, however, the Cu(1)-O(1) distance (1.795Å) becomes shorter than in Y 1-2-3. The green-colored Y_2BaCuO_5 semiconducting phase also has a layered structure of orthorhombic symmetry (space group Pbnm), in which edge-sharing trigonal prisms of oxygen atoms coordinating Y ions, form chains running along the b-axis. The chains are cross-linked by copper atoms in five-fold coordinated square pyramids of oxygen atoms. The Cu-O distances are all about 2.0 Å (ref. 18).

The tetragonal $Y_x(Ba_{2-x}Y_x)Cu_3O_{7+\delta}$ (Y 3-3-6) and orthorhombic $La_2CuO_{4-\delta}$ show filamentary superconductivity (ref. 13 and references therein) under particular annealing conditions. The Y 3-3-6 is isostructural with $YBa_2Cu_3O_6$ of space group P4/mmm but has a shorter c axis parameter (11.628Å compared with 11.8194Å for $YBa_2Cu_3O_6$). In addition, the Ba layer is partially doped with Y and the oxygens per formula unit exceed 7.0 (typically it is 7.1 to 7.3 when x = 0.5). The excess oxygen atoms go into the O(5) position between the chains to form disordered octahedra (Fig. 2). Unlike LaSrCu, $La_2CuO_{4-\delta}$ has the related orthorhombic structure of space group Pbma, which transforms to a tetragonal structure near ~170°C.

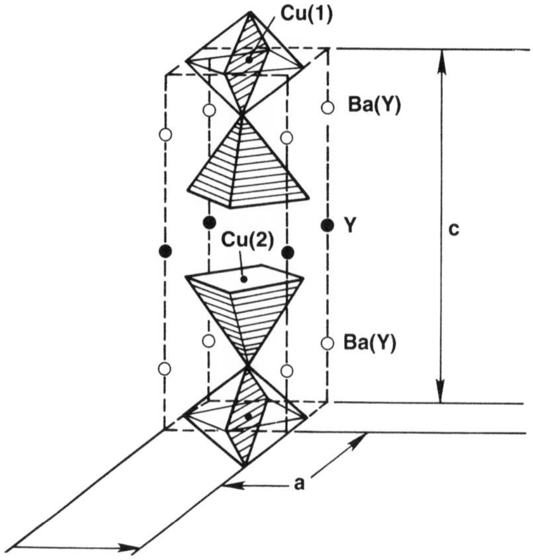

Fig. 2. Structure of the $Y(Ba_{2-x}Y_x)Cu_3O_{7-\delta}$(Y 3-3-6) phase.

The factor group analysis of the long wavelength (k ~ 0) copper-oxygen vibrations in the two superconducting structures under consideration is summarized in Table 1. The vibrational eigenvectors are localized on the oxygen atoms labelled in accordance with Fig. 1. Also shown are the symmetry distributions for Raman-active modes in which there is primary motion at the La and Cu(2) sites. These modes would undergo some mode mixing with oxygen vibrations of the same symmetry. Note that there is no Raman-activity involving the layer O(1) atoms in the tetragonal structure of LaSrCu. Similarly there is no Raman activity involving the chain O(4) atoms in the Y 1-2-3 structure. However, on going to the orthorhombic structure of La_2CuO_4 the layer oxygen mode becomes Raman-allowed in the K_2NiF_4-type structure.

Table. 1. Factor group analysis of the k ~ 0 Raman-active copper-oxygen and lanthanum vibrations of $La_{2-x}Sr_xCuO_4$ and $YBa_2Cu_3O_{7-\delta}$

Compound and Space Group	Atom	Mode Distribution
$La_{2-x}Sr_xCuO_4$	O(2)	$1\ A_{1g} + 1\ E_g$
(D_{4h}^{17} - I4mm)	La	$1\ A_{1g} + 1\ E_g$
$YBa_2Cu_3O_{7-\delta}$	O(1)	$1\ A_g + 1\ B_{2g} + 1\ B_{3g}$
(D_{2h}^{1} - Pmmm)	O(2)	$1\ A_g + 1\ B_{2g} + 1\ B_{3g}$
	O(3)	$1\ A_g + 1\ B_{2g} + 1\ B_{3g}$
	Cu(2)	$1\ A_g + 1\ B_{2g} + 1\ B_{3g}$

The local symmetries of the Cu-O modes in the octahedral environment of LaSrCu and the square planar environment of Y 1-2-3 are displayed in Fig. 3. In the octahedral environment the symmetric stretching mode corresponds to the O(2) stretching mode of A_{1g} symmetry. The asymmetric mode that corresponds to motion located at O(1) is Raman-inactive in the tetragonal unit cell as discussed before. The deformation mode which is localized at O(2) in the crystal corresponds to the doubly degenerate E_g mode listed in Table 1.

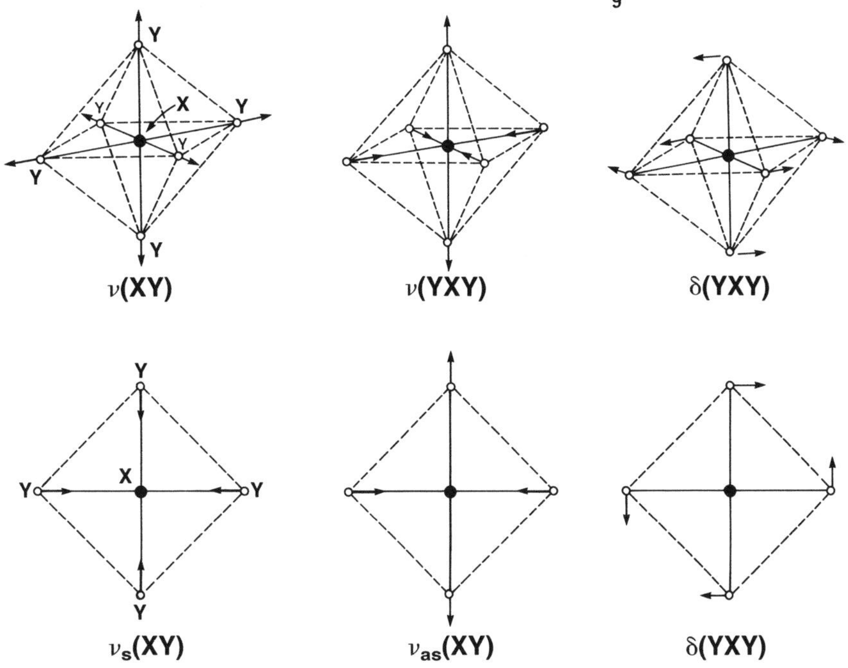

Fig. 3. Raman-active mode eigenvectors for octahedral and square planar local environments.

In the square planar environment of Y 1-2-3 at the Cu(1) site, the symmetric stretching local mode corresponds to the lattice A_g mode at O(1). In the puckered planar environment at Cu(2) the lattice A_g modes with eigenvectors at O(2) and O(3) are probably mixed with Ba^{2+} modes. The in-plane Cu-O deformations correspond to the B_{2g} and B_{3g} modes distributed between the two sites. The asymmetric stretching local mode would correspond to the chain O(4) mode in the crystal, which is Raman-inactive, and the B_{2g} and B_{3g} stretching modes at the O(2) and O(3) sites. In the O(4) depleted tetragonal lattice of $YBa_2Cu_3O_6$, the O(1) stretching mode remains intact, but the O(2) and O(3) modes evolve into a doubly degenerate vibration. However, in the tetragonal Y 3-3-6 structure the excess oxygen would induce local variations in the degree of ordering around Cu(1), which could activate the Cu(1) - O(4) vibration.

CALCULATED Cu-O MODE FREQUENCIES

A model calculation of the lattice dynamics of Y 1-2-3 has been preformed by Stavola et al (ref. 19) using values for force constants that are similar to an earlier model (ref. 20) for LaSr(Ba)Cu. For Y 1-2-3 the force constants are 120 Kdyn/cm for the Cu(1)-O(1) bond, 40 Kdyn/cm for the weak Cu(2)-O(1) bond, and 110 Kdyn/cm for the Cu(2)-O(3) and Cu(1)-O(4) bonds. The calculated Raman active, largely Cu-O stretching and deformation modes of A_g, B_{2g}, and B_{3g} symmetry, are listed in Table 2 based on the results of Stavola et al. The separation into stretching and bending modes in Table 2 is proposed on the basis of the discussion in the previous section.

Table. 2. Calculated Raman-active Cu-O stretching and bending vibrational modes frequencies (in cm^{-1}) in $YBa_2Cu_3O_{7-\delta}$ (after Stavola et al)

Mode Symmetry	Stretching	Deformation	
A_g	496	352	346
B_{2g}	599	339	314
B_{3g}	600	339	260

RESULTS AND DISCUSSION

$La_{1.8}Sr_{0.2}CuO_4$ and La_2CuO_4

Recent Raman data on $La_{1.8}Sr_{0.2}CuO_4$ have been published by Kourouklis et al (ref. 2), Blumenroeder et al (ref. 22) and Brun et al (ref. 23). The O(2) A_{1g} and E_g modes (Table 1 and Figs. 1 and 3) were observed at 380 cm^{-1} and 180 cm^{-1} by Blumenroeder et al. However, both Kourouklis et al and Brun et al observe a line near 430 cm^{-1} which they assign to the A_{1g} mode. The discrepancy between the results for the A_{1g} mode may be related to the exact oxygen content of the samples which were not reported. Careful Raman studies on single crystal La_2CuO_4 prepared by the flux method has also been reported by Kourouklis et al. A strong line at 526 cm^{-1} and a much weaker line at 426 cm^{-1} is observed in La_2CuO_4. The line at 536 cm^{-1} can be assigned to the stretching of the layer O(1) which is activated by reduction of the symmetry to orthorhombic as discussed above, while the line at 426 cm^{-1} corresponds to the motion of O(2). Even on heating the crystal above the orthorhombic-to-tetragonal phase transition, the line at 526 cm^{-1} (which is forbidden in the tetragonal K_2NiF_4 structure) persists. Kourouklis et al speculate that the presence of the 526 cm^{-1} line in the tetragonal phase is due to local symmetry breaking, possibly due to the antiferromagnetic interactions that have been shown to be present in La_2CuO_4. It is worth noting that the half-width of the 526 cm^{-1} line is ~20 cm^{-1} at 300K for La_2CuO_4. The relatively large linewidth may reflect static fluctuations of the oxygen ordering in those materials.

$YBa_2Cu_3O_{7-\delta}$

The Raman spectrum of polycrystalline $YBa_2Cu_3O_{7-\delta}$ has been reported by a number of workers. Here we will primarily discuss our results (ref. 11, 24) which are in close agreement with those reported by Kourouklis et al (ref. 25) and Cardona et al (ref. 26). Kourouklis et al and Cardona et al also report the spectra of isostructural $MBa_2Cu_3O_{7-\delta}$ compounds where M=Eu, Gd, Sm and Ho. Single crystal micro-Raman data were published very early in the field by Hemley and Mao (ref. 27), and results on a polycrystalline thin film of Y 1-2-3 on $SrTiO_3$ have been reported by Lyons et al (ref. 14).

Four features at 430, 497, 583 and 632 cm^{-1} are seen in most samples of Y 1-2-3 prepared in our laboratory via $BaCO_3$ (Fig. 4). Many samples also show the appearance of a line at 330 cm^{-1}. The linewidths at 300K are typically 20 to 30 cm^{-1}, which is similar to that observed in single crystal La_2CuO_4 (ref. 21) and a microcrystal of Y 1-2-3 (ref. 27). The linewidth does not show appreciable narrowing with decreasing temperature. In addition there is no evidence for scattering across the superconducting gap below 90K. The intensity of the line at 632 cm^{-1} varies from sample to sample. Carefully prepared samples show rather weak scattering at 632 cm^{-1}. Samples of Y 1-2-3 prepared via the citrate route (ref. 11) show complete absence of scattering at 632 cm^{-1}. The line at 330 cm^{-1} is, however, easily seen in single crystal samples (see below) and its absence in some data is due to the large elastic scattering in polycrystalline pellets.

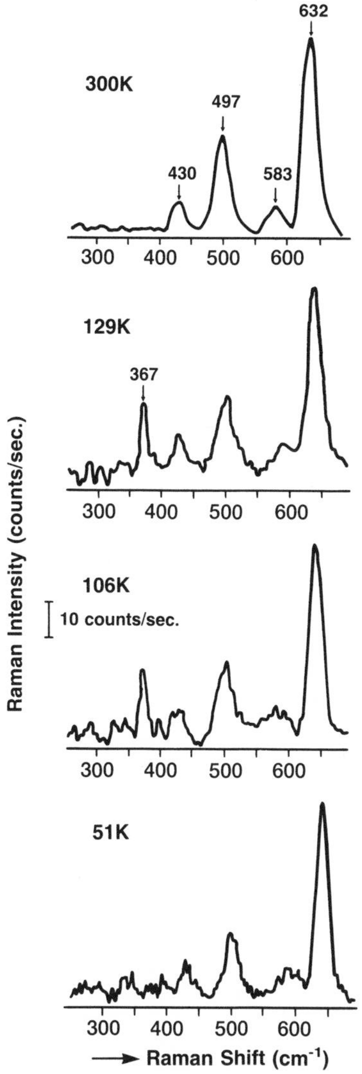

Fig. 4. Raman spectra at different temperatures of polcrystalline $YBa_2Cu_3O_{7-\delta}$ prepared via Ba-carbonate. Excitation is with 50mW 4880Å radiation.

The line at 632 cm^{-1} in Y 1-2-3 corresponds to a line at 626 cm^{-1} in $BaCuO_2$ of comparable cross section. X-ray diffraction, however, does not show greater than 1% by volume of $BaCuO_2$ as an impurity in these samples. This suggests that $BaCuO_2$-like defects are incorporated in Y 1-2-3 during certain synthesis protocols. The scattering at 632 cm^{-1} can, therefore, be assigned to Cu-O chain vibrations from the defects. Clearly these defects do not form when the Y 1-2-3 samples are produced via the atomically mixed citrate route.

The lines at 497 and 583 cm^{-1} in the polycrystalline samples can be assigned to the A_g

O(1) stretching and the B_{2g} and B_{3g} O(2) and O(4) stretching modes, respectively. This is consistent with the calculated values given in Table 2. The line at 330 cm^{-1} can also be assigned to a Cu-O deformation in agreement with the calculated values. However, the line near 430 cm^{-1} appears to be above the range of calculated values for the deformation modes and below the range for the calculated values for the stretching modes in the Y 1-2-3 structure.

Single crystal data on an as-grown Y 1-2-3 (001) plane platelet which shows ~10% flux exclusion below ~85K is shown in Fig. 5. XX scattering corresponding to A_{1g} modes is observed below 600 cm^{-1} at 490 and 556 cm^{-1}. The former line can be assigned to the O(1) stretching vibration. The shift to lower frequency of this mode is due to enhanced oxygen deficiency in the sample as indicated by a depressed T_c and decreased flux exclusion. The line at 556 cm^{-1} is likely to be the symmetric O(4) stretching mode which may be Raman-activated by the oxygen deficiency in the crystal. The 602 cm^{-1} line is associated with Y_2BaCuO_5 impurities which have been observed by X-ray diffraction in the CuO flux. The depolarized XY spectrum (which may contain other depolarized elements due to crystal misalignment) probably corresponds to modes of B_{2g} or B_{3g} symmetry with peaks at 330, 449 and 574 cm^{-1}, since modes of B_{1g} symmetry are forbidden according to the factor group analysis given in Table 1. The lines above 600 cm^{-1} in the XY spectrum are assignable to a $BaCuO_2$ phase which is also observed via X-ray diffraction in the CuO flux. The 574 cm^{-1} line can be assigned to the B_{2g} or B_{3g} symmetry stretching modes of the layer plane oxygen atoms, and the 330 cm^{-1} can be assigned to a Cu-O deformation mode. These assignments are consistent with the data on the polycrystalline material. Because of its relatively greater linewidth, particularly in the single crystal data, the line near 430 to 440 cm^{-1} could be a resonantly-enhanced second order band. Except for the absence of scattering around 440 cm^{-1} in the depolarized spectrum, the results described here are in agreement with the microcrystal Raman data published by Hemley and Mao (ref. 27). Note, however, that the single crystal examined here was independently checked for superconductivity.

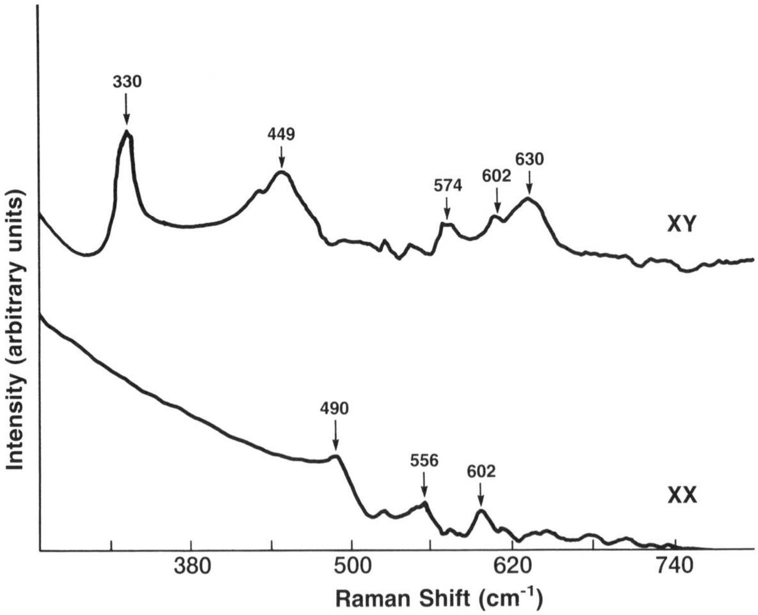

Fig. 5. Polarized back scattering Raman spectra of (001) plane single crystal $YBa_2Cu_3O_{7-\delta}$ at 300K. Excitation is with 20mW of 4880Å radiation.

Lyons et al have reported the Raman spectrum of a thin film on (100) $SrTiO_3$ using an iodine absorption cell, which allows reliable data to be obtained down to ~15 cm^{-1}. The lines observed at 80K are located at 338, 440 and 506 cm^{-1} in agreement with the polycrystalline and single crystal data discussed above. A sharp line at 142 cm^{-1} reported by Hemley and Mao in their microcrystal data has not, however, been observed in any of the measurements discussed above.

$\underline{YBa_2CuO_3O_6}$ and $\underline{Y(Ba_{2-x}Y_x)Cu_3O_{7-\delta}}$

In this section we will discuss the Raman spectra from the semiconducting Y_2BaCuO_5, tetragonal $YBa_2CuO_3O_6$ and $Y(Ba_{2-x}Y_x)Cu_3O_{7+\delta}$(Y 3-3-6) phases. Note that the Y 3-3-6 phase shows filamentary superconductivity when annealed near 520°C under 3 atmospheres of oxygen (ref. 13).

The Raman spectra of Y_2BaCuO_5 and $YBa_2Cu_3O_6$ recorded under exactly the same conditions, are shown together with that of the superconducting Y 1-2-3 phase in Fig. 6. We will not

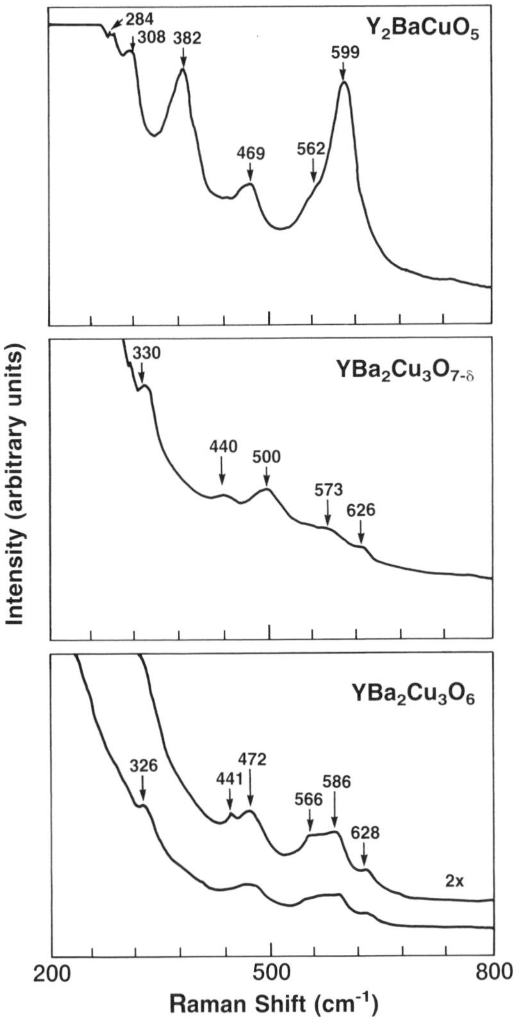

Fig. 6. Raman spectra of Y_2BaCuO_5, $YBa_2Cu_3O_{7-\delta}$ and $YBa_2Cu_3O_6$ at 300K recorded under the same conditions. Excitation is with 50mW 4880Å radiation.

analyze the vibrations of the Y_2BaCuO_5 phase here but only point out that the stretching modes ranging between 450 and 600 cm^{-1} and the bending modes below 400 cm^{-1} are comparable to those observed in the Y 1-2-3 phase. The tetragonal YBa_2CuO_6 structure shows relatively strong scattering at 566 and 586 cm^{-1}. The line at 586 cm^{-1} corresponds to the layer stretching mode in Y 1-2-3. The appearance of the line at 566 cm^{-1} suggests that as in the case of the Y 1-2-3 single crystal, the O(4) mode is activated by the presence of short-to medium-range oxygen disorder. This disorder is due to the presence of Cu-oxygen clusters in the chain direction although the neutron results, which give the average picture, does not show this. The 440 and 330 cm^{-1} lines remain essentially unchanged as one goes from Y 1-2-3 to $YBa_2Cu_3O_6$. However the line near 500 cm^{-1} shifts down in frequency due to c-axis expansion with increasing oxygen depletion.

The Raman spectra of a variety of Y 3-3-6 samples are shown in Fig. 7. Focussing on the parent Y 3-3-6 x = 0.5 phase, the Raman spectrum appears to be similar to $YBa_2Cu_3O_6$ except

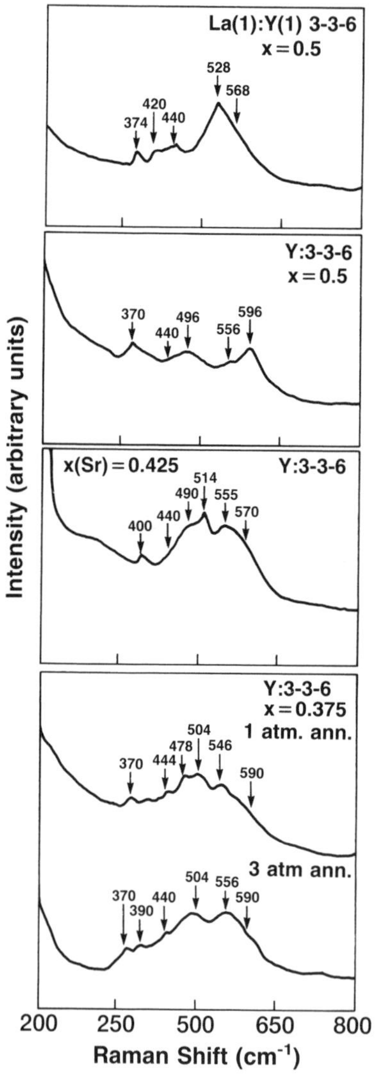

Fig. 7. Raman spectra of different Y 3-3-6 samples at 300K. The sample details are indicated on the figures. Excitation of the spectra is with 50mW 4880Å radiation.

that the Cu-O(1) stretching mode frequency remains near its value in Y 1-2-3, since the c-axis parameter in Y 3-3-6 is comparable to that in Y 1-2-3. In addition, the bending mode at 330 cm^{-1} in Y 1-2-3 shifts up in frequency to 370 cm^{-1} in Y 3-3-6. With decreasing x the relative intensity of the line near 550 cm^{-1} increases. This may reflect a higher degree of oxygen disorder since the oxygen content decreases with decreasing x (ref. 13). These results combined with those on Y 1-2-3 and $YBa_2Cu_3O_6$ indicate that careful studies of the Raman cross sections around 550 cm^{-1} may provide useful information regarding short-to medium-range order in these materials. More detailed studies of this are now in progress in this laboratory.

On mixing Y 3-3-6 with La in a 1:1 ratio we observe a decrease in the stretching frequencies which may be a reflection of the increase in the c-axis parameter with insertion of the larger La^{3+} ion. In addition, the La mixed sample shows more intense scattering at 526 cm^{-1} which corresponds to the line near 550 cm^{-1} in the pure Y 3-3-6 samples. This would indicate enhanced oxygen disorder in the La mixed sample relative to the pure Y 3-3-6 sample at x = 0.5.

In summary, I have presented and reviewed a range of Raman scattering data on a series of Y,La-Ba,Sr-Cu oxide superconducting and non-superconducting materials. The key copper-oxygen vibrations have been identified and a method of sampling the degree of short-to medium-range oxygen order in these materials is pointed out.

Acknowledgements - I would like to acknowledge the collaboration of H. Eckhardt, A. Bose, F. Reidinger, B. Chai and B.L. Ramakrishna (Arizona State University) in transport, magnetic, diffraction and crystal growth experiments on these exciting new materials. I would also like to thank Ray H. Baughman for his encouragement and support.

REFERENCES

1. G. Bednorz and K.A. Muller, Z. Phys. B64, 18 (1986).
2. R.J. Cava, R.B. van Dover, B. Batlogg and E.A. Rietman, Phys. Rev. Lett. 58, 408 (1987).
3. M.K. Wu, J.R. Ashburn, C.J. Torng, P.H. Hor, R.L. Meng, L. Gao, Z.J. Huang, Y.Q. Wang and C.W. Chu, Phys. Rev. Lett. 58, 908 (1967).
4. R.J. Cava, B. Batlogg, R.B. van Dover, D.W. Murphy, S. Sunshine, T. Siegrist, J.P. Remeika, E.A. Rietman, S. Zahurak and G.P. Espinosa, Phys. Rev. Lett. 58, 1676 (1987).
5. D.U. Gubser and M. Schluter (Editors) High Temperature Superconductors, Materials Research Society Symposium, Anaheim (1987).
6. J. Bardeen, L.N. Cooper and J.R. Schrieffer, Phys. Rev. 106, 162 (1957).
7. Z. Iqbal, S.W. Steinhauser, A. Bose, N. Cipollini, F. Reidinger and H. Eckhardt, Proc. SPIE Intl. Conf. on Raman and Luminescence Spectroscopy in Technology, Vol. 822 - in press.
8. B. Batlogg, R.J. Cava, A. Jayaraman, R.B. van Dover, G.A. Kourouklis, S. Sunshine, D.W. Murphy, L.W. Rupp, H.S. Chen, A. White, K.T. Short, A.M. Mujsee and E.A. Rietman, Phys. Rev. Lett. 58, 2333 (1987).
9. L.C. Bourne, M.F. Crommie, A. Zettl, H.C. zur Loye, S.W. Keller, K.L. Leary, A. Stacy, K.J. Chang, M.L. Cohen and D.E. Morris, Phys. Rev. Lett. 58, 2337 (1987).

10. L. Ran, R. Merlin, M. Cardona, Hj. Mattausch, W. Bauhofer, A. Simon, F. Garcia-Alvarado, E. Moran, M. Vallet, J.M. Gonzalez-Calbet and M.A. Alario, Solid State Commun. 63, 839 (1987).
11. Z. Iqbal, S.W. Steinhauser, A. Bose, N. Cipollini and H. Eckhardt, Phys. Rev. B36, 2283 (1987).
12. D.L. Kaiser, F. Holtzberg, B.A. Scott and T.R. McGuire, Appl. Phys. Lett. 51, 1040 (1987).
13. Z. Iqbal, F. Reidinger, A. Bose, N. Cipollini, H. Eckhardt, B. L. Ramakrishna and E.W. Ong, Proc. Symp. High T_c Superconductors, MRS, Boston (1987) - in press.
14. K.B. Lyons, S.H. Liou, M. Hong, H.S. Chen, J. Kwo and T.J. Negran, Phys. Rev. B36, 5592 (1987).
15. M. Onoda, S. Shamoto, M. Sato and S. Hosoya, Jap Jour. Appl. Phys. 26, L363 (1987).
16. F. Beech, S. Miraglia, A. Santoro and R.S. Roth, Phys. Rev. B35, 8778 (1987).
17. A. Santoro, S. Miraglia, F. Beech, S.A. Sunshine, D.W. Murphy, L.F. Schneemeyer and J.V. Waszczak, Mat. Res. Bull. 22, 1007 (1987).
18. R.M. Hazen, L.W. Finger, R.J. Angel, C.T. Prewitt, N.L. Ross, H.K. Mao, C.G. Hadidiacos, P.H. Hor, R.L. Meng and C.W. Chu, Phys. Rev. B35, 7238 (1987).
19. M. Stavola, D.M. Krol, W. Weber, S.A. Sunshine, A. Jayaraman, G.A. Kourouklis, R.J. Cava and E.A. Rietman, Phys. Rev. B36, 850 (1987).
20. W. Weber, Phys. Rev. Lett. 58, 1371 (1987).
21. G.A. Kourouklis, A. Jayaraman, W. Weber, J.P. Remeika, G.P. Espinosa, A.S. Cooper and R.G. Maines Sr. - preprint.
22. S. Blumenroeder, E. Zirngiebl, J.D. Thompson, P. Killough, J.L. Smith and Z. Fisk, Phys. Rev. B35, 8840 (1987).
23. T. Brun, M. Grimsditch, K.E. Gray, R. Bhadra, V. Maroni and C.K. Loong, Phys. Rev. B35, 8837 (1987).
24. Z. Iqbal, B. Chai, B.L. Ramakrishna and E.W. Ong - to be published.
25. G.A. Kourouklis, A. Jayaraman, B. Batlogg, R.J. Cava, M. Stavola, D.M. Krol, E.A. Rietman and L.F. Schneemeyer - preprint.
26. M. Cardona, L. Genzel, R. Liv, A. Wittlin, Hj. Mattausch, F. Garcia-Alvarado and E. Garcia-Gonzalez, Solid State Commun. - in press.
27. R.J. Hemley and M.K. Mao, Phys. Rev. Lett. 58, 2340 (1987).

A simple method to prepare high T_c superconducting thin films: pulsed laser deposition

X. D. Wu[1], T. Venkatesan[2]

[1]Department of Physics, Rutgers University, Piscataway, NJ 08854, USA
[2]Bell Communications Research, Red Bank, NJ 07701, USA

Abstract-We have used a laser deposition method to prepare high T_c superconducting thin films of $RBa_2Cu_3O_{7-x}$ (R=Y, Gd, Eu) from a single high T_c target. The composition of the films was very close to that of the target, with uniformity over a deposition area of >1 cm^2. The method is simple, easy and inexpensive compared to other techniques. After a proper annealing in oxygen the films show superconductivity with zero resistance over liquid nitrogen temperatures and transition widths as narrow as 2 K. The film on (100) $SrTiO_3$ has a preferential orientation with the c-axis of the film normal to the substrate surface. The properties of the films were dependent on the substrates on which the films were deposited.

INTRODUCTION

Since the discovery of superconductivity at 30 K in the Ba-La-Cu-O system (ref. 1) by Bednorz and Muller and at 90 K in the multiphase Y-Ba-Cu-O system (ref. 2), tremendous effort has been made by scientists all over the world to study the physics and the technology of these high transition temperature superconductors. A large fraction of the technological thrust since the discovery of the new high T_c materials has been in the fabrication of wires to be used in power lines and magnets and thin films for superconducting devices. Besides practical applications, these films are also useful for basic physical measurements such as critical current density, flux-pinning etc.. Superconducting thin films have been successfully prepared by *e*-beam evaporation (ref. 3,4), sputtering (ref. 5), multilayer evaporation (ref. 6), molecular beam deposition (ref. 7) and laser deposition (ref. 8). All these methods could be generically divided into two catagories: multiple-source and single source deposition. To get uniform composition over a large area of the film is difficult by the multiple source methods , since the overlap of the trajectoriers of the metal elements in the deposition plane is a problem. In addition, the relative deposition rate of each source has to be maintained constant during the entire process to get the correct film composition. Hence, in the long run single source systems would be the preferred choice. Sputtering is an example of one such process. But the sputtered film composition will not be the same as the target due to different sticking coefficients for the metal elements in addition to their different sputtering rates. In this paper, we report a far simpler approach to prepare these high T_c superconducting thin films. The method is pulsed laser deposition from a single target, which is simple, easy, and inexpensive compared to most of the other techniques.

The laser evaporation method has been used to deposit thin films of semiconductors (ref. 9) and dielectrics (ref. 10). Even superconducting films of $BaPb_{1-x}Bi_xO_3$ have been made by using pulsed Ruby laser (ref. 11) deposition from a single target; the best films after thermal annealing showed zero resistance temperatures at 5 K compared to the bulk transition temperature of 14 K. We have used the same method to prepare the new high T_c superconducting thin films. Under practical conditions the laser enables us to get films with stoichiometry close to that of the target. After annealing in oxygen the films show superconductivity with zero resistance at 86 K on $SrTiO_3$, 82 K on ZrO and 75 K on Al_2O_3. For all the films, the superconducting onset temperature was always about 93 K, similar to the bulk material.

EXPERIMENT

The detailed description of the deposition system has been published earlier (ref. 8). A small 'O' ring sealed vacuum system with a base pressure of $5x10^{-7}$ Torr was used. High T_c pellets of $RBa_2Cu_3O_{7-x}$ (R= Y, Eu, Gd) with zero resistance temperature about 93 K as targets were prepared using standard ceramic technique (ref. 12). The target was mounted about 3 *cm* from the substrate holder which was heated to 450 C during deposition, and irradiated by a KrF excimer laser (2480 Å, 30 *ns*) at 45° angle of incidence through a quartz window (Fig. 1). The typical energy density on the target was about 2 J/cm^2. To remove the material uniformly, the target was slowly rotated during deposition. The typical deposition rate was about 1 Å/pulse. Films with thickness of 2000 to 7000 Å were deposited on substrates of (100) and (110) $SrTiO_3$, ZrO and Al_2O_3. The film composition was examined by Rutherford backscattering spectrometry (RBS) and Auger electron spectrometry (AES). The surface morphology was studied by scanning electron microscopy (SEM). The structure in the films was determined by X-ray and transmission electron microscopy (TEM). Electrical measurements were made using a standard four probe technique. Contacts were made by indium solder or silver ink. Typical contact resistance was about 30 to 50 ohms. The applied current was 1 to 10 μA.

RESULTS AND DISCUSSION

X-ray and electron diffraction revealed that the as-deposited films were disordered. TEM studies on the films indicated the crystal size to be < 10 Å. The films did not show any measurable room temperature conductivity. It was therefore necessary to subsequently anneal the films in oxygen. The typical annealing condition was: insert at 650 C , hold for 10 m, then raise to 900 C within 15 m, hold for 1 h and finally slowly cool to room temperature in a few hours.

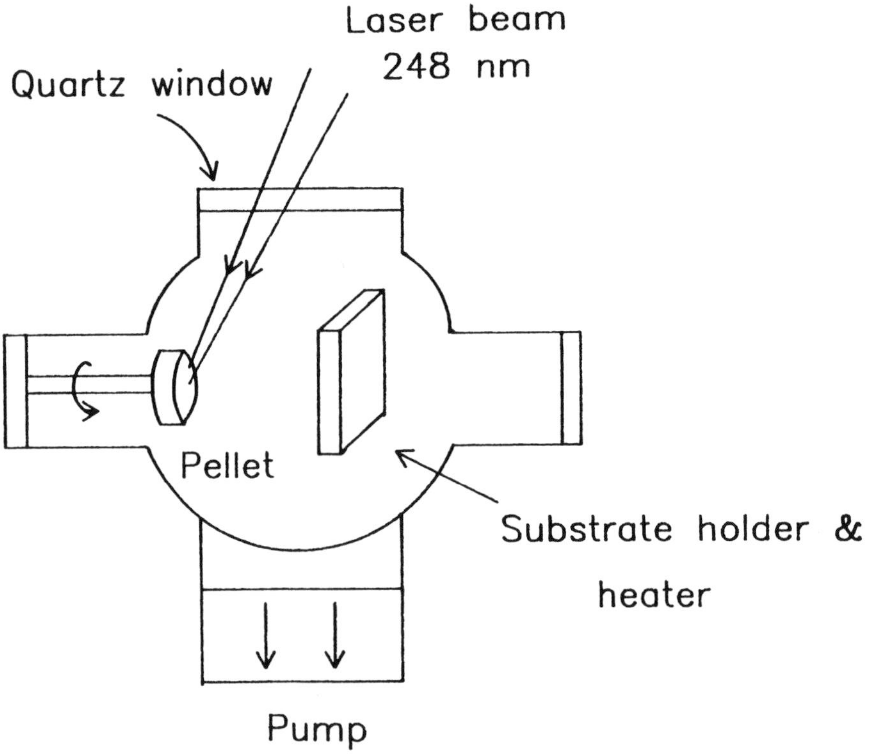

Fig. 1. Experimental setup.

The first experiment was performed in air. A target of Y-Ba-Cu-Oxide and a carbon substrate were mounted in a V-shaped holder. After 1000 laser shots, the substrate was examined by RBS with the results shown in Fig. 2. The solid line is the experimental data and the dashed line is a simulation assuming the Y, Ba, Cu ratio of 1:2:3 (The simulation was made by RUMP program (ref. 13)). It is clear from the simulation that the film deposited in air has all the three metal elements with a composition very close to that of the bulk target. The deposition rate was low since the mean free path of the elements in air was very short. Subsequently, all the depositions were performed in a vacuum system. Figure 3 is an RBS spectrum of a film on a sapphire substrate made in vacuum. The composition of the films was close to $YBa_2Ba_3O_6$ (within 10%) as shown in the simulation. Although there is no oxygen flow in the system, the as-deposited films had a large amount of oxygen . That is probably the reason why our as-deposited films are much more stable compared to those made by *e*-beam evaporation (ref. 3). So far, the precise measurement of the oxygen content in the films is still a problem.

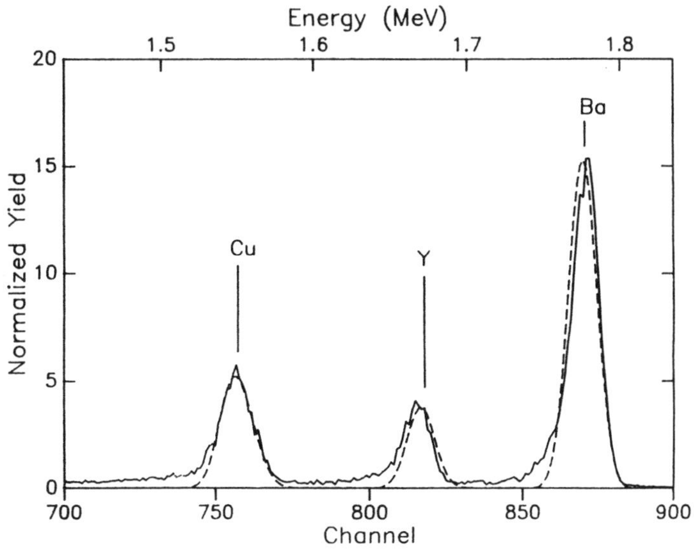

Fig. 2. RBS-spectrum (2 Mev He^+) of a very thin (about 100 Å) as-deposited film made in air (solid line); the dashed line is the simulation of $Y_1Ba_2Cu_3O_6$.

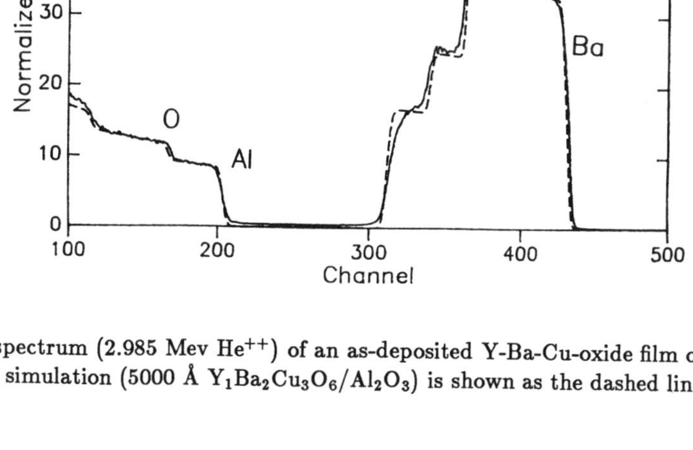

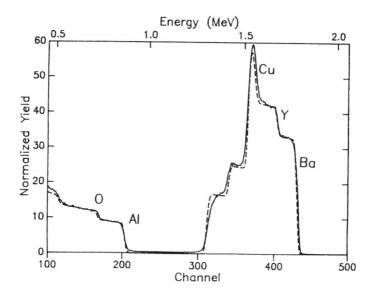

Fig. 3. RBS-spectrum (2.985 Mev He^{++}) of an as-deposited Y-Ba-Cu-oxide film on sapphire in vacuum. The simulation (5000 Å $Y_1Ba_2Cu_3O_6/Al_2O_3$) is shown as the dashed line.

Substrate temperature during the deposition was about 450 C. The films deposited at lower temperature do not generally adhere well to the substrates. The deposition temperature effects were studied by processing two films identically but for two different deposition temperatures. The results of the electrical measurements are shown in Fig. 4. Both films were deposited on Al_2O_3 and annealed at 850 C for 1 h. The temperature dependence of the resistances were measured only down to 77 K. The onsets of superconductivity for both films were close (at 93 K); but the ratio of the onset resistance to the room temperature value is much larger for the film made at lower deposition temperature.

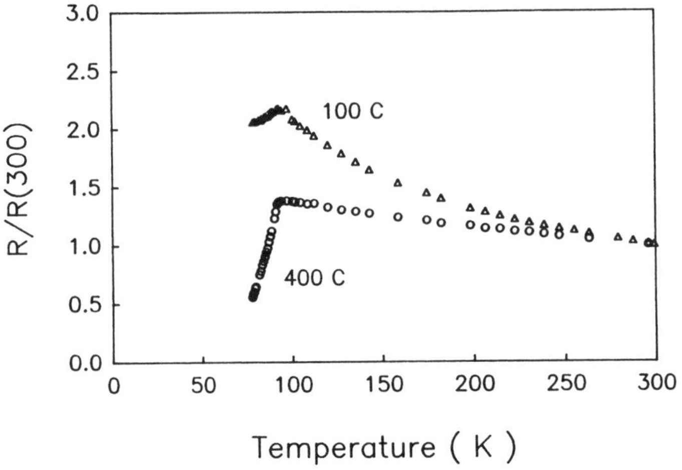

Fig. 4. Resistance vs. temperature measurements for two annealed films on sapphire deposited at different substrate temperatures indicated in the figure.

We have also studied the annealing conditions to obtain reproducible, good superconducting films. The annealing parameters included the initial insertion temperature, ramp up time, final temperature and duration and cooling rate. It was found that some of these were very important to achieve superconductivity at high temperatures. For example, slow cooling from high temperature to room temperature over a few hours is essential. The air quenched films showed almost no zero resistance temperature all the way down to 8 K. But after annealing over 800 C in oxygen for more than 30 m, the films always exhibited an onset of about 93 K. Figure 5 shows a sharp superconducting transition of a film on (100) $SrTiO_3$. The zero resistance temperature is 84 K and transition width (90%-10%) is about 4 K for this film. The critical current density on this films was about $1x10^4$ A/cm^2 at 77 K, and $1x10^6$ A/cm^2 at 4 K. The measurement was made by a four probe technique on a 1 mm wide strip with a thickness of 3000 Å.

X-ray diffraction was used to determine the phases in the films. Details of the x-ray results will be published elsewhere (ref. 14). Figure 6 shows the diffraction pattern for a film deposited on Al_2O_3 with zero resistance temperature of 55 K. The dominant phase in the film is orthorhombic as in the high T_c bulk material.

From the study, we found that under identical processing conditions the films on various substrates showed different electrical characteristics. It was concluded that the substrate influences the film properties primarily in three ways (ref. 15): the thermal expansion mismatch introduces cracks in the film, the interface reaction changes the film composition and the lattice match with substrate results in different crystallite sizes. Since the films have to be annealed at high temperatures to get superconductivity, the cracks formed during the thermal cycle destroy the percolation path for superconductivity in the film. Furthermore, the interdiffusion between the film and substrate creates a dead layer at the interface. We have shown that films of two different thickness on (100) $SrTiO_3$ under identical processing conditions exhibited different superconducting behavior (ref. 16). The thick (3500 Å) and thin (1200 Å) films had zero resistance temperatures of 85 K and about 8 K respectively although the onsets were same. Therefore, low temperature processing is necessary to obtain high zero resistance temperatures on most substrates and for preparation of very thin ($<$ 1000 Å) films. From x-ray and ion channeling studies, the films on $SrTiO_3$ were oriented

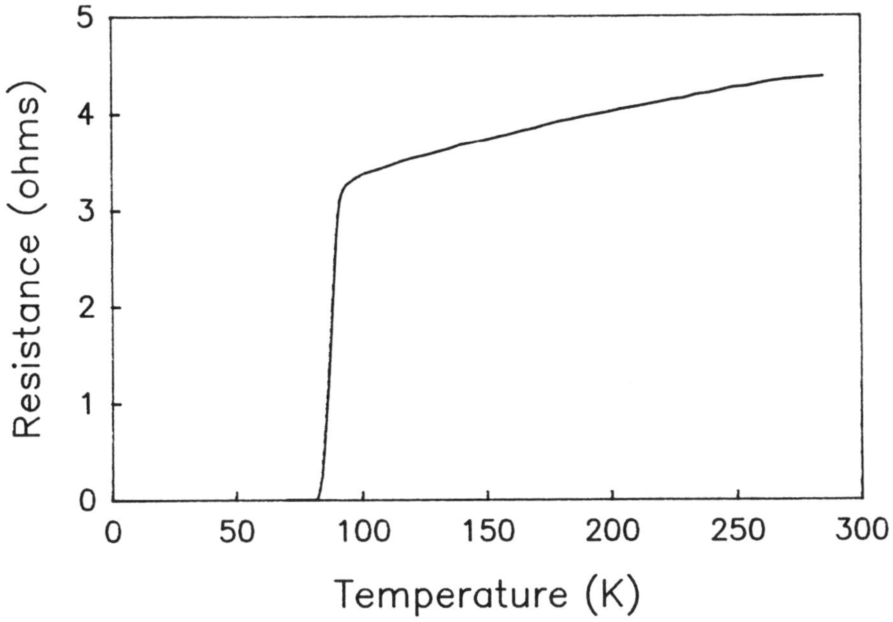

Fig. 5. Resistance vs. temperature for an annealed Y-Ba-Cu-Oxide film on (100) $SrTiO_3$.

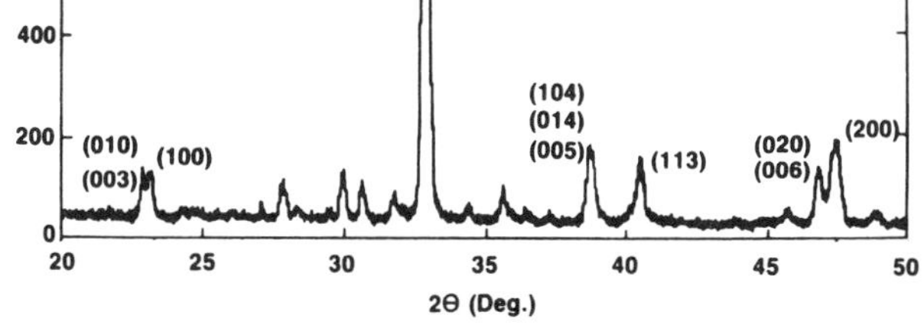

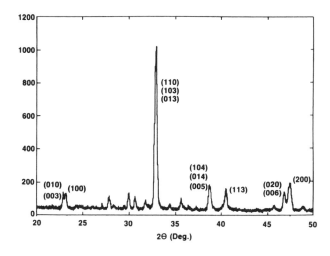

Fig. 6. X-diffraction pattern for an annealed film on sapphire with random orientation.

with respect to the substrate since the lattice match between the film and substrate is within a few percent. We found that the c-axis of the film was normal to the substrate plane. Orientation of the c-axis of the film both in and normal to the plane on (100) $SrTiO_3$ were observed in the films prepared by other methods (ref. 3,4,7). We believe that the orientation after annealing is dependent on the starting film composition. Due to the solid phase expitaxy, the cystallite size in the films on (100) $SrTiO_3$ is relative larger, about 3000 to 5000 Å, compared to 500 Å on Al_2O_3 from TEM studies (ref. 15).

We have also prepared the superconducting thin films of Eu- and Gd-Ba-Cu-oxide by simply changing the target. Figure 7 and 8 show the RBS spectra of Eu and Gd substituted films respectively. The dashed lines again are simulations with compositions close to those in the targets. After annealing in oxygen, zero resistance temperature over 77 K was also achieved in these films (ref. 17). It is very clear that the laser deposition method is relatively simple and general since for all the other techniques a change in the source composition would imply developement of a new process recipe.

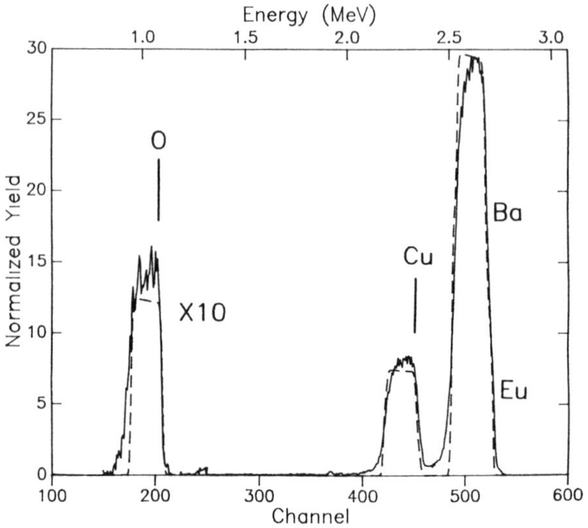

Fig. 7. RBS-spectrum (3 Mev He^{++}) of an as-deposited Eu-Ba-Cu-Oxide film on a carbon substrate. The dashed line is a simulation of $Eu_1Ba_2Cu_3O_6$.

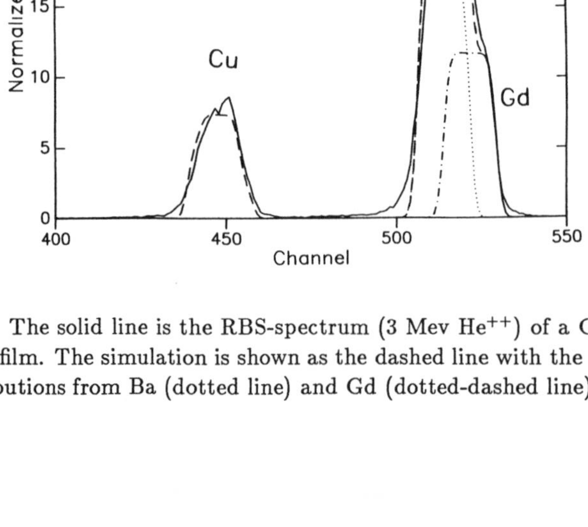

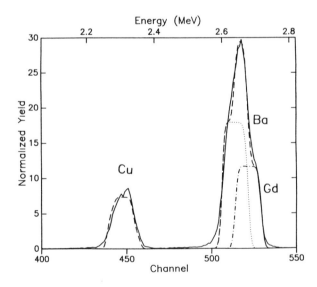

Fig. 8. The solid line is the RBS-spectrum (3 Mev He^{++}) of a Gd-Ba-Cu-Oxide film. The simulation is shown as the dashed line with the individual contributions from Ba (dotted line) and Gd (dotted-dashed line).

The laser we are using is operating at 2480 Å with a 30 *ns* pulse width. It was shown that pulsed Ruby laser (6943 Å) could be used to get superconducting films of $BaPb_{1-x}Bi_xO_3$ (ref. 11). A YAG laser (10600 Å) was used to get 56 K superconducting thin films of Y-Ba-Cu-oxide (ref. 18). So we believe that pulsed lasers from deep-UV to near infra-red could be utilized to prepare the new superconducting films.

In summary, we have shown the pulsed laser deposition method to be a simple, versatile technique for preparing the high T_c superconducting thin films. The technique seems viable for commercial upscaling.

The authors wish to thank D. Dijkkamp, now at Philips Research Laboratories, Eindhoven; S. B. Ogale at Poona University, India; E. W. Chase, C. C. Chang, D. M. Hwang, P. F. Miceli, S. A. Schwarz, L. A. Farrow, J. M. Tarascon, N. Stoffel, B. Wilkens, P. Barboux, L. H. Greene and J. M. Rowell at Bell Communications Research; A. Inam, S. A. Shaheen, N. Jisrawi, K. Marcantonio, W. L. McLean and M. Croft at Rutgers University for various help and discussions.

REFERENCES

[1] J. G. Bednorz and K. A. Muller, Z. Phys. **B64**, 189(1986).

[2] M. K. Wu, J. R. Ashburn, C. T. Torng, P. H. Hor, R. L. Meng, L. Gao, Z. J. Huang, Y. Q. Wang and C. W. Chu, Phys. Rev. Lett. **58**, 908(1987).

[3] R. B. Laibowitz, R. H. Koch, P. Chaudhari and R. J. Gambino, Phys. Rev. **B35** , 8821(1987). P. Chaudhari, R. H. Koch, R. B. Laibowitz, T. R. Mcguire, and R. J. Gambino, Phys. Rev. Lett. **58**, 2684(1987).

[4] M. Naito, R. H. Hammond, B. Oh, M. R. Hahn, J. W. P. Hsu, P. Rosenthal, A. Marshall, M. R. Beasley, T. H. Geballe, and A. Kapitulnik, (unpublished).

[5] Y. Enomoto, T. Murakami, M. Suzuki, and K. Moriwaki, Jpn. J. Appl. Phys, **26**, L1248(1987).

[6] Z. L. Bao, F. R. Wang, Q. P. Jiang, S. Z. Wang, Z. Y. Ye, K. Wu, C. Y. Li, and D. L. Yin, Appl. Phys. Lett. **51**, 946(1987).

[7] J. Kwo, T. C. Hsieh, R. M. Fleming, M. Hong, S. H. Liou, B. A. Davidson, and L. C. Feldman, Phys. Rev. **36**, 4039(1987).

[8] D. Dijkkamp, T. Venkatesan, X. D. Wu, S. A. Shaheen, N. Jisrawi, Y. H. Min-Lee, W. L. McLean, and M. Croft, Appl. Phys. Lett. **51**, 619(1987).

[9] J. T. Cheung and T. Magee, J. Vac. Sci. Technol. **A1**, 1604(1983).

[10] H. Sankur, in *Laser Contolled Chemical Processing of Surfaces*, edited by A. W. Johnson, D. J. Ehrlich, and H. R. Schlossberg (North-Holland, New York, 1984), p.373.

[11] S. V. Zaitsev-Zotov, A. N. Martynyuk, and E. A. Protasov, Sov. Phys. Solid State, **25**, 100(1983).

[12] J. M. Tarascon, W. R. McKinnon, L. H. Greene, G. W. Hull, and E. W. Vogel, Phys. Rev. **B36**, 226(1987).

[13] L. R. Doolittle, Nucl. Instr. and Meth. in Phys. Res. **B9**,344(1985).

[14] P. F. Miceli (unpublished).

[15] T. Venkatesan, C. C. Chang, D. Dijkkamp, S. B. Ogale, E. W. Chase, L. A. Farrow, D. M. Hwang, P. F. Miceli, S. A. Schwarz, J. M. Tarascon, X. D. Wu, and A. Inam, J. Appl. Phys., to be published.

[16] X. D. Wu, D. Dijkkamp, S. B. Ogale, A. Inam, E. W. Chase, P. F. Miceli, C. C. Chang, J. M. Tarascon, and T. Venkatesan, Appl. Phys. Lett. **51**, 861(1987).

[17] X. D. Wu *et al.*, (unpublished).

[18] Y. H. Shen, (Fudan University, China, private communication).

High-temperature superconductivity in bismuth and thallium cuprates

C.N.R. Rao, L. Ganapathi, A.K. Ganguli, R.A. Mohan Ram, R. Vijayaraghavan, A.M. Umarji, G.N. Subbanna and P. Somasundaram

Solid State and Structural Chemistry Unit and Materials Research Laboratory, Indian Institute of Science, Bangalore-560012, India.

Abstract Structure, stoichiometry and superconductivity of oxides of the Bi-Ca-Sr-Cu-O and related systems are discussed. Various compositions of the Tl-Ca-Ba-Cu-O system (2212, 2122, 2223 etc.) show onset of superconductivity around 120K. In the Bi and Tl cuprates, although some of the Cu is nominally to be in the 3+ state, only Cu^{+} and Cu^{2+} states are actually present, indicating the possible role of oxygen holes. These cuprates absorb microwave radiation in the superconducting state. Both Bi and Tl cuprates possess structures analogous to those of the Aurivillius family of bismuth oxides containing perovskite layers. Electron microscopy has been employed extensively to study these cuprates. Evidence is found for the presence of dislocations and intergrowths. It is pointed out that intergrowths may be playing an important role in the superconductivity of these cuprates and provides a possible means of increasing the T_c to much higher temperatures.

INTRODUCTION

High-temperature superconductivity in cuprates of K_2NiF_4 structure ($T_c \approx 30 \pm 10K$) and of the 123 variety ($T_c \approx 90K$) has been the subject of intense investigation since January 1987. In the last few weeks, there is much excitement in studying oxides containing bismuth, alkaline earth metal and copper or thallium, alkaline earth and copper for high-temperature superconductivity above liquid nitrogen temperature. Present indications are that T_cs around 100K or higher are possible with these oxides (see High T_c update, Vol. 2, February 15, 1988). Michel et al (ref. 1) showed recently that an oxide of the Bi-Sr-Cu-O family becomes superconducting in the 7-22K range. The oxide was of the composition $Bi_2Sr_2Cu_2O_{7+\delta}$ with a perovskite related structure. Maeda et al (ref. 2) found onset of superconductivity around 105K in $BiCaSrCu_2O_x$ as determined by resistivity measurements, but with a low-temperature tail extending down to ~ 80K. Chu et al (ref. 3) have reported onset temperatures upto 120K in multiphasic Bi-Ca-Sr-Cu-O and Bi-Al-Ca-Sr-Cu-O systems. All the bismuth cuprates possess structures akin to those of the Aurivillius family of oxides (ref. 4), $(Bi_2O_2)^{2+}$ $(A_{n-1}B_nO_{3n+1})^{2-}$. Several groups of workers have initiated studies on the bismuth cuprates. Although there has been some success in identifying one or two of the superconducting phases, in the Bi-Ca-Sr-Cu-O system, there is still uncertainty with regard to the exact description of the phases which give T_cs in the 80K region and above 100K; no obvious difference in the structures has been noticed between these two phases. Several superconducting compositions have been identified in the Tl-Ca-Ba-Cu-O system. It seems that at least part of the difficulty with these new oxide systems may be due to the presence of unusual defects and considerable disorder. Such defects and disorder are characteristic of the Aurivillius family of oxides. It was shown quite early (ref. 5) that the Aurivillius family of oxides, exhibit extensive dislocations and disorder, side-stepping of Bi_2O_2 layers being commonly encountered. More importantly, these oxides form disordered as well as recurrent intergrowths (ref. 6-8). Presence of such intergrowths and/or stacking disorder could indeed affect the superconducting transition temperatures in the Bi and Tl cuprates. In this article, we shall discuss the structure, stoichiometry and superconducting properties of the Bi-M-Cu-O (M = alkaline earth) and Tl-M-Cu-O families of high-temperature superconductors.

SUPERCONDUCTING PROPERTIES OF $Bi_mM_nCu_pO_x$

Some of the samples of Chu et al (ref. 3) showed a superconducting transition with a midpoint temperature around 90K, while some others showed steps in both resistivity and magnetic measurements corresponding to a $T_c \approx 115K$. Temperature variation of the resistivity and the d.c. magnetization (ref. 9) of two samples of nominal composition $BiCaSrCuO_x$ (a and b) and one sample of $BiCaSrCu_3O_x$(c) subjected to different heat treat-

---

*Contribution No. 523 from the Solid State & Structural Chemistry Unit.

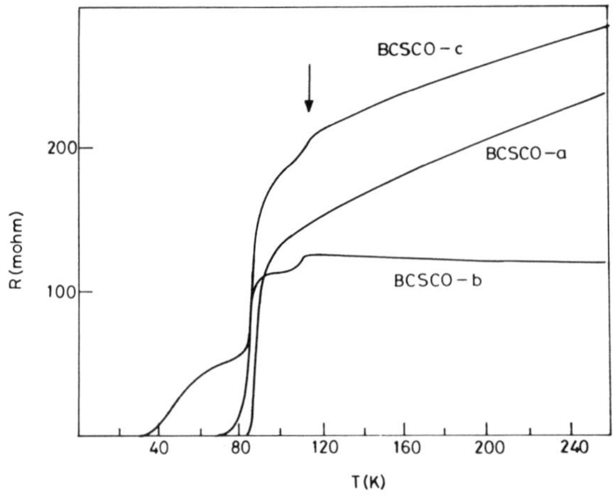

Fig. 1. Temperature variation of the resistance of Bi-Ca-Sr-Cu oxide superconductors. (a) and (b) have the composition $BiCaSrCuO_x$ and (c) has the composition $BiCaSrCu_3O_x$. (From Hazen et al, ref. 9.).

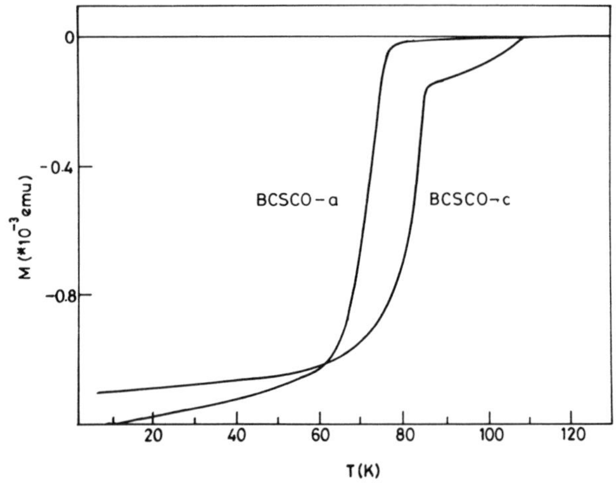

Fig. 2. Temperature variation of the d.c. magnetization of Bi-Ca-Sr-Cu oxide superconductors, (a) and (c) are same as in Fig. 1. (From Hazen et al, ref. 9).

ments are shown in Figs. 1 and 2. Sample (a) of $BiCaSrCuO_x$ shows only one superconducting transition, but sample (b) shows two just as $BiCaSrCu_3O_x$ (sample c). Nearly 12% Meissner effect is associated with the 115K transition, the total effect being around 40%. Even higher Meissner effects upto 60% or more have also been detected in some of the samples. It is believed that much of the Meissner effect is associated with the 115K transition. The nature of heat treatment of the sample seems to be crucial. In Figs. 3 and 4 we show the resistivity data of several Bi-M-Cu-O (M = Mg, Ca, Sr, Ba) oxides of the general formula $Bi_mM_nCu_pO_x$ (see Table 1) obtained in this laboratory (ref. 10). Heating the samples over long periods just below melting improves the superconducting properties and sharper transitions than in Figs. 3 and 4 are obtained. Once melted, the properties are badly affected. The resistivity drops (or steps) around 200K found in some of the samples (Fig. 3 and Table 1) are noteworthy. Comparing these steps with the 115K step in Fig. 1, one is tempted to believe that they may signify superconducting transitions. We shall return to this aspect later in this article. For purpose of information, we have listed in Table 2 bismuth cuprates (prepared in this laboratory) which have not shown superconductivity.

The importance of the heat treatment of the samples is clearly brought out by the data of Tarascon et al (ref. 11). In Fig. 5 we show the resistivity data of $Bi_2(Ca, Sr)_3Cu_3O_x$ for a polycrystalline sintered sample (I) and for another sample after annealing in oxygen at 1070K (8 hr.) and further vacuum annealing for 24 hr. (II). Sample I exhi-

Table 1

Superconducting properties of $Bi_mM_nCu_pO_x$

Composition	Preparative conditions (temperature in K)	Onset temperature (K)	Zero resistance temperature (K)
$Bi_2CaSrCuO_{6+\delta}$	1100-24h/air +1115-34h/air	85	25
$Bi_2CaSrCu_2O_{7+\delta}$	1100-32h/air +1100-18h/oxygen annealed	77 80 (steps at 220 and 60)	20 25
$Bi_2CaSr_2Cu_2O_{8+\delta}$	(a) 1100-32h/air +1100-18h/oxygen annealed	70 (step at 55)	28
	(b) 1100-48h/air +1160-12h/oxygen quenched	100	--
	(c) 1100-32h/air +1115-30h/air	85 (step at 65)	45
$Bi_2Ca_{1.5}Sr_{1.5}Cu_2O_{8+\delta}$	1100-24h/air +1115-42h/air	100	43
$Bi_{2.1}Ca_{1.15}Sr_{1.75}Cu_2O_{8+\delta}$	1050-12h/air +1100/12h/air +1115-12h/air	110	60
$Bi_{2.1}Ca_{1.45}Sr_{1.45}Cu_2O_{8+\delta}$	- " -	110	53
$Bi_{2.15}Ca_{1.17}Sr_{1.68}Cu_2O_{8+\delta}$	(a) 1050-12h/air 1100-12h/air 1115-12h/air	100	20
$Bi_2Ca_2SrCu_2O_{8+\delta}$	1100-48h/air +1160-12h/oxygen quenched	90	-
$Bi_{2.2}Ca_{1.2}Sr_{1.6}Cu_2O_{8+\delta}$	1070-12h/air 1150-24h/oxygen quenched	100	60
$Bi_{2.25}Ca_{1.2}Sr_{1.55}Cu_2O_{8+\delta}$	- " -	90 (steps at 280 and 65)	49
$Bi_2CaMgSrCu_2O_{8+\delta}$	1100-32h/air	100 (steps at 200 and 65)	34
$Bi_2CaMgSr_2Cu_2O_x$	1100-32h/air	80	--
$Bi_2Ca_2Sr_2Cu_3O_x$	1100-48h/air +1160-6h/oxygen quenched	95	60
$Bi_2Ba_2Sr_2Cu_3O_x$	1100-32h/air +1100-18h/oxygen annealed	65 94	28 30
$Bi_3CaSr_2Cu_3O_x$	1100/32h/air +1100-18h/oxygen annealed	70 100 (step at 70)	-- --
$Bi_3Ca_2SrCu_3O_x$	1100-48h/air quenched	73	--

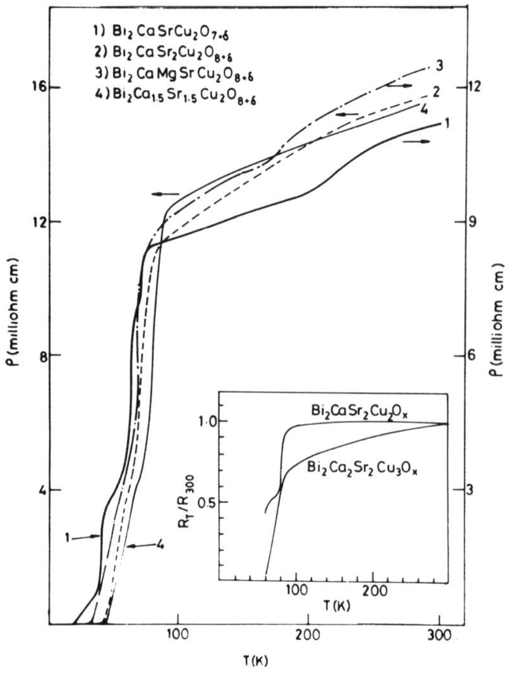

Fig. 3. Resistivity behaviour of superconducting $Bi_mM_nCu_pO_x$ compositions. Behaviour of two preparations of $BiCaSr_2Cu_2O_x$ is shown (ref. 10).

Table 2

$Bi_mM_nCu_pO_x$ compositions not yet found to be superconducting

Composition	Preparative conditions (temperatures in K)	Resistivity behaviour
$Bi_2BaSrCu_2O_x$	1110-36h/air +1110-18h/oxygen	insulating
$Bi_2BaSr_2Cu_2O_x$	1110-36h/air +1110-13h/oxygen	semiconducting
$Bi_2Ba_2SrCu_2O_x$	1145-36h/air +1145-12h/oxygen	semiconducting
$Bi_3CaSrCu_2O_x$	1110-36h/air +1110-18h/oxygen	metallic
$Bi_3Sr_2Cu_2O_x$	1110-36h/air +1110-18h/oxygen	semiconducting
$Bi_3BaSrCu_2O_x$	1110-36h/air +1110-18h/oxygen	insulating
$Bi_2Ba_2SrCu_3O_x$	1110-36h/air +1110-18h/oxygen	semiconducting

bits a T_c at 85K; in sample II the T_c is shifted to low temperatures. AC susceptibility measurements also show a T_c around 85K for sample I (Fig. 6). Resistivity and susceptibility data of $Bi_2(Ca, Sr)_3Cu_3O_x$ which was annealed at 1150K for 48 hrs. (sample III) is shown along with the data of $Bi_2(Ca, Sr)_3Cu_2O_y$ (samfple IV) annealed similarly. We readily see that these samples show a T_c around 110K. Sample III also shows linear resistivity below 200K exhibiting a negative curvature down to T_c ; $YBa_2Cu_3O_{7-\delta}$, on the other hand, shows linear resistivity with temperature from room temperature down to T_c. What is also rather intriguing is the long resistivity tail exhibited by the 110K T_c samples after the sharp resistivity drop. It should be noted that the total Meissner effect of sample IV is 50% with 26% and 24% corresponding to the 110K and 85K transitions.

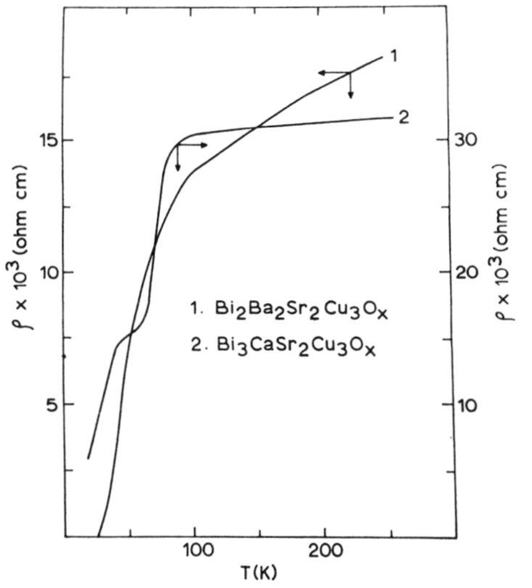

Fig. 4. Resistivity behaviour of superconducting $Bi_mM_nCu_pO_x$ compositions (ref. 10).

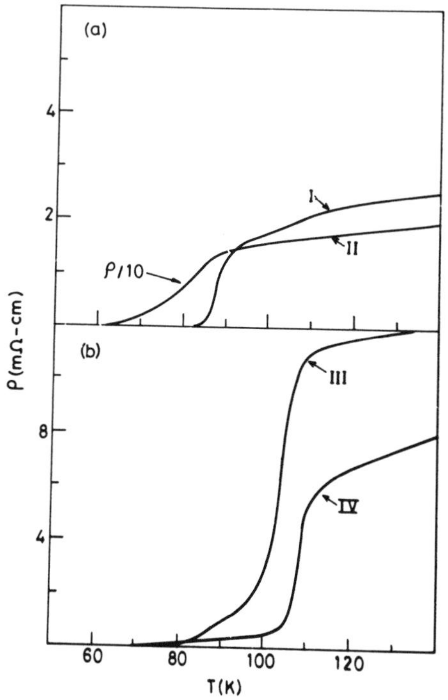

Fig. 5. Temperature variation of the resistivity of $Bi_2(Ca, Sr)_3Cu_3O_x$ superconductors (From Tarascon et al, ref. 11). For description of samples, see text.

The composition of the important superconducting phase in the Bi-Ca-Sr-Cu-O system seems to be Bi_2 (Ca, Sr)$_3$ Cu_2 O_x. X-ray diffraction patterns of the 85 and 110K phases seem to show little or no differences. This composition suggests a similarity of the structure to that of the Aurivillius family of oxides (ref. 4) of the general formula $(Bi_2O_2)^{2+}$ $(A_{n-1}B_nO_{3n+1})^{2-}$. In Fig. 7, we show the similarity between the Aurivillius phase $Bi_4Ti_3O_{12}$ (orthorhombic) and Bi_2 (Ca, Sr)$_3$ Cu_2 O_9. The c-parameters of these two oxides would be comparable (31 ± 1 Å). We can have lower members of the Bi-Ca-Sr-Cu-O family with c-parameters of around 24 and 18A. For example, Bi_2(Ca, Sr)$_2$ CuO_x would have a similar structure except that there would be two (Ca, Sr)O layers and one CuO_2 layer and the c-parameter would be ~ 24A. Such a layered atomic arrangement with alternating bismuth oxide double layers and pairs of $(CuO_2)_\infty$ layers seperated by the alkaline earth cations has been proposed for $Bi_2CaSrCu_2O_{8+\delta}$ (ref. 12). Hazen et al (ref. 9) have pro-

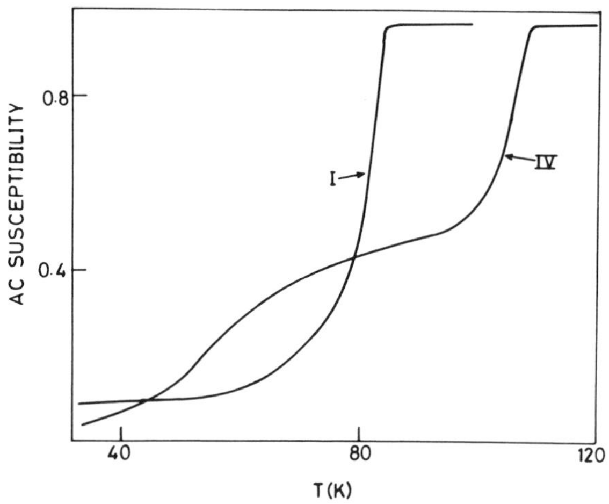

Fig. 6.   Temperature variation of the ac susceptibility of $Bi_2(Ca, Sr)_3Cu_3O_x$ superconductors (From Tarascon et al, ref. 11).

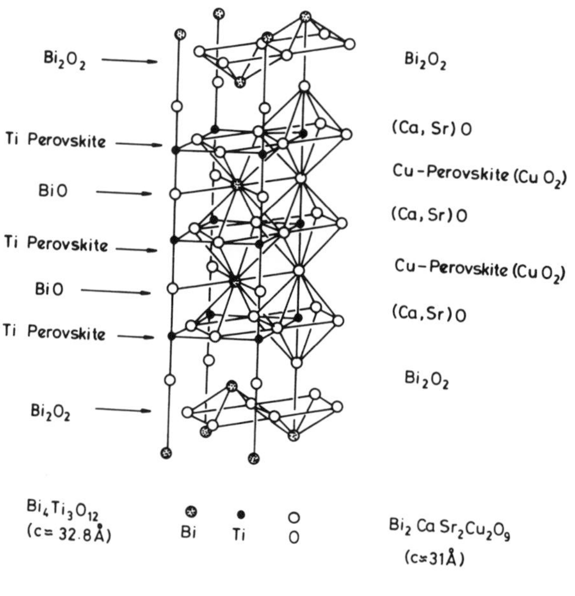

Fig. 7.   Structures of $Bi_4Ti_3O_{12}$ and $Bi_2CaSr_2Cu_2O_9$ (ref. 10).

posed an A-centered orthorhombic unit subcell 5.41 x 5.44 x 30.78Å for Bi_2Ca, Sr_2Cu_2
$O_{8+\delta}$. A space group of P̲n̲n̲n̲ corresponding to a primitive orthorhombic unit cell has
been assigned to this oxide (ref. 13). A pseudotetragonal cell has also been assigned
with a = 3.817Å and c̲ = 30.6Å (ref. 11, 14). The space groups I̲4̲m̲m̲ and I̲4̲/m̲m̲n̲ have been
suggested, the latter being more likely for a high-temperature oxide. The tetragonal
structure of Tarascon et al (ref. 11) is shown in Fig. 8. Knowing the Aurivillius phases,
the author has the feeling that these Bi-Ca-Sr-Cu-O oxides may have an orthorhombic
or monoclinic distortion.

High-resolution electron microscopy of the Bi-Ca-Sr-Cu-O oxides show the expected diffraction patterns and lattice images. We show the lattice image of $Bi_2CaSrCu_2O_x$ in Fig.
9, with an ordered arrangement of lattice fringes perpendicular to the c̲-axis as expected. $Bi_2(CaSr)_3Cu_2O_x$, on the other hand shows 15Å fringes corresponding to a c̲-parameter
of ~31Å (Fig. 10); the dark bands correspond to the bismuth layers. These lattice images
are similar to those of the Aurivillius family of oxides studied extensively earlier
(ref. 5). As expected of these oxides, considerable disorder (stacking faults) and other
defects are seen in the electron micrographs. The Aurivillius family oxides show side-
stepping of Bi_2O_2 layers and extensive dislocations. We see evidence for extensive dislocations as well as other defects in the electron micrographs of $Bi_2(Ca, Sr)_2Cu_2O_x$

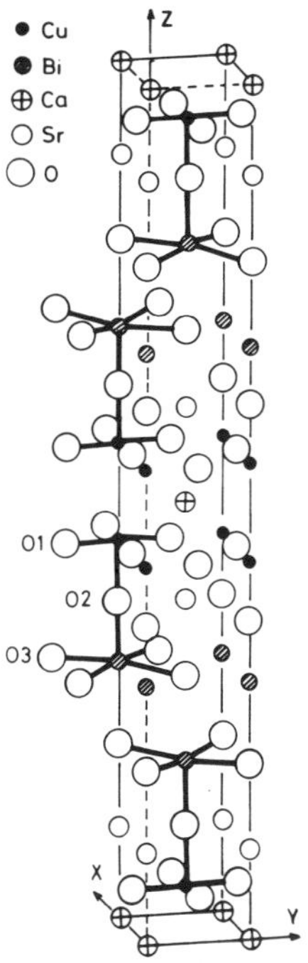

Fig. 8. Unit cell of $Bi_2CaSr_2Cu_2O_8$. The O_4 is located directly below 01 midway between the Bi planes (From Tarascon et al, ref. 11).

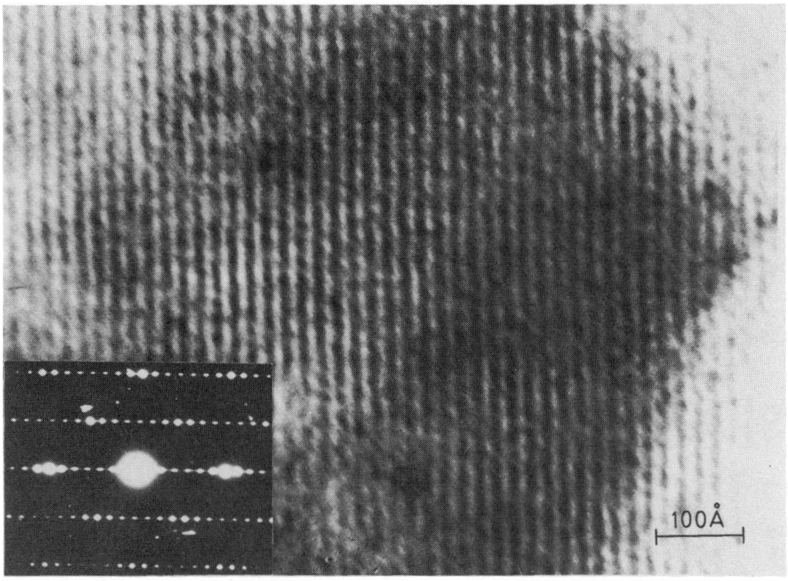

Fig. 9. Electron diffraction pattern and lattice image of $Bi_2CaSrCu_2O_x$ perpendicular to the c-axis showing 21A fringes (ref. 10). This oxide mage be similar to the phase described by Michel et al (ref. 1).

and related oxides. In Fig. 11 we show edge dislocations in $Bi_2CaSr_2Cu_2O_{8+\delta}$. The dislocations are likely to be associated with the facile side-stepping of the bismuth layers.

In Fig. 12 we show the x-ray diffraction patterns of several Bi-Ca-Sr-Cu-O oxides along with the onset temperatures to show their structural similarity. By varying the Ca/Sr ratio in $Bi_2(Ca, Sr)_3Cu_2O_8$, we have found that compositions on the calcium-excess side form pure phase. Thus $Bi_2Ca_2SrCu_2O_x$ and $Bi_2Ca_{1.5}Sr_{1.5}Cu_2O_x$ show better monophasic x-ray patterns. In the case of $Bi_2Ca_{1.5}Sr_{1.5}Cu_2O_x$, the x-ray pattern shows a mixture of two phases with c = ~24A and ~31A when first prepared; if the sample is heated for a longer

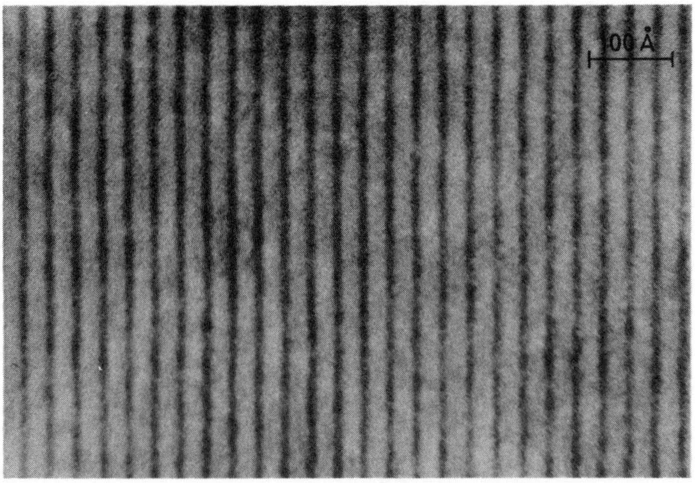

Fig. 10. Lattice image of $Bi_2Ca_{1.5}Sr_{1.5}O_{8.1}$ (from this laboratory).

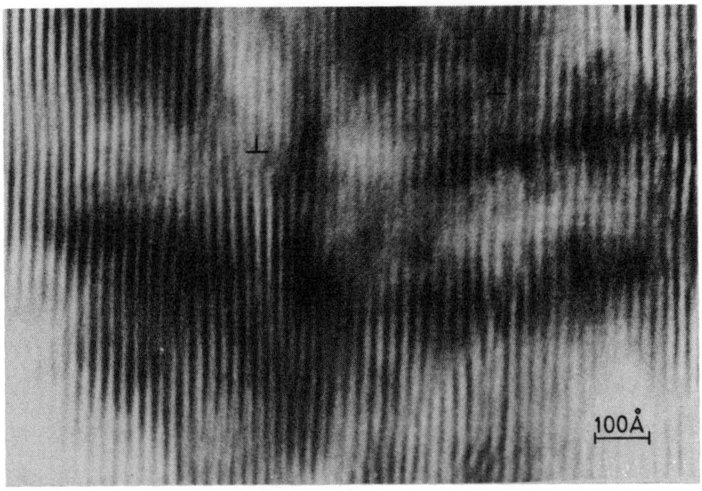

Fig. 11. Edge dislocations found in the lattice image of $Bi_2CaSr_2O_{8+\delta}$ (From this laboratory).

period at 1130K, the pattern shows only the ~ 31Å phase exactly expected for this composition (Fig. 13). We have found a similar change from the $\underline{c} \approx$ 24Å phase to the $\underline{c} \approx$ 31Å phase when a mixture of oxide with Bi-Ca-Sr-Cu ratio of 4.234 is heated over long periods at 1120K (Fig. 14). Such changes may be related to the presence of two T_cs (85 and 110K in such oxides), the higher T_c manifesting itself after prolonged heating.

Copper in the Bi-Ca-Sr-Cu-O oxides is formally mixed valent and the oxygen content of the $Bi_2(Ca, Sr)_2Cu_2O_x$ as determined by TGA (Fig. 15) and other methods is 8.3 ± 0.15 and the formula may therefore be represented as $Bi_2(Ca, Sr)_2Cu_2O_{8+\delta}$. The actual stoichiometry and cation distribution in this oxide is interesting. The Ca/Sr ratio is variable in this oxide and the composition $Bi_2Ca_{1.5}Sr_{1.5}Cu_2O_{8+\delta}$ seems to be closer to that of the superconducting phase. The actual contribution seems to be $Bi_{2.15}Ca_{1.17}Sr_{1.68}Cu_2O_{8+\delta}$. We have made different compositions of Bi-Ca-Sr-Cu oxides with a slight Bi-excess ($\geq$ 2.0), but all of them show a major superconducting transition around 100K. The composition $Bi_{2.25}Ca_{1.2}Sr_{1.55}Cu_2O_x$ showed some unusual features with a possible very high T_c phase, although it was a mixture of phases as prepared.

XPS and Auger studies show the bismuth cuprates to have Cu mainly in the 1+ and 2+ states. There is no evidence for Cu^{3+}. There appears to be some evidence for the presence of O^{1-} (oxygen holes) besides O^{2-}.

THALLIUM CUPRATES

Thallium forms oxides with copper and alkaline earth metals and these oxides are analogous to those of bismuth discussed earlier. Superconducting Tl-Ba-Cu-O and Tl-Ca-Ba-Cu-O

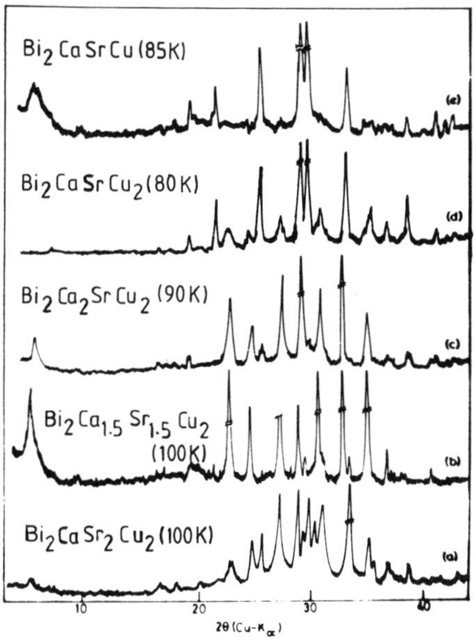

Fig. 12. X-ray diffraction patterns of different compositions of Bi-Ca-Sr-Cu oxides (From this laboratory).

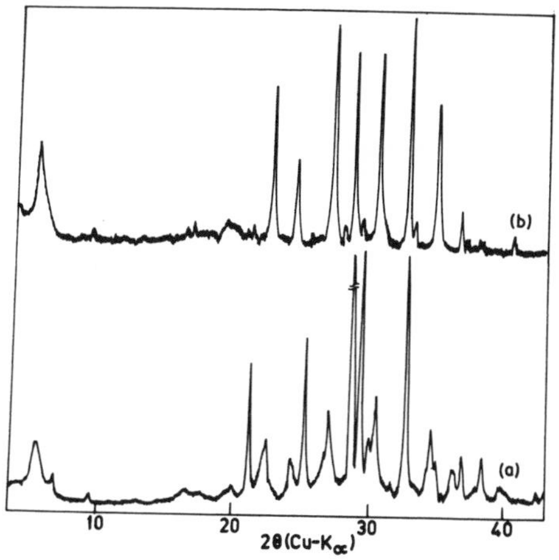

Fig. 13. X-ray diffraction patterns of $Bi_2Ca_{1.5}Sr_{1.5}Cu_2O_{8+\delta}$ before (a) and after (b) heating at 1130K (From this laboratory).

oxides ($T_c \sim$ 100K) have been synthesized (ref. 15). $Tl_2Ca_2Ba_2Cu_3O_{10+\delta}$ and $Tl_2CaBa_2Cu_2O_{8+\delta}$, both with onset T of 120K and zero resistance of 100K, have been isolated (ref. 16). Both are pseudo-tetragonal with 5.40 x 5.40 x 36.3A and 5.44 x 5.44 x 29.6A pseudo-tetragonal cells. The structures are better described as orthorhombic. As we would expect from Fig. 7, $Tl_2Ca_2Ba_2Cu_3O_{10+\delta}$ would have additional alkaline earth oxide and CuO_2 layers compared to $Tl_2CaBa_2Cu_2O_{8+\delta}$. Electrical resistivity data of Tl-Ca-Ba-Cu oxides of 2213 and 2223 compositions are shown in Fig. 16 to illustrate the high T_c of these materials. In Fig. 17 we show the resistivity behaviour of 2122, 2212, 2213 and 2223 members of the Tl-Ca-Ba-Cu-O system found in this laboratory (ref. 17); sharp transitions with onset temperatures of $\sim$ 120K are exhibited by these samples. AC susceptibility data of two of the samples (Fig. 18) show large Meissner effect. The 2223 sample of Fig. 16 and Fig. 17 has a small proportion of 2122 as well. Nearly pure 2223 with an onset temperature of $\sim$ 125K can be prepared starting from 1313. In Table 3, we have listed various other thallium cuprates which we have studied.

The thallium cuprates are truly remarkable in that they show very sharp superconducting transitions. They are prepared readily (by heating for a very short period in a preheated furnace at 1170K), although care has to be taken to avoid evaporation of Tl_2O_3. In Fig. 19 we show the x-ray diffraction patterns of the various members of the Tl-Ca-Ba-Cu-O

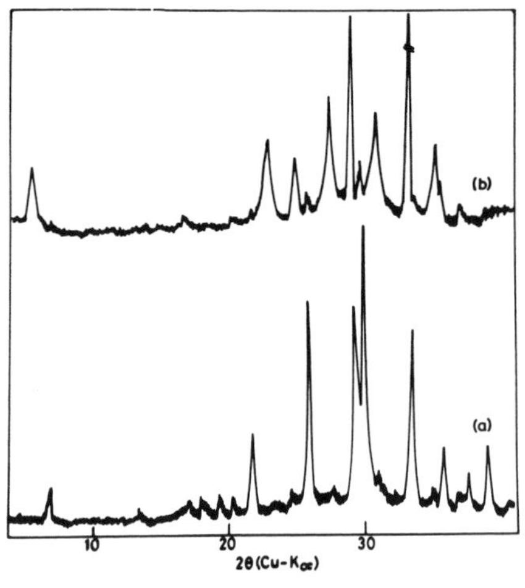

Fig. 14. X-ray diffraction patterns of the $Bi_4Ca_2Sr_3Cu_4O_x$ mixture before and after heating at 1120K (From this laboratory).

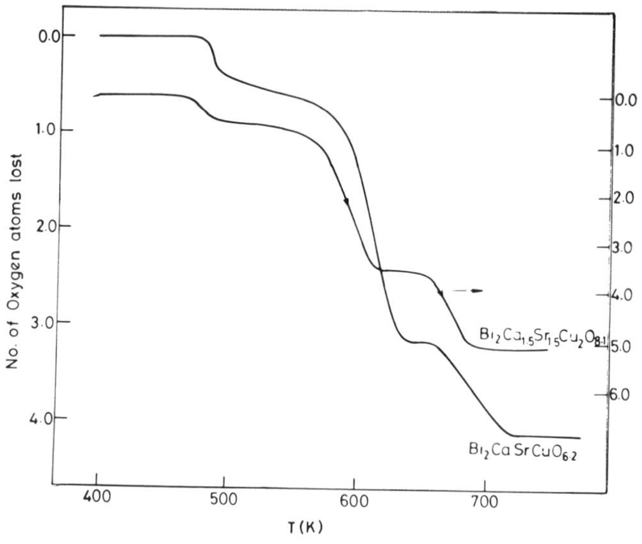

Fig. 15. TGA of $Bi_2Ca_{1.5}Sr_{1.5}Cu_2O_x$ and $Bi_2CaSrCuO_y$ where x and y are shown to be close to 8 and 6 respectively (From this laboratory).

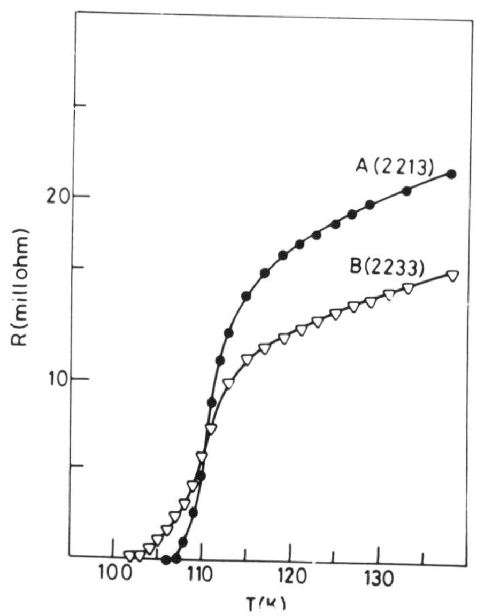

Fig. 16. Temperature variation of resistance of (A)$Tl_2Ca_2BaCu_3O_x$ and (B) $Tl_2Ca_2Ba_2Cu_3O_x$ superconductors (From Hazen et al, ref. 16).

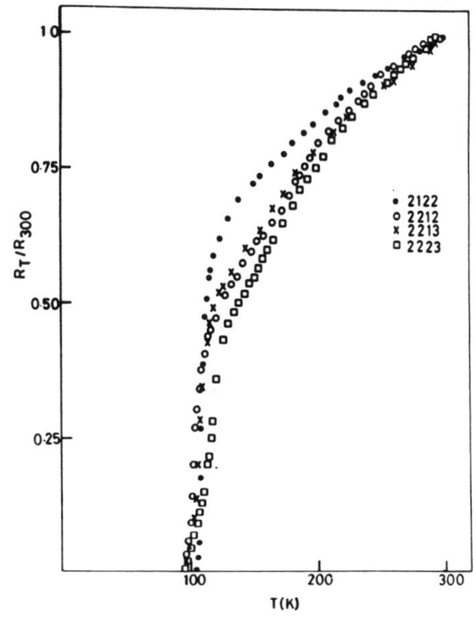

Fig. 17. Resistivity behaviour of 2122, 2212, 2213 and 2223 members of the Tl-Ca-Ba-Cu-O system (ref. 17).

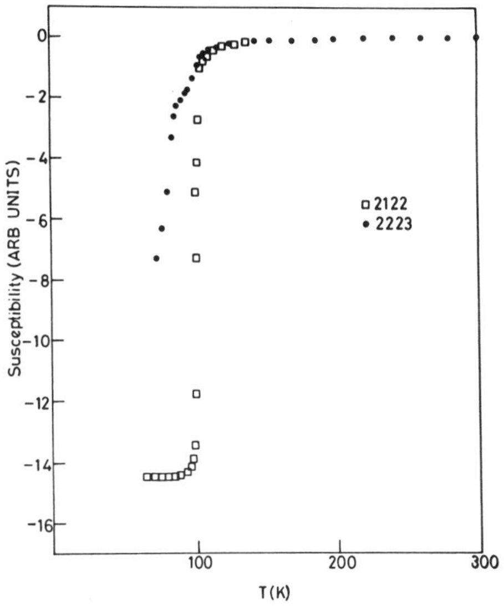

Fig. 18. AC susceptibility data of 2122 and 2223 members (ref. 17).

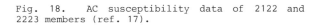

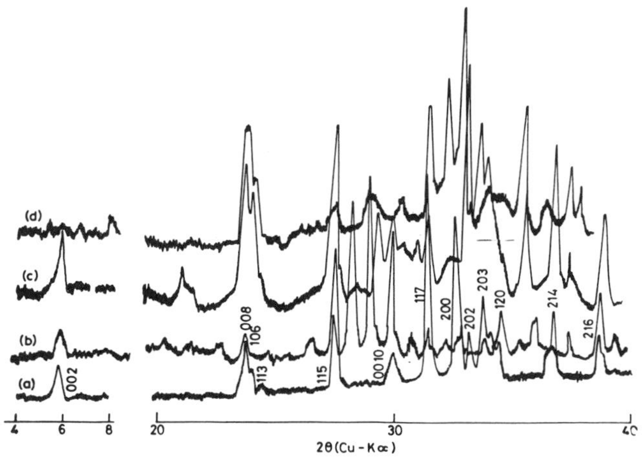

Fig. 19. X-ray diffraction patterns of (a) 2122 as prepared, (b) 2122 heated further, (c) 2212 and (d) 2213 (ref. 17).

Table 3

Electrical properties of a few thallium cuprates

Nominal composition	Electrical property[a]
$Tl_2Pb_3Cu_2O_x$	Metallic
$Tl_2CaBaPbCu_2O_x$	Metallic
$Tl_2MgBa_2Cu_2O_x$	Metallic
$TlBiCa_2BaCu_3O_x$	High resistivity
$TlBiCaBa_2Cu_2O_x$	T_c (onset) = 102K
$Tl_2Ca_3Cu_2O_x$	Metallic
$Tl_2Sr_3Cu_2O_x$	High resistivity
$TlCa_3BaCu_3O_x$	High resistivity
$Tl_2CaSr_2Cu_2O_x$	High resistivity
$Tl_2CaBa_{2.25}Cu_2O_x$	Metallic
$Tl_{2.15}CaBa_2Cu_2O_x$	T_c (zero) = 85K
$Tl_2Ca_{0.75}Ba_{2.25}Cu_2O_x$	T_c (zero) = 72K
$Tl_3Ca_4Ba_3Cu_5O_x$	T_c (zero) = 97K
$Tl_4Ca_3Ba_2Sr_2Cu_4O_x$	T_c (onset) = 108K
$TlCa_4BaCu_4O_x$	T_c (onset) = 94K
$Tl_2Ca_3Ba_2Cu_4O_x$	T_c (onset) = 112K
$Tl_4Ca_3Ba_4Cu_5O_x$	T_c (zero) = 100K
$Tl_4Ca_3Ba_3Cu_5O_x$	T_c (zero) = 95K

[a] resistivity measured down to 60K only.

system. Many of the samples contain $BaCuO_2$ impurity; extensive heating increases this impurity concentration. The x-ray pattern of the 2122 phase is identical to that of $Bi_2CaSr_2Cu_2O_{8+\delta}$ (Table 4). Electron microscopy of the thallium cuprates show some evidence for their layered morphology (ref. 17).

In Fig. 20 we show the ordered lattice image of the 2122 oxide alongwith the electron diffraction pattern. The image shows the expected 15A fringes. The thallium cuprates also show the presence of extensive dislocations just as in the bismuth cuprates

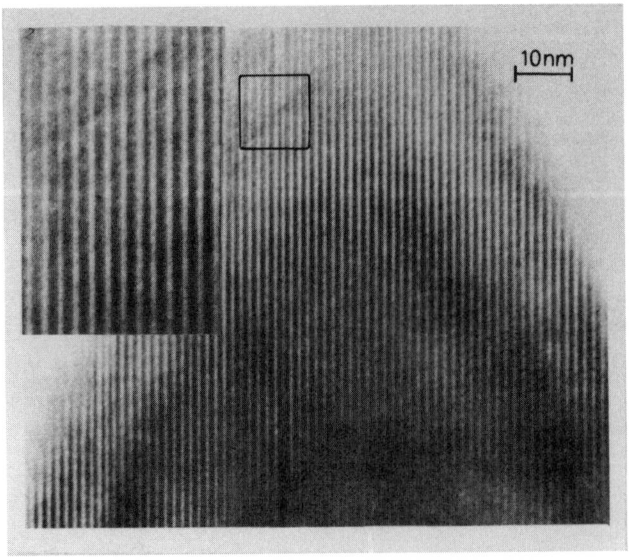

Fig. 20. Lattice image of $Tl_2CaBa_2Cu_2O_{8+\delta}$ (ref. 17) showing 15Å fringes.

Table 4

Powder X-ray diffraction data of $Tl_2CaBa_2Cu_2O_x$*

h k l	d_{obs}(Å)	d_{calc}(Å)	I/I_o
0 0 2	14.979	14.900	10
0 0 4	7.406	7.450	2
0 0 8	3.746	3.725	25
1 0 6	3.700	3.684	15
1 1 3	3.619	3.613	4
1 1 5	3.243	3.251	60
0 0 10	2.980	2.980	15
1 1 7	2.858	2.867	100
2 0 0	2.743	2.740	85
2 0 2	2.698	2.698	15
2 0 3	2.652	2.647	10
-	2.600	-	8
1 2 0	2.455	2.455	20
2 1 4	2.332	2.329	12
2 1 6	2.201	2.198	15
2 0 8			
1 1 12	2.103	2.094	10
0 2 10	2.027	2.020	8
2 2 0	1.939	1.940	40
3 0 0	1.829	1.826	25
3 1 5	1.664	1.664	20
2 2 10	1.632	1.626	5
3 1 7		1.605	
	1.601		25
1 1 17		1.597	
0 0 19	1.576	1.568	10
2 2 12	1.520	1.529	14
2 2 14	1.435	1.429	8
2 0 18			
	1.423		4
0 0 21		1.419	
4 0 0	1.372	1.370	8

*Indexed on an orthorhombic cell with a = 5.479, b = 5.493 and c = 29.80Å.

and in the Aurivillius family of oxides. In Fig. 21 we show an electron micrograph of the 2122 oxides with a large number of edge dislocations.

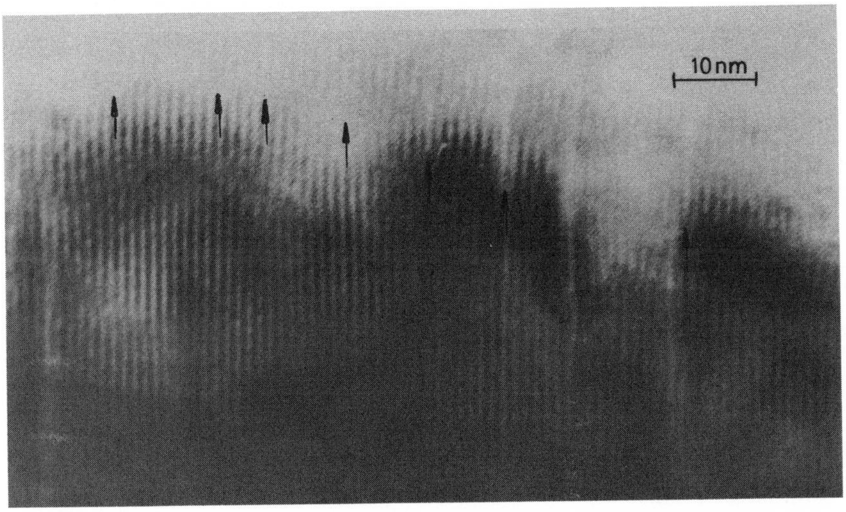

Fig. 21. Lattice image of the 2122 oxide showing edge dislocations (From this laboratory).

INTERGROWTH STRUCTURES AND SUPERCONDUCTIVITY

An increasing number of metal oxides forming long-period structures due to the recurrent intergrowth of two chemically distinct but structurally related units are getting to known in recent years (ref. 6,7,13). These novel structures have given rise to new chemistry at solid-solid interfaces. Besides such ordered intergrowth structures, many oxides with random intergrowth (similar to stacking faults in polytypes) are known. The Aurivillius family of oxides are some of the best known examples of systems forming intergrowth structures (ref. 8,19). In Fig. 22, we show schematically, the first three

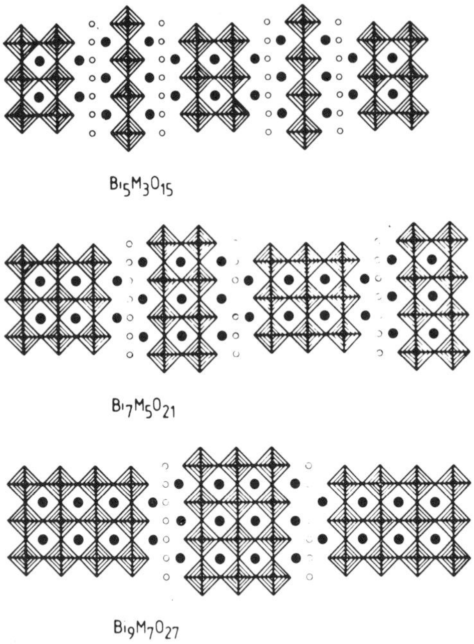

Fig. 22. Homologous series of ordered intergrowths formed by the Aurivillius family of bismuth oxides (From Rao, ref. 7).

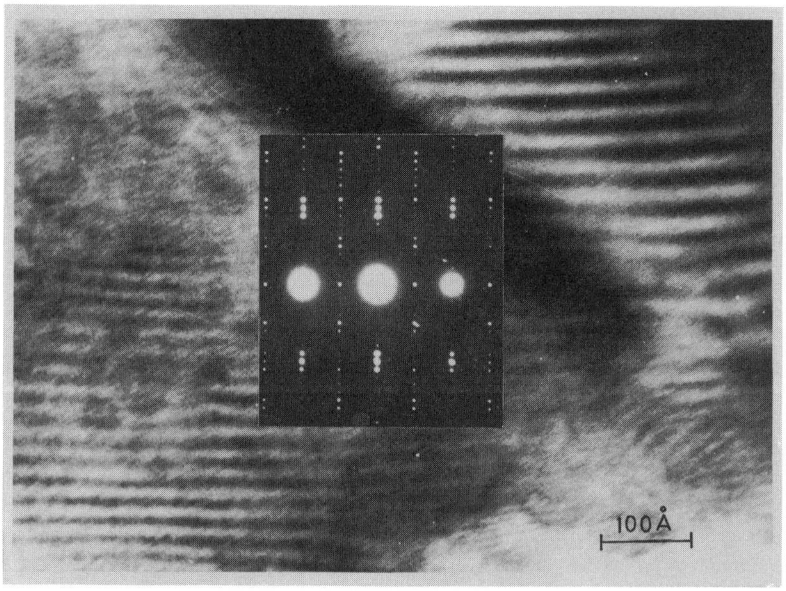

Fig. 23. Lattice image showing intergrowth of two Bi-Ca-Sr-Cu-O phases with two different c - parameters (ref. 10).

members of the homologous series of oxides of the general formula $Bi_4A_{m+n-2}B_{m+n}O_{3(m+n)+6}$ formed by the Aurivillius family of oxides.

Preliminary studies of certain compositions of the type $Bi_{m_1+m_2}M_{n_1+n_2}Cu_{p_1+p_2}O_{x_1+x_2}$ formed by two cuprates of the type $Bi_mM_nCu_pO_x$ have yielded some encouraging results.

Accordingly, lattice images of the composition $Bi_4Ca_2Sr_3Cu_3O_y$ (supposedly formed by $Bi_2CaSr_2Cu_2O_x$ and $Bi_2CaSrCuO_x$) gave evidence for the presence of two different <u>c</u> - spacings in the lattice images (Fig. 23). Similar intergrowths were seen in the lattice images of Bi Ca Sr Cu O as well (Fig. 24). These lattice images are interesting although they do not show the presence of ordered intergrowths. Resistivity measurements of such compositions showed clear drops or steps around 200K or higher (Fig. 25) suggesting that they may be due to intergrowths. Magnetic susceptibility also shows evidence for Meissner effect around these temperatures. We have some reason to believe that "intergrowth mechanism" may provide a way of increasing T_cs of these layered materials.

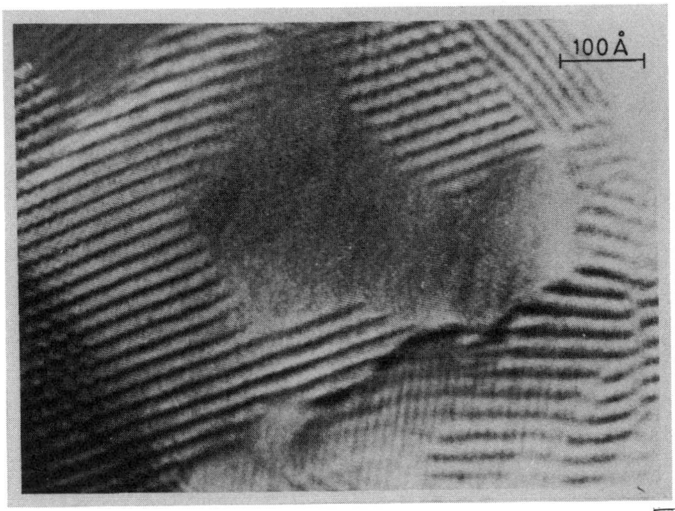

Fig. 24. Lattice image of $Bi_4Ca_2Sr_3Cu_4$ showing phases with different <u>c</u> - parameters (From this laboratory).

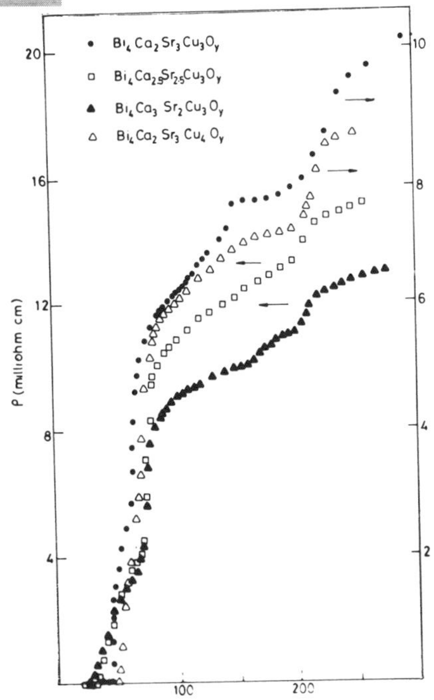

Fig. 25. Electrical resistivity data of certain Bi-Ca-Sr-Cu-O compositions showing drops around 200K besides the major superconducting transition (From this laboratory). Note that the resistivity drop at ~ 200K is more marked.

Preliminary studies on the Tl-Ca-Ba-Cu-O system have shown the presence of intergrowths (Fig. 26). It is possible that intergrowths play a role in the superconducting of these oxides as well.

CONCLUDING REMARKS

The bismuth and thallium cuprates constitute two of the rare-earth-free high T_c oxide systems. They are easy to prepare, although the exact conditions for obtaining the best superconducting behaviour need to be established. Oxygen seems to be more stable in

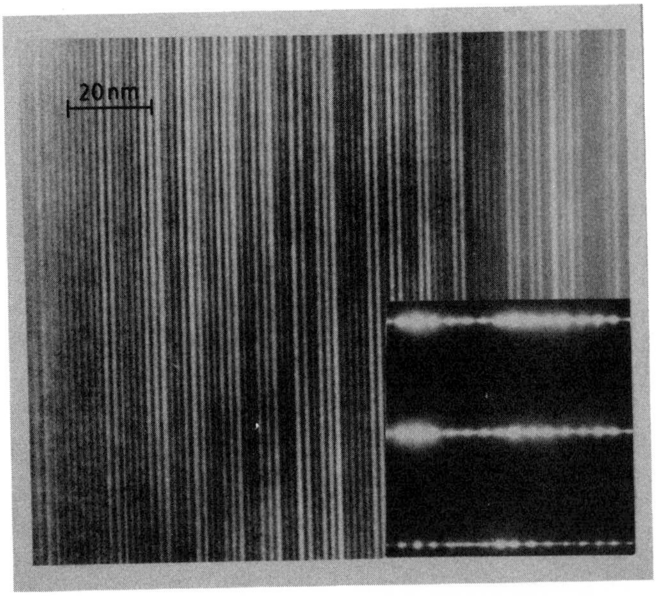

Fig. 26. Intergrowths seen in the lattice image of $Tl_2CaBa_2Cu_2O_{8+\delta}$ (from this laboratory).

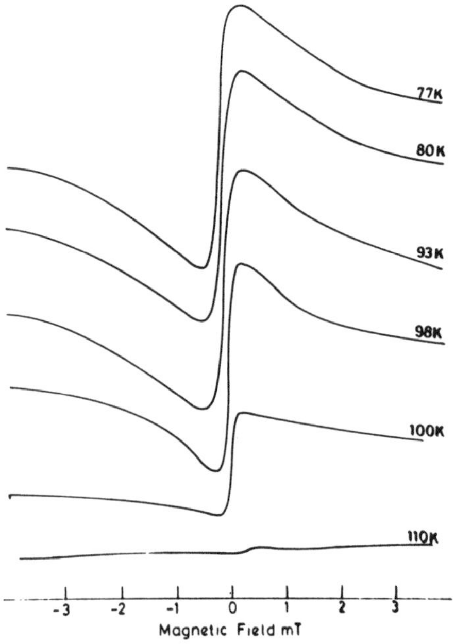

Fig. 27. Derivatives of absorption signals of $Tl_2CaBa_2Cu_2O_{8+\delta}$ at various temperatures recorded at 9.11 GHz (ref. 17).

these cuprates, suggesting greater stability of these materials.

The Bi and Tl cuprates also have a nominal mixed valency with copper, in 2+ and 3+ states, but Cu is present only in 1+ and 2+ states. It is possible that oxygen holes will play an important role in the mechanism of superconductivity of these cuprates.

The Bi and Tl cuprates possess structures containing perovskite layers similar to the Aurivillius family of bismuth oxides. They have low-dimensional features due to the presence of copper-oxygen sheets, but there are no Cu-O chains.

The bismuth and thallium cuprates appear to be granular type II superconductors just as the 123 oxides. Both the Bi and Tl cuprates absorb microwaves in the superconducting state (ref. 17) as shown in Fig. 27, the intensity going to zero just above the superconducting transition temperature. Microwave absorption can be used to characterize these oxide superconductors; the absorption is sensitive to the sample quality as well as ambient atmosphere, and shows hysteresis just as $YBa_2Cu_3O_7$ (ref. 20).

It is relatively easy to grow crystals of the bismuth cuprates. We have had some success in this effort as well.

REFERENCES

1. C. Michel, M. Herview, M.M. Borel, A. Grandin, F. Deslandes, J. Provost and B. Raveau, Z. Physik., B68, 421 (1987).
2. H. Maeda, Y. Tanaka, M. Fukutomi and T. Asano, Jpn. J. Appl. Phys. to be published (1988).
3. C.W. Chu, J. Bechtold, L. Gao, P.H. Hor, Z.J. Huang, R.L. Meng, Y.Y. Sun, Y.Q. Wang and Y.Y. Xue, Phys. Rev. Lett., to be published (1988).
4. B. Aurivillius, Arkiv Kemi, 1, 463, 499 (1950); 2, 519 (1950).
5. J.L. Hutchison, J.S. Anderson and C.N.R. Rao, Proc. Roy. Soc. London, A355, 301 (1977).
6. C.N.R. Rao and J. Gopalakrishnan, "New Directions in Solid State Chemistry", Cambridge University Press, 1986.
7. C.N.R. Rao, Bull. Mat. Sci., 7, 155 (1985).
8. J. Gopalakrishnan, A. Ramanan, Rao, C.N.R. and D.A. Jefferson, J. Solid State Chem., 55, 101 (1984).
9. R.M. Hazen, C.T. Prewitt, R.J. Angel, N.L. Ross, L.W. Finger, C.G. Hadidiacos, D.R. Veblen, P.J. Heaney, P.H. Hor, R.L. Meng, Y.Y. Sun, Y.Q. Wang, Y.Y. Xue, Z.J. Huang, L. Gao, J. Bechtold and C.W. Chu, Phys. Rev. Lett., to be published (1988).
10. C.N.R. Rao, A.M. Umarji, R.A. Mohan Ram, R. Vijayaraghavan, K.S. Nanjundaswamy, P. Somasundaram and L. Ganapathi, Pramana-J. Phys., to be published (1988), April issue.
11. J.M. Tarascon, Y. Le Page, P. Barboux, B.G. Bagley, L.H. Greene, W.R. McKinnon, G.W. Hull, M. Giroud and D.M. Hwang, to be published.
12. M.A. Subramanian, C.C. Torardi, J.C. Calabrese, J. Gopalakrishnan, K.J. Morrissey, T.R. Askew, R.B. Flippen, U. Chowdhry and A.W. Sleight, Science, to be published (1988).
13. D.R. Veblen, P.J. Heaney, R.J. Angel, L.W. Finger, R.M. Hazen, C.T. Prewitt, N.L. Ross, C.W. Chu, P.H. Hor, R.L. Meng, Nature, (to be published).
14. E.T. Muromachi, Y. Uchida, A. Ono, F. Izumi, M. Onoda, Y. Matsui, K. Kosuda, S. Takekawa and K. Kato, Jpn. J. Appl. Phys., submitted.
15. Z.Z. Sheng and A.M. Hermann, Nature 33, 55, 138 (1988) and other papers to be published.
16. R.M. Hazen, L.W. Finger, R.J. Angel, C.T. Prewitt, N.L. Ross, C.G. Hadidiacos, P.J. Heaney, D.R. Veblen, Z.Z. Sheng, A.E. Ali and A.M. Hermann, to be published.
17. A.K. Ganguli, G.N. Subbanna, A.M. Umarji, S.V. Bhat and C.N.R. Rao, Pramana-J.Phys., May issue (1988).
18. C.N.R. Rao and J.M. Thomas, Acc. Chem. Res., 18, 113 (1985).
19. D.A. Jefferson, M.K. Uppal and C.N.R. Rao, Mat. Res. Bull., 19, 1403 (1984).20. S.V. Bhat, P. Ganguly, T.V. Ramakrishnan and C.N.R. Rao, J. Phys. C. Solid State, 20, L559 (1987).